I0074253

Topological Phase Transitions
and
New Developments

Topological Phase Transitions
and
New Developments

Editors

Lars Brink
Chalmers University of Technology, Sweden

Mike Gunn
University of Birmingham, UK

Jorge V José
Indiana University, USA

John Michael Kosterlitz
Brown University

Kok Khoo Phua
Nanyang Technological University, Singapore

World Scientific

NEW JERSEY • LONDON • SINGAPORE • BEIJING • SHANGHAI • HONG KONG • TAIPEI • CHENNAI • TOKYO

Published by

World Scientific Publishing Co. Pte. Ltd.

5 Toh Tuck Link, Singapore 596224

USA office: 27 Warren Street, Suite 401-402, Hackensack, NJ 07601

UK office: 57 Shelton Street, Covent Garden, London WC2H 9HE

British Library Cataloguing-in-Publication Data
A catalogue record for this book is available from the British Library.

TOPOLOGICAL PHASE TRANSITIONS AND NEW DEVELOPMENTS

Copyright © 2019 by World Scientific Publishing Co. Pte. Ltd.

All rights reserved. This book, or parts thereof, may not be reproduced in any form or by any means, electronic or mechanical, including photocopying, recording or any information storage and retrieval system now known or to be invented, without written permission from the publisher.

For photocopying of material in this volume, please pay a copying fee through the Copyright Clearance Center, Inc., 222 Rosewood Drive, Danvers, MA 01923, USA. In this case permission to photocopy is not required from the publisher.

ISBN 978-981-3271-33-3

For any available supplementary material, please visit
https://www.worldscientific.com/worldscibooks/10.1142/11016#t=suppl

Preface

In 2016 the physics Nobel Selection Committee announced that **Duncan Haldane, Michael Kosterlitz and David Thouless** were the winners of this pinnacle cherished prize. They stated that *"This year's Laureates opened the door on an unknown world where matter can assume strange states. They have used advanced mathematical methods to study unusual phases or states of matter, such as superconductors, superfluids or thin magnetic films. Thanks to their pioneering work, the hunt is now on for new and exotic phases of matter. Many people are hopeful of future applications in both materials science and electronics."* At the initiative of Professor Kok Khoo Phua, a workshop on Topological Phase Transitions and New Developments was co-organized by us and held at the Institute of Advance Studies, Nanyang Technological University, June 5–8, 2017. The two leading lectures during the meeting were delivered by the 2016 Nobel Laureates, Haldane and Kosterlitz. There were also lectures presented by leading invited speakers, from Asia, Europe and America. They covered a wide range of subjects related to the subject matters of the awards. Most of the chapters in this book are articles describing material presented in lectures at the meeting, or related extensions, enhancing the quality of this book. The book also includes the abstracts of presentations made at the meeting that do not have a paper contribution.

We feel that this book gathers in a single convenient place material which will be very useful to researchers who want to get a glimpse of new developments, extensions and questions related to the seminal ideas introduced by Haldane, Kosterlitz and Thouless. It also makes clear why the three physicists fully deserved the recognition bestowed on them by the 2016 Nobel Physics Prize selection committee.

Sincerely,

Lars Brink, Mike Gunn, Jorge V. José,
John Michael Kosterlitz and Kok Khoo Phua

Contents

*Abstract.

---

*Abstract.

---

*Abstract.

Topological defects and phase transitions

J. Michael Kosterlitz

Brown University
E-mail: J_Kosterlitz@brown.edu

This talk reviews some of the applications of topology and topological defects in phase transitions in two-dimensional systems for which Kosterlitz and Thouless split half the 2016 Physics Nobel Prize. The theoretical predictions and experimental verification in two-dimensional superfluids, superconductors and crystals will be reviewed as they provide very convincing quantitative agreement with topological defect theories.

Full paper found at https://www.nobelprize.org/nobel_prizes/physics/laureates/2016/kosterlitz-lecture.pdf.

References

1. J. M. Kosterlitz, *Int. J. Mod. Phys. B*, **32**, 1830005 (2018) https://doi.org/10.1142/S02179792/8300050.

Geometry of flux attachment in the fractional quantum hall effect states

F. Duncan M. Haldane

Princeton University
E-mail: haldane@princeton.edu

The unexpected experimental discovery of the topologically-ordered Fractional Quantum Hall (FQH) states showed that the powerful diagrammatic perturbation theoretic methods of the time were only useful for a subclass of problems adiabatically related to free-particle problems, and instead, Laughlin's discovery of a model state that describes "flux attachment" to form composite particles has been the source of most subsequent understanding of the effect. In recent years, it has become apparent that "flux attachment" has important sort-distance geometrical properties as well as long-distance topological entanglement properties. I will describe geometric analogies between the unit cell of a solid and the "composite boson" which is the elementary unit of incompressible FQH liquids, and the place for "composite fermions" in their description.

The attraction between antiferromagnetic quantum vortices as origin of superconductivity in cuprates

P. A. Marchetti

Dipartimento di Fisica e Astronomia, Università di Padova, INFN,
Padova, I-35131, Italy
E-mail: marchetti@pd.infn.it

We propose as key of superconductivity in (hole-doped) cuprates a novel excitation of magnetic origin, characteristic of two-dimensions and of purely quantum nature: the antiferromagnetic spin vortices. In this formalism the charge pairing arises from a Kosterlitz-Thouless-like attraction between such vortices centered on opposite Néel sublattices. This charge pairing induces also the spin pairing through the action of a gauge force generated by the no-double occupation constraint imposed in the *t-J* model of the CuO planes of the cuprates. Superconductivity arises from coherence of pairs of excitations describing Zhang-Rice singlets and it is not of standard BCS type. We show that many experimental features of the cuprates can find a natural explanation in this formalism.

Keywords: Superconductivity, Cuprates, Vortices, Gauge field theories.

1. Introduction

Thirty years after the discovery of the first high-T_c superconducting cuprate[1], the microscopical mechanism behind superconductivity in this class of materials is still not understood, despite constant experimental advances. It is commonly believed that antiferromagnetism (AF) is a key ingredient for the superconductivity in cuprates, then a natural pairing glue would be provided by the spin fluctuations, i.e. antiferromagnetic spin-waves (see e.g. Ref. 2). Their action would be enhanced by nesting of the Fermi surface (FS), but evidence for this is not so clear. We propose as pairing glue another excitation still emerging from AF, but of purely quantum origin: antiferromagnetic spin vortices. In the antiferromagnetic phases the spin group $SU(2)$ is broken to $U(1)$, the quotient $SU(2)/U(1)$ is isomorphic to the 2-sphere S^2 whose points label the directions of the magnetization. Their fluctuations are described by spin waves. The unbroken $U(1)$ group describes unphysical gauge fluctuations. However in two dimensions (2D) one can consider vortices of Aharonov-Bohm type in this $U(1)$; due to AF the vortices have opposite chirality when centered in two different Néel sublattices, hence we dub them antiferromagnetic spin vortices. Lowering the temperature such gas of vortices in 2D undergoes a Kosterlitz-Thouless-like transition, with the formation of a finite density of vortex-antivortex pairs. If the vortices are centered on charges, this induces a new form of charge-pairing, again due to AF, but different from the spin-fluctuation pairing. As discussed later this pairing finally leads to superconductivity.

To present how this is realized in the (hole-doped) cuprates is the fundamental aim of this paper. One discovers that through this key idea many structural features of the phase diagram of the cuprates could be understood and many physical properties successfully computed. The derivations are based on well-defined conjectures and approximations and many experimental consequences are consistent with availabla data, as sketched in the final section. Let us stress that, due to their structure, the antiferromagnetic vortices are specific to quasi-2D doped antiferromagnets and some phenomena they give rise to, like the charge pairing discussed above, the induced spin-pairing and the metal-insulator crossover of in-plane resistivity, are peculiar to this approach and do not appear in this form in other approaches to the physics of cuprates.

Whenever we succeed we present the intuitive ideas at the beginning of the section, giving later on a brief summary of their formal implementation and deferring to the references for the explicit proofs.

This paper reviews the results on a mechanism for superconductivity arising from a long joint project on a spin-charge gauge approach to the physics of cuprates initiated with Z.-B. Su and L. Yu, to whom I express a deep gratitude for all the knowledge and ideas shared with me. Along the way we profit of fundamental contributions from many researcher, among which a leading role have been played by F. Ye and G. Bighin, but it is also a pleasure to cite as co-authors J.-H. Dai, L. De Leo, G. Orso, A. Ambrosetti, M. Gambaccini and last but not least to gratefully acknowledge the original insight by J. Froehlich.

2. Model and phase diagram

The high T_c cuprates share a layered structure incorporating one or more copper–oxygen (CuO) planes; an excellent recent summary of their properties can be found in Ref. 3. There is no consensus yet on the theoretical interpretation of their low-energy physics, but an agreement has been achieved that superconductivity is due to formation of Cooper pairs with principal locus the CuO planes and that the order parameter is a spin singlet with orbital symmetry $d_{x^2-y^2}$. Still a large consensus have the ideas that the formation of Cooper pairs takes place independently within different CuO multilayers and that repulsive electron-electron interaction plays a key role, the electron-phonon interaction being not the principal mechanism of their formation. However, even with these restrictions several quite different proposal for the pairing mechanism leading to superconductivity have been made. A recent, partial comparison of different theories with some experimental data can be found in Ref. 4.

Most of the researchers believe that for the hole-doped cuprates, considered in this paper, the key actors in the superconducting transition are the so-called Zhang-Rice singlets[5]. Let us sketch what they are and why they naturally lead to a t-J model-like description of the CuO planes. This basic structural unit is formed by a square lattice of Cu atoms with oxygen atoms in the middle of the sides of the square. In the undoped materials the relevant orbitals are the $3d_{x^2-y^2}$ orbitals of the Cu, containing one electron per site with a large Coulomb repulsion prohibiting a low-energy double occupation, and the p orbitals of the O, oriented along the sides of the square and completely filled. The spins of the Cu electrons at low temperature are antiferromagnetically ordered. The Cu and O orbitals have similar energies and they strongly hybridize. If a doping hole is introduced, it goes predominantly in the combinations of four oxygen p orbitals centered around a copper site forming a spin-singlet with the spin of the corresponding Cu site. These are the Zhang-Rice singlets and, from the point of view of the square lattice of the Cu, the corresponding site has 0 charge and 0 spin. One can thus describe effectively the low-energy of the CuO plane in terms of a square lattice, corresponding to the positions of Cu atoms, whose sites are either singly occupied or empty. Furthermore every oxygen orbital contributing to the hybridization is shared by two Cu, so that these singlets, described by empty sites, can hop between the sites of the above square lattice. Therefore a reasonable Hamiltonian H describing the low-energy physics of the CuO planes is given by the t-J model

$$H_{t-J} = \sum_{\langle i,j \rangle} P_G \left[-t c_{i\alpha}^* c_{j\alpha} + J \vec{S}_i \cdot \vec{S}_j \right] P_G, \tag{1}$$

where i, j denote nearest neighbour (nn) sites of the lattice, c the hole field operator, P_G the Gutzwiller projection eliminating double occupation, α the spin indices and summation over repeated spin (and vector) indices is understood, here and in the following. A typical value for the nn hopping is $t \approx 0.3\,\text{eV}$ and for the anti-ferromagnetic Heisenberg coupling $J \approx 0.1\,\text{eV}$. (These are the values used in the numerical computations reported in the last section.) Furthermore to get a reasonable shape of the Fermi surface in the tight binding approximation one adds to the nn hopping term at least a nnn term with coefficient t', $|t'| \sim 0.03 - 0.1\,\text{eV}$, strongly material dependent.

Let us briefly sketch the structure of the phase diagram of the (hole-doped) cuprates; for a very recent review see Ref. 6. In the ground state of the undoped materials all the Cu sites are singly occupied, thus realizing an antiferromagnetic Mott insulator, due to the strong on-site Coulomb repulsion. Long-range antiferromagnetism disappears when the doping concentration, δ, in a CuO plane exceeds a critical value ($\delta \approx 0.03$ at $T = 0$), through a Néel transition, but still a "charge-transfer" gap exists between the Zhang-Rice band and the upper Hubbard band of the Cu. Superconductivity (SC) appears when δ reaches ≈ 0.05 and as a function of doping the superconducting temperature T_c exhibits a dome shape. The doping concentration corresponding to the highest T_c is called optimal doping; the cuprates with lower doping are called underdoped and those with a higher one overdoped. In a V-shaped region in the δ-T plane around $\delta \approx 0.19$, called "Strange Metal" (SM), ARPES experiments reveal a Fermi surface roughly consistent with the band calculations, but the quasi-particle peak is broad and the incoherent tail anomalously large. Also the transport presents strange features, like the linear in

T behaviour of in-plane resistivity. In the underdoping region above T_c and below a temperature T^* indicating a crossover (or transition) to the SM there is a region called Pseudogap (PG), characterized by a suppressed density of states near the Fermi level. The current picture of the pseudogap opening is the following (see e.g. Ref. 7): in the large Fermi surface seen in the SM as the temperature decreases below T^*, the pseudogap first opens near $(0,\pm\pi)$ and $(\pm\pi,0)$ (antinodal region), it then gradually "eats" up the original FS, converting it into a "pseudogapped" part, eventually leaving only short disconnected arcs around $(\pm\pi/2,\pm\pi/2)$ (nodal region). Finally, these arcs shrink abruptly to nodal points of the superconducting gap function at the superconducting transition, maintaining the d-wave symmetry. Actually some experimentalists and theorists suggest the existence of two "pseudo-gap temperatures" (see e.g. Refs. 8, 9), a "high" one that joins the superconducting dome in the overdoped side and a "low" one crossing the superconducting dome and reaching $T = 0$ near $\delta \approx 0.19$ if superconductivity is suppressed. The first one corresponds to the emergence of the "pseudogap" phenomenology on the FS described above, but also e.g. to a decrease of the Knight shift or a deviation from linearity of in-plane resistivity. The second one corresponds e.g. to an inflection point of the resistivity curve and to a broad peak in the specific heat coefficient. Another crossover line dome-shaped above the dome of the superconducting transition signals the onset of a Nernst signal due to magnetic vortices and diamagnetism[16]. We call "Nernst" (N) the non-superconducting region below this crossover. Both the "low" pseudogap and the Nernst crossover line seem to be universal in the cuprates, whereas somewhat material-dependent appear the "high" pseudogap crossover and even more the superconducting transition. Finally, there is a crossover between AF and SC below which there is a "spin-glass" region and recently it has been firmly established[11] a dome-shaped crossover, quite distinct from the superconducting dome, around $\delta \approx 1/8$ below which one finds evidence of charge-density waves.

3. Hints from 1D

Two natural questions arising from the previous sections are: how the antiferromagnetic vortices appear in the t-J model and how much they explain of the above sketched phenomenology of the cuprates. Let us come back to the t-J model. Inspired by an idea pioneered by Anderson[12] and Kivelson[13], one can get rid of the no-double occupation constraint imposed by the Gutzwiller projector P_G by rewriting the hole field c_α as a product of a charge 1 spinless fermion field h, the holon, and a neutral spin 1/2 boson field $\tilde{s}_\alpha$, the spinon, as $c_\alpha = h^*\tilde{s}_\alpha$. (We introduced the tilde because we will denote by s_α a slightly different field). The holon h being spinless implements exactly the Gutzwiller constraint by the Pauli principle. Furthermore if the constraint $\tilde{s}_\alpha^*\tilde{s}_\alpha = 1$ is imposed, since $c_\alpha^*c_\alpha = 1 - h^*h$ we see that h^*h is just the density of empty sites in the model. However, if we treat the holon in mean-field, precisely because it is spinless one doesn't get a reasonable Fermi surface. Therefore we need to "dress" it, and actually we will see that also the spinon needs a "dress". A hint on how to proceed comes from the one-dimensional t-J model, which in some limit is exactly solvable, thus allowing a check. Let us here give the intuitive idea, then we briefly sketch its formal implementation in the next section.

Consider the Heisenberg spin $1/2$ chain, describing the 1D t-J model in the limit of zero doping and take as reference state that with the spin antiferromagnetically ordered, the semi-classical ground state mimicking the Néel order in 2D. Now let us insert a dopant by removing the spin from a site; as a consequence the two neighbouring spins will be ferromagnetically aligned. Let then the empty site to hop by a simultaneous opposite hopping of the spin, then we get two separate excitations. There is an empty site, but with neighbouring spins antiferromagnetically aligned, thus carrying charge but not spin; the corresponding excitation is the holon. There is another site where there is a domain wall between two different Néel sublattices, hence carrying spin $1/2$ but neutral; the corresponding excitation is the spinon (see e.g. Ref. 14). Notice however that attached to the site with a spin mismatch of $1/2$ there is a string of spins flipped w.r.t. the reference state from that site to the holon position, which is integral part of the spinon excitation. It turns out that, due to this "spin string", interchanging two spinon fields one gets a phase factor $e^{\pm i\pi/2}$, hence the spinon is a semion. A semion, in fact, is a particle excitations obeying braid statistics, which can be characterized precisely by the phase factor $e^{\pm i\pi/2}$ acquired by the many-body wave-function or by the product of two fields operators when two semions are exchanged (see e.g. Ref. 15), the sign depending on the relative orientation of the exchange w.r.t. the space. Naively one can describe the holon as a spinless fermion h, but to maintain the fermion statistics of the hole one must add a "charge string" turning it into a semion too. The charge string has an additional effect: to modify the Haldane (exclusion) statistics[17] of the holons assigning to them an exclusion statistics parameter $1/2$. A way of characterizing the exclusion statistics parameter g is that, setting the Fermi energy, the maximal particle number N_g for given g is determined by $N_0 = N_g(1 - g)$, where N_0 is the maximal fermion filling number for $g = 0$, the exclusion statistics parameter for fermions[18]. Therefore for the holon the maximum occupation at fixed momentum is 2 and the Fermi momenta of the spinless semionic holon is the same of the Fermi momenta of a spin $1/2$ fermion. Hence, since the spinon has no chemical potential, the composite hole, product of spinon and holon, satisfies the Luttinger theorem. This was exactly what we were looking for in 2D. Therefore, if we understand how to implement mathematically this picture, this will be a guide for tackling the 2D problem. Although the unphysical mathematical detour in 1D of the next section will be somewhat long, we think it is useful since many ideas there can be explicitly tested by comparison with known results and they will be taken up for the 2D treatment.

4. The spin-charge gauge formalism

Let us sketch how the picture described in the previous section can be implemented mathematically in the path-integral formalism using Chern-Simons gauge fields, in the so-called spin-charge gauge approach; for details see Ref. 19. We start coming back to 2D and basing our theoretical treatment of the t-J model on the following

Theorem 4.1.[20, 21]

We embed the lattice of the 2D t-J model in a 2-dimensional space, denoting by $x = (x^0, x^1, x^2)$ the coordinates of the corresponding 2+1 space-time, x^0 being the euclidean time. We couple the fermions of the t-J model to a $U(1)$ gauge field, B^μ, gauging the global charge symmetry, and to an $SU(2)$ gauge field, V^μ, gauging the global spin symmetry of the model, and we assume that the dynamics of the gauge fields is described by the Chern-Simons actions:

$$S_{c.s.}(B) = -\frac{1}{2\pi} \int d^3x \epsilon_{\mu\nu\rho} B^\mu \partial^\nu B^\rho(x),$$

$$S_{c.s.}(V) = \frac{1}{4\pi} \int d^3x \text{Tr} \epsilon_{\mu\nu\rho} [V^\mu \partial^\nu V^\rho + \frac{2}{3} V^\mu V^\nu V^\rho](x), \qquad (2)$$

where $\epsilon_{\mu\nu\rho}$ is the Levi-Civita anti-symmetric tensor in 3D. Then the spin-charge (or $SU(2) \times U(1)$) gauged model so obtained is exactly equivalent to the original t-J model. In particular the spin and charge invariant correlation functions of the fermions fields $c_{j\alpha}$ of the t-J model are exactly equal to the correlation functions of the fields $\exp[i \int_{\gamma_j} B] P(\exp[i \int_{\gamma_j} V])_{\alpha\beta} c_{j\beta}$, where c denotes now the fermion field of the gauged model, γ_j a string at constant euclidean time connecting the point j to infinity and $P(\cdot)$ the path-ordering, which amounts to the usual time ordering $T(\cdot)$, when "time" is used to parametrize the curve along which one integrates.

(For a careful discussion of boundary conditions and further details, see Ref. 20.)

A good feature of introducing the above gauge fields is that they allow a more flexible treatment of charge and spin responses within a spin-charge decomposition scheme; we have argued above that spin and charge are quite independent in 1D and this turns out to be partially true even in 2D. A comment on the notations: for lattice fields space coordinates are denoted by sub-indices and time coordinate as a variable (e.g. $c_j(x^0)$), but often omitted for simplicity; for continuum fields space-time coordinates are denoted as variables without indices and often space coordinates with the vector symbol (e.g. $x = (x^0, \vec{x})$).

Let us give the key ideas of the proof of the above theorem for the partition function. We represent the partition function of the gauged model in the first-quantized formalism in terms of the worldlines of fermions. After integrating out the gauge fields, due to the chosen coefficients of the Chern-Simon action, the effect of the coupling to $B_\mu(V_\mu)$ is only to give a factor $e^{\pm i\frac{\pi}{2}} (e^{\mp i\frac{\pi}{2}})$ for any single exchange of the fermion worldlines, so the two effects cancel each other exactly. The two sign in the phases correspond to the two possibilities of under-crossing or over-crossing of the worldlines; due to the Gutzwiller projection there are no intersections between the worldlines, so the crossing are well-defined. We just mention that the role of the strings γ, appearing in the representation of the fields in the theorem, is to "close" the worldlines emerging when a fermion is created and annihilated, to guarantee gauge-invariance also in the correlators.

To discuss the t-J model in 1D we simply restrict the lattice to 1D and choose the strings γ parallel to the lattice, whose space direction we label with the index 1, towards $+\infty$ but infinitesimally shifted in the orthogonal direction, labelled by the index 2, to avoid the intersections with the worldlines of the fermions.

We apply to the fermion of the gauged model the spin-charge decomposition discussed in Sec. 3, but now we identify $\exp[i \int_{\gamma_j} B] h_j$ as the holon and $P(\exp[i \int_{\gamma_j} V])_{\alpha\beta} \tilde{s}_{j\beta}$ as the spinon fields. The Chern-Simons coupling automatically ensures that the corresponding field operators obey both semionic braid statistics. As we will see the two strings introduced above are precisely the "dressing" alluded in Sec. 2, reproducing in fact the correct result in 1D.

To understand this result we start noticing that in absence of holons the spin of the Heisenberg model is written in terms of $\tilde{s}_\alpha$ as $\vec{S} = \frac{1}{2}\tilde{s}_\alpha^* \vec{\sigma}_{\alpha\beta} \tilde{s}_\beta$, where $\vec{\sigma}$ denote the Pauli matrices. Therefore, to come closer to the picture discussed in the previous section we first gauge-fix the $SU(2)$ gauge invariance imposing (even in the presence of holons) the requirement that the spins are antiferromagnetically ordered, by setting

$$\tilde{s}_j = \sigma_x^{|j|} \begin{pmatrix} 1 \\ 0 \end{pmatrix}, \tag{3}$$

where $|x| = x^1 + x^2$. (In 1D for j a lattice site we have simply $|j| = j$, but this notation is useful in 2D.)

After the holon-spinon decomposition introduced above the action of the gauged t-J model at doping δ is given by[22]:

$$S(h, h^*, B, V) = \int_0^\beta dx^0 \sum_j h_j^* [\partial_0 - iB_0(j) - \delta] h_j + iB_0(j) + i(1 - \tag{4}$$

$$h_j^* h_j)(\sigma_x^{|j|} V_0(j)\sigma_x^{|j|})_{11} - \sum_{\langle i,j \rangle} t h_j^* \exp[i \int_{\langle i,j \rangle} B] h_i (\sigma_x^{|i|} P(\exp[i \int_{\langle i,j \rangle} V])\sigma_x^{|j|})_{11}]$$

$$+ \frac{J}{2}(1 - h_j^* h_j)(1 - h_i^* h_i)[|(\sigma_x^{|i|} P(\exp[i \int_{\langle i,j \rangle} V])\sigma_x^{|j|})_{11}|^2 - \frac{1}{2}] + S_{c.s.}(B) + S_{c.s.}(V).$$

The two-point correlation function of the hole can be written as:

$$\langle c_{x\alpha}^*(x^0) c_{y\alpha}(y^0) \rangle = \int \mathcal{D}h\mathcal{D}h^*\mathcal{D}B\mathcal{D}V \exp[-S(h, h^*, B, V)] h_x^* \exp[-i \int_{\gamma_x} B](x^0)$$

$$\exp[i \int_{\gamma_y} B] h_y(y^0)(\sigma_x^{|x|} P(\exp[i \int_{\gamma_x} V]))_{1\alpha}(x^0)(P(\exp[i \int_{\gamma_y} V])\sigma_x^{|y|})_{\alpha 1}(y^0)$$

$$[\int \mathcal{D}h\mathcal{D}h^*\mathcal{D}B\mathcal{D}V \exp[-S(h, h^*, B, V)]]^{-1}, \tag{5}$$

where x, y are lattice sites and x^0, y^0 are euclidean times.

Let us discuss the case of 1D. Since the 0-component of B appears linearly in (5), we can safely integrate it out getting for the numerator of (5) the constraint (with $\mu, \nu = 1, 2, z \in \mathbf{R}^3$):

$$\epsilon_{\mu\nu}\partial^\mu B^\nu(z) = \pi\delta(z^2)[\sum_j \delta(z^1 - j)(1 - h_j^* h_j)(z^0)]. \tag{6}$$

By imposing the gauge-fixing $B^2 = 0$ one finally gets

$$B^1(z) = \frac{\pi}{2}\text{sgn}(z^2)[\sum_j \delta(z^1 - j)(1 - h_j^* h_j)(z^0))], \tag{7}$$

so that the holon field turns out to be given by

$$\exp[i \int_{\gamma_j} B]h_j = \exp[i\frac{\pi}{2}\sum_{\ell>j}(1 - h_\ell^* h_\ell)]h_j. \tag{8}$$

If the doping is δ, the fermion c_α of the 1D t-J model in the tight-binding approximation has a Fermi momentum $\pi(1-\delta)/2$ since two fermions with opposite spin can have the same momenta. For the spinless fermion h^* the Fermi momentum would be $\pi(1 - \delta)$ since only one spinless fermion con have a fixed momenta. However, if we consider the phase string attached to the spinless fermion in (8) we see that, since the expectation value of $h_\ell^* h_\ell$ is δ, it contributes to the Fermi momentum a term $-\pi(1 - \delta)/2$, so that the semionic holon of (8) obeys an Haldane statistics of parameter $1/2$. Hence when the electron is reconstructed combining the holon with the spinon, it has the same Fermi surface (in the leading approximation) of the tight-binding approximation for the fermion. With the due changes the same mechanism will be advocated in 2D.

The treatment of the spinon is slightly more involved. Since we have already gauge-fixed the $SU(2)$ gauge-invariance in (3), the gauge field V has to be integrated without gauge-fixing. Then we split the integration over V into an integration over a field $\bar{V}$, satisfying the gauge-fixing condition $\bar{V}^2 = 0$ and its gauge transformations expressed in terms of an $SU(2)$-valued scalar field g, i.e., $V^\mu = g^\dagger \bar{V}^\mu g + g^\dagger \partial^\mu g$. Notice that $P(\exp[i \int_x^y V]) = g_y^\dagger P(\exp[i \int_x^y \bar{V}])g_x$. We now find the configuration of g that optimize the partition function of holons in a fixed, but holon-dependent, background. It is rigorously proved in Ref. 21 that such configuration is given by $g_j^m = \exp[-i\frac{\pi}{2}\sigma_x \sum_{\ell>j} h_\ell^* h_\ell]$. This is precisely the "string" of spin flipping that we have encountered in the preliminary qualitative discussion in the previous section. Finally we set $g = Ug^m$, with $U \in SU(2)$ describing the fluctuations around the optimal configuration. We write

$$U = \begin{pmatrix} s_1 & -s_2^* \\ s_2 & s_1^* \end{pmatrix}, \tag{9}$$

and we will call in the following the s_α again "spinons". One such spinon can be identified with the 1/2-spin local configuration in the previous section. As a "mean field" approximation we neglect the fluctuations U in the calculation of $\bar{V}$. Proceeding now as done for the B field and integrating over $\bar{V}^0$ one finds that $\bar{V}^1$ is diagonal and one of the two components of $P(\exp[i \int \bar{V}])$ acts erasing the pure phase factor of g^m, the other acts doubling this phase factor. One can then show that in this way one can reproduce[21] the correct longwavelength limit of the correlation functions of the 1D t-J.

Let us remark an interesting feature[19] of the above approach: thanks to the holon-depending spin flips, since the holon in a oriented hopping link is at its end, $(\sigma_x^{|i|} P(\exp[i \int_{\langle i,j \rangle} V] \sigma_x^{|j|})_{11}]$ in the t-term of (4) equals $s_{\alpha i}^* s_{\alpha j}$, whereas in the J-term, where there are no holons, it equals $\epsilon_{\alpha\beta} s_{\alpha i} s_{\beta j}$. But under the constraint $s_{\alpha j}^* s_{\alpha j} = 1$ the following identity holds: $|s_{\alpha i}^* s_{\alpha j}|^2 + |\epsilon_{\alpha\beta} s_{\alpha i} s_{\beta j}|^2 = 1$ so that if one optimize the t-term choosing $|s_{\alpha i}^* s_{\alpha j}| = 1$, simultaneously one optimize also the J-term. This optimizing property is somewhat strange, since the same expression in terms of V gives rise to different expressions in terms of the spinons s in the t- and the J-terms. It is intrinsically due to the $SU(2)$ gauge degrees of freedom, absent in the more standard gauge approach of the slave boson formalism, which involves only $U(1)$ gauge degrees of freedom. Let us finally briefly comment on why in the picture of Sec. 3 the position of the spinon, corresponding to a spin kink, is different from that of the holon. The reason is that if we eliminate the holons in the remaining Heisenberg chain there is a gas of spin kink-antikinks that interpolate between two semiclassical realizations of the Néel order, related by parity in 1D. However, since as well known there is no Néel order in 1D, this gas is dense and e.g. an anti-kink of this gas annihilate a spin kink generated by the holon leaving a spin kink arbitrarily far from the holon position. Carrying the know-how gained in 1D we come back now to the physical 2D model.

5. The slave-particle gauge field

Let us try naively to export to 2D the string mechanism discussed in 1D in Sec. 3. We immediately find a big difference: when there is a string of spin flips in a chain the two adjacent chains will have the spin with the same orientation of those of the string and being the spin interaction antiferromagnetic this configuration costs an energy proportional to the length of the string, i.e. holon and spinon connected by a string of spin flips are confined! Here we see the appearance of a new actor that in our discussion in 1D we have overlooked: a gauge force between spinon and holon. Mathematically its origin comes from the spin-charge decomposition of the fermion of the (gauged) t-J model: $c_{j\alpha} = \exp[i \int_{\gamma_j} B] h_j P(\exp[-i \int_{\gamma_j} V])_{\alpha\beta} \tilde{s}_{j\beta}^*$. This decomposition is clearly invariant under the $U(1)$ gauge transformation: $h_j \longrightarrow h_j e^{i\Lambda_j}, \tilde{s}_{j\alpha} \longrightarrow s_{j\alpha} e^{i\Lambda_j}$ (we can also replace $\tilde{s}$ by s), with Λ a real lattice function. We call this emergent symmetry slave-particle (or h/s) gauge symmetry, not to be confused with the charge gauge symmetry whose gauge field is B. At least in the longwavelength continuum limit one can make this symmetry explicit by introducing the related gauge field that we denote by a^μ. This can be done in absence of holons simply by assuming a continuum limit of the spinon field s of the form $s_{x\alpha}(x^0) \longrightarrow s_\alpha(x) + (-1)^{|x|} p_\alpha(x)$, where s and p in the r.h.s. are continuum fields. Integrating out the ferromagnetic component p, as shown in Ref. 23, both in

1D and 2D we obtain a low-energy spinon Lagrangian in the form of a CP^1 model:

$$\mathcal{L}_s = \frac{1}{g}\left[v_s^{-2} |(\partial_0 - ia_0)\, s_\alpha|^2 + \left|\left(\vec{\nabla} - i\vec{a}\right) s_\alpha\right|^2 \right], \tag{10}$$

where $g \sim \epsilon^{D-1}J^{-1}$ and $v_s \sim J\epsilon$ is the spinon velocity, with ϵ the lattice spacing, in the following set to 1. Furthermore $a^\mu(x) = s_\alpha^*(x)\partial^\mu s_\alpha(x)$, with μ a space-time index, and the implicit constraint $s_\alpha^* s_\alpha = 1$ is understood. However, in 1D there is an additional $\Theta = \pi$ term

$$\mathcal{L}_\Theta = i\frac{\Theta}{2\pi}\epsilon_{\mu\nu}\partial^\mu a^\nu. \tag{11}$$

(A hedhodge gas discussed in Ref. 23 would appear in 2D instead of the Θ-term, but it disappears when we add holons and therefore here will not be considered.) Here we see at a deeper level the difference between 2D and 1D discussed at the beginning of this section: precisely at the value $\Theta = \pi$ in 1D there are two degenerate ground states connected by a parity transformation (see e.g. Ref. 25) as in the Heisenberg chain. Kinks interpolating between the two destroy the Coulomb attraction generated by a^μ, which is massless since the SU(2) symmetry is unbroken, and the spinons are deconfined. The absence of this Θ-term in 2D makes the slave-particle gauge field confining in the phase with unbroken symmetry[24] and as a result the spinon s and the anti-spinon s^* are bound together into a spin 1 spin-wave. Also in the broken-symmetry antiferromagnetic phase in 2D the excitations are spin waves, but now they appear as Goldstone bosons. When we introduce holons, as we will see, we go from long-range to short-range AF in the CP^1 model, and in this approach the main actors are the antiferromagnetic spin-vortices, at last! Confinement is destroyed by the vacuum polarization of holons, but there remains a short-range attraction between holon and spinon and between spinon and anti-spinon mediated by a which has no analogue in 1D. In fact, due to the finite density of holons, the propagator of the transverse component of the slave-particle gauge field $a_\perp$, absent in 1D, is of the form

$$\langle a_\perp a_\perp\rangle(\omega, \vec{k}) \sim (-\chi|\vec{k}|^2 + i\kappa\frac{\omega}{|\vec{k}|})^{-1}, \tag{12}$$

for $\omega, |\vec{k}|, \omega/|\vec{k}| \sim 0$, where χ is the diamagnetic susceptibility and κ the Landau damping. By plugging as typical energy scale T we see that there is a typical momentum scale, the so-called Reizer[26] momentum, $Q(T) = (\kappa T/\chi)^{1/3}$. Below this momentum scale, or alternatively beyond the inverse momentum scale in space, there is an effective attraction mediated by $a_\perp$ between opposite charges relative to the slave-particle $U(1)$ group. Since we are in 2D we expect that an attractive interaction binds together excitations with opposite charges. In fact a weak binding is precisely the result obtained in 2D within some approximations in Refs. 27, 28 for the hole as composite of spinon and holon and for the magnon as a composite of spinon and anti-spinon. They exhibit a "small" resonance life-time, T-dependent, originated from the slave-particle gauge coupling, leading to a behaviour of these excitations less coherent than in a standard Fermi-liquid. For some considerations supporting a gauge-induced composite structure of the hole, see Ref. 29.

6. The charge- and antiferromagnetic spin-vortices

If we accept the suggestion coming from the 1D model for the statistics of holon and spinon, one should search for a semionic representation of the hole field also in 2D. However in 2D the typical topological excitation is the vortex, not the kink, characteristic of 1D. Therefore the analogy with 1D suggests the appearance of a charge-vortex attached to the holon and a spin-vortex attached to the spinon and this is indeed what happens. These vortices are somewhat analogous to those introduced by Laughlin in the FQHE and in fact a semionic representation of the hole was advocated by him[30] quite soon after the discovery of high T_c.

We now make these ideas precise in the spin-charge gauge formalism. Let us come back to the action of the 2D gauged t-J model in the form of (4). As in 1D one can integrate out B^0; we obtain the constraint (with $\mu, \nu = 1, 2, z \in \mathbf{R}^3$):

$$\epsilon_{\mu\nu}\partial^\mu B^\nu(z) = \pi[\sum_j (1 - h_j^* h_j)(z^0)\delta^{(2)}(\vec{z} - j)]. \tag{13}$$

Imposing the Coulomb gauge-fixing $\partial_\mu B^\mu = 0$ one gets $B^\mu(z) = \bar{B}^\mu + b^\mu(z)$ where $\bar{B}^\mu$ introduces a π-flux phase, i.e. $\exp[\int_{\partial p} \bar{B}] = -1$ for every plaquette p and

$$b^\mu(z) = -\frac{1}{2}[\sum_j \partial^\mu \arg(\vec{z} - j)(h_j^* h_j)(z^0))]. \tag{14}$$

We can recognize $\partial^\mu \arg(\vec{z} - j)$ as the vector potential of a vortex centered on the holon position j, i.e. centered on an empty site of the t-J model. We already see here that strings in 1D are replaced in 2D by vortices. The vortices in (14) appear in the charge $U(1)$ group but their spin companion will be finally the antiferromagnetic vortices we are looking for the pairing. Following an argument given in Ref. 31 we conjecture that also in 2D, due to the π-flux, the holon field $\exp[i \int_{\gamma_j} b]h_j$ obeys an exclusion statistics with parameter $1/2$. A complete proof of this conjecture is still lacking but is presently under investigation and preliminary results are encouraging. In the following we take this conjecture for granted.

We turn to the spin degrees of freedom and proceed as in 1D: we rewrite $V^\mu = g^\dagger \bar{V}^\mu g + g^\dagger \partial^\mu g, \mu = 0, 1, 2, g \in SU(2)$ with $\bar{V}$ satisfying the Coulomb condition $\partial_\nu \bar{V}^\nu = 0, \nu = 1, 2$. Then ideally one would search a configuration of g, depending on the holon configuration, optimizing the holon-partition function in that g background, and around that configuration, g^m, one would consider spinon fluctuations. We didn't succeeded to rigorously found such configuration as in 1D, but we still found a configuration optimal "on average" by expanding the holon partition function in terms of holon worldlines in the first-quantization formalism as done in in one-dimension. Looking at (4) we see that on links appearing in the J-term, thus not containing holons, the optimization requires $|(\sigma_x^{|i|} P(\exp[i \int_{(i,j)} V])\sigma_x^{|j|})_{11}| = 0$.

In the links of the hopping term the holons are coupled to the complex abelian gauge field:

$$X_{\langle i,j \rangle} = \exp[i \int_{\langle i,j \rangle} B][(\sigma_x^{|i|} P(\exp[i \int_{\langle i,j \rangle} V]) \sigma_x^{|j|})_{11}]. \tag{15}$$

In Ref. 22 it is shown that the optimal modulus of X is 1 and Lieb[32] has proved rigorously that at half-filling ($\delta = 0$) the optimal configuration for an abelian gauge field on a square lattice in 2D has a flux π per plaquette at arbitrary temperature. On the basis also of the results of Refs. 33 and of a numerical simulation (L. Qin, unpublished) we conjecture that at least on average the optimal flux per plaquette is $\pi(1 - \delta)$ for sufficiently small doping and temperature. On the other hand, it is well known that at high doping or temperature a vanishing flux per plaquette is favored. Therefore we expect a crossover corresponding to the melting of the π-flux lattice and we propose to identify such crossover, which we denote by T^*, with the experimental "low-pseudogap" crossover in the cuprates on the basis of comparison between theoretical curves and experimental data, in particular of in-plane resistivity as discussed in Sec. 10. Consistently with the terminology used in Sec. 2 for the phenomenology of the cuprates, we call PG also for the t-J model in our approach the region below the $\pi \to 0$ crossover and we call SM the region above. We see from (6) that the term $\exp[i \int_{\langle i,j \rangle} B]$ has in average a flux $\pi(1 - \delta)$ per plaquette, hence in PG on the links derived from the hopping term we impose $(\sigma_x^{|i|} g_j^\dagger P(\exp[i \int_{\langle i,j \rangle} \bar{V}]) g_i \sigma_x^{|j|})_{11} = 1$. We find that g^m involves a spin flip at the holon position as in 1D. Hence writing $g = U g^m$ with U as in (9), the way in which the s spinons appear in the t and J terms is the same seen in 1D. Therefore the considerations on the simultaneous optimization both of the t- and the J-term made at the end of sect. 4 apply, in approximate form, also to 2D and this motivates "a priori" this spin-charge gauge approach. Recalling that in the continuum limit the lattice field s generates its continuum version as discussed in the previous section, one realizes that $s_{\alpha i}^* s_{\alpha j}$ appearing in the t-term of (4) generates in the continuum limit a coupling of the holon with the slave-particle gauge field $a^\mu \approx s_\alpha^* \partial^\mu s_\alpha$. In SM the optimal configuration g^m contains also a phase factor deleting the contribution of $\bar{B}$ in the loops of hopping links of the holons, in order to get effectively an approximately 0 flux. Forgetting at first the spinon fluctuations and assuming an exclusion Haldane statistics with parameter $1/2$ for the holon, in SM one then recover a "large FS" roughly consistent with band calculations[34]. The effect of the field $\bar{B}$ is then to introduce an unbalanced π-flux in PG. As discussed in the slave-boson approach in Ref. 35, through Hofstadter mechanism the π-flux converts the holon of SM with dispersion, in this approximation, $\omega_h \sim 2t[(\cos k_x + \cos k_y) - \delta]$ into a pair of "Dirac fields", with pseudospin index corresponding to the two Néel sublattices and with dispersion: $\omega_h \sim 2t[\sqrt{\cos^2 k_x + \cos^2 k_y} - \delta]$ restricted to the magnetic Brillouin zone (BZ). One thus obtain two "small FS" centred at $(\pm\frac{\pi}{2}, \pm\frac{\pi}{2})$ with Fermi momenta $k_F \sim \delta$. When the holon is combined with the spinon to reconstruct the hole with dispersion in the full BZ, the Dirac structure of the holon suppresses the spectral weight outside of the magnetic BZ[28]. Neglecting, as in 1D, the spinon fluctuations U in the computation of $\bar{V}$, one gets (with $\mu = 1, 2$):

$$\bar{V}^\mu(z) = -\frac{1}{2}\sum_j (-1)^{|j|}\partial^\mu \arg(\vec{z} - j)h_j^* h_j(z^0)\sigma_z. \tag{16}$$

We recognize in the term $(-1)^{|j|}\partial^\mu \arg(\vec{z} - j)$ the vector potential of a vortex centered on the holon position j, with opposite vorticity (or chirality) for the center in opposite Néel sublattices. These are finally the antiferromagnetic spin vortices alluded in the beginning of the paper. As one see they are along the spin direction, z, of the magnetization; they are the topological excitations of the $U(1)$ subgroup of the $SU(2)$ spin group unbroken in the antiferromagnetic phase. These vortices are of purely quantum origin, since of Aharonov-Bohm type, inducing a topological effect far away from the position of the holon itself, where their classically visible field strength is supported. Hence in this approach the empty sites of the 2D t-J model, mimicking the Zhang-Rice singlets and corresponding to the holon locations, are the cores of spin vortices, a quantum distortion of the antiferromagnetic spin background, recording in their vorticity the Néel structure of the lattice. Therefore they are still a peculiar manifestation of the antiferromagnetic interaction, like the more standard antiferromagnetic spin waves. We now show that these vortices are responsible both for short-range AF and for charge-pairing.

7. Short-range AF and charge pairing

The effect of the spin vortices appeared in the last section is twofold: acting as impurities on the gapless spin waves of the Heisenberg model describing undoped cuprates they "localize" them, more precisely they make them gapped. Furthermore being a 2D gas of vortices with both chiralities (corresponding to center in the two Néel sublattices) they undergo a Berezinski-Kosterlitz-Thouless (BKT) transition. Since they are centered on holons the KT attractive pairing induces a charge pairing.

Mathematically the presence of the antiferromagnetic spin vortices is recorded in the field $\bar{V}$ of (16). In the low-energy continuum limit one can see from (4) that in the t-term of the action $\bar{V}$ appears linearly; since its spatial average vanishes in a "mean field" treatment we ignore it. In the J-term instead it appears quadratically and assuming self-consistently that for sufficiently large δ the full $SU(2)$ symmetry is restored at large scales, treating the spinons in "mean field" one finds an interaction between vortices and spinons proportional to:

$$\int d^3x (\bar{V}^\mu \bar{V}_\mu)(x)s_\alpha^* s_\alpha(x). \tag{17}$$

A quenched average, $\langle \cdot \rangle$, over the positions of the center of spin-vortices yields the following estimate[22]: $\langle \bar{V}^\mu \bar{V}_\mu \rangle \approx \delta |\log \delta|$. Thus the term (17) provides a mass-gap to the spinons. Leaving aside for the moment the monomial quartic in the holons of the J-term in (4), we treat in mean field the monomial quadratic in the holons, so that the antiferromagnetic coupling is renormalized to $J(1 - 2\delta)$. We see here the effect of strong reduction of antiferromagnetism due to the increase of the density of empty sites, corresponding to Zhang-Rice singlets in the cuprates. As a result one finds that the continuum limit of the term

$$\int dx^0 \sum_j i(\sigma_x^{|j|} V_0(x^0, j) \sigma_x^{|j|})_{11}$$

$$+ \sum_{\langle i,j \rangle} \frac{J}{2}(1 - h_j^* h_j - h_i^* h_i)[|(\sigma_x^{|i|} P(\exp[i \int_{\langle i,j \rangle} V]) \sigma_x^{|j|})_{11}|^2 - \frac{1}{2}](x^0) \quad (18)$$

within the above approximations is given by

$$\int d^3x (1 - 2\delta) \frac{1}{g} \left[v_s^{-2} |(\partial_0 - i a_0) s_\alpha|^2 + \left| \left(\vec{\nabla} - i\vec{a} \right) s_\alpha \right|^2 + m_s^2 s_\alpha^* s_\alpha \right](x), \quad (19)$$

where $m_s \approx \sqrt{\delta |\log \delta|}$. Hence the gapless spinons s forming the spin waves of the CP^1 (or equivalently $O(3)$) model describing the undoped system, traveling in a gas of spin vortices centered on holons acquire a gap $\approx J(1 - 2\delta)\sqrt{\delta |\log \delta|}$, converting the long-range AF of the undoped model in the short-range AF when doping exceeds a critical value. This is selfconsistent with the previous assumption of $SU(2)$ symmetry restoration at large scales. We thus see that in this approach the transition corresponding to the onset of short-range AF is indeed due to the antiferromagnetic spin vortices. We now show that the same term (17) describing the interaction of spinons with spin vortices generates also the charge pairing, by treating in mean-field $s_\alpha^* s_\alpha(x)$ instead of $\bar{V}^\mu \bar{V}_\mu(x)$. This averaging produces the term

$$< s_\alpha^* s_\alpha > \sum_{i,j} (-1)^{|i|+|j|} \Delta^{-1}(i - j) h_i^* h_i h_j^* h_j, \quad (20)$$

where Δ is the 2D Laplacian. In the static approximation for holons (20) describes a 2D lattice Coulomb gas with charges ± 1 depending on the Néel sublattice. In particular the interaction is attractive between holons in opposite Néel sublattices, with maximal strength for nearest neighbour sites, along the lattice directions with a d-wave symmetry. Putting back the coefficients one finds that the coupling constant of this interaction is $J_{eff} = J(1 - 2\delta) < s_\alpha^* s_\alpha >$, thus decreasing with doping. For 2D Coulomb gases with the above parameters, pairing appears below a temperature $T_{ph} \sim J_{eff}$ which turns out to be inside the SM "phase". Comparing the effect of this pairing on the spectral weight of the hole with ARPES data we propose to identify this crossover temperature with the experimental "high" pseudogap in the cuprates, discussed in Sec. 2. The formation of holon pairs, in fact, induces a reduction of the spectral weight of the hole, starting from the antinodal region, which will be discussed in Sec. 9. Inserting the nnn hopping term, T_{ph} strongly depends on its coefficient t', contrary to the "low" pseudogap T^*. To identify more precisely the crossover temperature T_{ph} we use a continuum approximation for the effective interaction, valid in the large-scale limit, taking into account the

Coulomb screening effect, with a screening length $\ell \sim 1/\sqrt{k_F}$ in the Thomas-Fermi approximation, where k_F is the average fermi momenta of the FS of the holon. Then we treat the resulting Coulomb-screened attractive interaction in the BCS approximation obtaining a d-wave order parameter for the holons[36][37]. The d-wave nature of the order parameter is a natural consequence of the D_2 (or $Z_2 \times Z_2$) symmetry of the FS of holons due to the "charge" π flux and the structure of the holon pairing (20), both arising from AF. The modulus of the order parameter, Δ^h, on the FS turn out to be $\sim J_{eff}\sqrt{k_F}\exp[-\text{const.}\ t/J_{eff}]$. As expected the holon-pair order parameter decreases with doping and its scale is proportional to J and not to t, as natural due to its magnetic origin, but reduced by the effect of the empty sites. In the next section we show that this charge pairing, due to the antiferromagnetic spin vortices, indeed leads to superconductivity.

8. Spin pairing and superconductivity

The charge pairing alone does not yet lead to hole-pairing, since the spins are still unpaired. It is the slave-particle gauge attraction between holon and spinon that induces the formation of short-range spin-singlet (RVB) spinon pairs at a lower temperature T_{ps}, in a sense, using the holon-pairs as sources of attraction. More in detail: the spinon gap is due to the presence of a gas of unpaired spin vortices and when the charge-pairing occurs the tight vortex-antivortex pairs don't contribute anymore to the spinon gap. Hence charge-pairing leads to a lowering of the spinon energy proportional to the density of spinon pairs. Further, the monomial quartic in the holons of the J term of the action in (4) which involves the RVB spinon-pair field is repulsive and its condensation energy is negative and decreased by the formation of spinon pairs. However fortunately the lowering of the spinon gap implies an enhancement of the contribution of the vacuum energy of the slave-particle gauge field. The competition between these two effects finally produces a saddle point with finite density of RVB spinon-pairs. From the combined charge- and spin-pairs we get a gas of incoherent spin-singlet hole-pairs. Finally, at a even lower temperature, the superconducting transition temperature T_c, the hole pairs become coherent and a d-wave hole condensate (in BCS approximation) appears, leading to superconductivity. The appearance of two temperatures, one for pair formation and a lower one for pair condensation, is typical of a BEC-BCS crossover regime for a fermion system with attractive interaction[38]. In this sense the incoherent hole pairs discussed above play a role analogous to that of the "preformed pairs" considered e.g. in Refs. 39. However, due to the composite spin-charge structure of the hole we have here two distinct temperatures: for charge-pair formation, T_{ph}, and for spin-pair formation, T_{ps}. Let us look how the above ideas are implemented mathematically.

As remarked above in this approach a key ingredient of the spinon pairing is the monomial quartic in the holons of the J term in (4) that we have so far ignored. Rewritten in terms of the spinon fluctuations s and neglecting the subleading contribution of $\bar{V}$ this Heisenberg term is given by: $\frac{J}{2}h_j^*h_jh_i^*h_i|\epsilon_{\alpha\beta}s_{\alpha i}s_{\beta j}|^2$. This interaction term is clearly positive, hence repulsive, due to the semionic mean-field approach, contrary to the similar term in the slave-boson approach. It can be reasonably neglected in the normal state, since in mean field it is of $O(\delta^2)$, but as soon as holon-pairing becomes relevant the quartic holon monomial may become in mean-field $O(\delta)$ and it cannot be neglected anymore. To investigate its effect we apply a Hubbard-Stratonovic (HS) transformation and go to the continuum limit. Denote in this limit the HS field by $\Delta_s^\mu, \mu = 1, 2$ and treat the holon pairs in mean field. Then the additional term that one obtains to add to the spinon lagrangian in (19) takes the form $-2|\Delta_s(x)^\mu|^2/(J|\langle h_jh_i\rangle|^2)+\Delta_s(x)^\mu\epsilon_{\alpha\beta}s_\alpha\partial_\mu s_\beta(x)+h.c.$ Treating the HS field in mean-field one finds[36] that a non-vanishing expectation value, Δ_s, of $\Delta_s(x)$ modifies the positive branch of the spinon dispersion, $\omega_s(\vec{k}) = \sqrt{m_s^2 + \vec{k}^2}$, replacing it with two branches:

$$\omega_{s\pm}(\vec{k}) = \sqrt{(m_s^2 - |\Delta_s|^2) + (|\vec{k}| \pm |\Delta_s|)^2}. \tag{21}$$

This doubling arises because for a finite density of spinon pairs there are two (positive energy) excitations, with different energies, but the same spin and momenta. They are given, e.g., by creating a spinon up in the presence of the pairs and by destructing a spinon down in one of the RVB spin-pairs. The lower branch exhibits a minimum with an energy lower than m_s, analogous to the one appearing in a plasma of relativistic fermions[40]; it implies a backflow of the gas of spinon-pairs dressing the "bare" spinon. Hence RVB pairing lowers the spinon kinetic energy. However, to get a non-vanishing Δ_s in mean-field we needs to compete with the spinon repulsion generated by the Heisenberg term discussed above; after the HS transformation the corresponding relevant lagrangian term is $\sim -|\Delta_s|^2/|\langle h_jh_i\rangle|^2$. For this competition the slave-particle gauge field is the relevant actor. In fact, since the spinon is gapped in absence of RVB pairing its contribution to the gauge effective action is a Maxwell term of the form $(\partial^\mu a^\nu - \partial^\nu a^\mu)^2/m_s$. When spinon pairing occurs we see from (21) that the mass is reduced to $M_s \approx m_s - |\Delta_s|^2/m_s$ (for small $|\Delta_s|/m_s$), hence integrating over the gauge field we get a lagrangian contribution $\sim |\Delta_s|^2/m_s^2$ from the ground state variation. If it overcomes the previous term one finds a non-vanishing saddle point in Δ_s. In the underdoped region this occurs only beyond a critical concentration, since $|\langle h_jh_i\rangle|^2 \sim O(\delta)$ but $m_s^2 \sim O(\delta|\log \delta|)$. More precisely, the resulting "gap equation" has the form:

$$\frac{3}{2m_s} - \frac{1}{||\langle h_ih_j\rangle|^2} = \frac{1}{2|\Delta_s|V}\sum_{\vec{k}}[\frac{k}{\omega_{s-}\tanh\frac{\omega_{s-}}{2T}} - \frac{k}{\omega_{s+}\tanh\frac{\omega_{s+}}{2T}}]. \tag{22}$$

As soon as the RVB pairs in mean-field condense at T_{ps} the slave-particle global gauge symmetry is broken from $U(1)$ to Z_2 in mean-field. The Anderson-Higgs mechanism then implies a gap for the gauge field a^μ. However if we reinsert the phase fluctuations of the holon-pair and spinon-pair order parameters, the true superconducting transition is shifted below T_{ps}, which remains just as a crossover.

In fact below T_{ps} the low-energy effective action obtained integrating out spinons is a Maxwell-gauged 3D XY model, where the angle-field ϕ of the XY model is the "square root" of the phase of the long-wave limit of the hole-pair field and the gauge field is the slave-particle gauge field (suitably dressed). The holon contribution is subleading and will be ignored here. The euclidean effective action is then given by

$$S(\phi, a_\mu) \approx \frac{1}{6\pi M_s} \int d^3x [(\partial_\mu a_\nu - \partial_\nu a_\mu)^2 + 2|\Delta_s|^2(\partial_0 \phi - a_0)^2 + |\Delta_s|^2(\vec{\nabla}\phi - \vec{a})^2] \tag{23}$$

where, since in the range considered $v_s/T \sim J/T \ll 1$, we extended the integration to the full Euclidean space-time $\mathbf{R}^3$, but retaining the temperature dependence of the coefficients in the action. The gauged XY model has two phases: Coulomb and Higgs. If the coefficient, $\sim |\Delta_s|^2/M_s$, of the Anderson-Higgs mass term for a is sufficiently small, the phase field ϕ fluctuates so strongly that actually no mass gap is generated. This is the Coulomb phase, where there is a plasma of standard slave-particle "magnetic" (not spin!) vortices-antivortices. In the presence of a temperature gradient a perpendicular external magnetic field induces an imbalance between vortices and antivortices, giving rise to a Nernst signal, even if the hole pairs are not condensed yet. Therefore we conjecture that T_{ps} corresponds in the phase diagram of cuprates to the onset of non-superconducting diamagnetic and vortex Nernst signals. Since it is dominated by spinons physics, this crossover is essentially independent of details of the Fermi surface, hence material-independent. For a sufficiently large coefficient of the mass term for a the gauged XY model is in the broken-symmetry phase: the fluctuations of ϕ are exponentially suppressed and we have a quasi-condensation (power-law decaying order parameter) at $T > 0$; accordingly the "magnetic" vortex-antivortex pairs become dilute, so the slave-particle gauge field is gapped and hence suppressed in the long wavelength limit. At the same time the holon, and hence the hole, acquires the gap outside the nodes of the d-wave order parameter; this is the superconducting phase. Therefore, the superconducting transition is almost of classical 3D XY type, driven by condensation of "magnetic" vortices and related to phase coherence as in BEC systems, in spite of the BCS-like charge-pairing discussed above.

9. Phase fluctuations of charge-pairs and Fermi arcs

When we reinsert the phase fluctuations of the charge-pair order parameter we are going beyond the BCS approximation. The main effect is to reproduce between T_{ph} and T_c the phenomenology of Fermi arcs coexisting with gap in the antinodal region discussed in Sec. 2. Indeed when a gas of incoherent holon pairs is present, the scale of the inverse correlation length ("mass" m_{ph}) of the quanta of the phase of the holon-pair field separates self-consistently low energy modes with a Fermi liquid behavior from high energy modes with a d-wave superconducting behavior. More precisely an approximate evaluation of self-energy correction for the holon, Σ, due to these phase fluctuations in the longwavelength continuum limit gives[37]:

$$\Sigma(\omega, \vec{k}) = \frac{|\Delta_h(\vec{k})|^2}{(i\omega + \omega_h(\vec{k}))}[1 - \frac{m_{ph}}{\sqrt{\omega^2 + \omega_h(\vec{k})^2 + m_{ph}^2}}], \tag{24}$$

where $\Delta_h(\vec{k})$ is the d-wave holon-pair order parameter and $\omega_h(\vec{k})$ the holon dispersion. Clearly if $m_{ph} = 0$ one goes back to the BCS approximation and in the opposite limit $m_{ph} \to \infty$ the correction vanishes. The value of m_{ph} decreases with T, thus lowering T one finds a gradual reduction of the spectral weight on the FS at small frequency as we move away from the diagonals of the Brillouin zone, due to the d-wave structure of $\Delta_h(\vec{k})$. Simultaneously at larger frequencies we have the formation and increase of two peaks of intensity precursors of the excitations in the superconducting phase. The sketched mechanism of spectral weight suppression on the FS exhibits the fingerprint of the presence of the slave-particle gauge field, because the smooth interpolation between Fermi liquid and superconducting behaviour discussed above is actually due to the interaction of the phase of the holon pairs with the gauge field[37]. Without the gauge field the superconducting-like peaks in the spectral weight in the normal phase are almost absent, as in BCS. The physical hole is obtained as a holon-spinon bound state produced by the gauge attraction and it inherits the above holon features, but with a strongly enhanced scattering rate, due to the spinon contribution. Notice that in PG the suppression of the spectral weight discussed above sums up with the antinodal gap produced by the "charge" π flux, leaving only small isolated segments of FS for the hole. This seems in agreement with recent experiments[41]. Below T_{ps} beyond mean-field we have to add the phase fluctuations also of the spin-pairs which combine with phase fluctuations of the charge-pairs to give rise to the hole-phase field ϕ introduced in the previous section.

10. Brief comparison with experiments

In this final section we compare some experimental data with theoretical computations performed within the above sketched approach to the t-(t')-J model, with some experimental inputs for one doping to fix parameters that are then used consistently in all the calculations. So one may verify that many experimental features of the low-energy physics of hole-doped cuprates can find a natural explanation within the formalism presented here.

– Phase diagram. In Fig. 1 we present the δ-T phase diagram theoretically derived[45] compared with some experimental data. We find that the general structure of the experimental crossovers is well reproduced, with even a semi quantitative agreement for T^* with the "low pseudogap" and T_{ps} with the boundary of "Nernst phase". The only crossover that is completely missed is the onset of charge-density waves around $\delta \approx 1/8$; we conjecture that this is due to a contribution of the oxygen orbitals not taken into account by the t-(t')-J model.

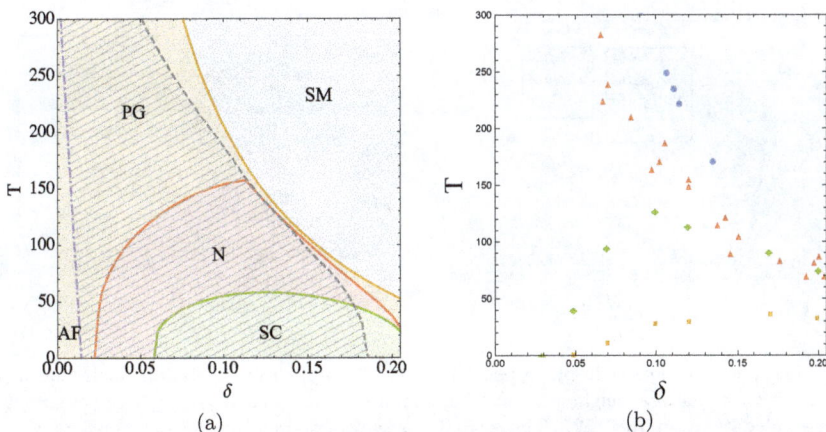

Fig. 1. (a) Theoretically derived phase diagram from Ref. 45: holon pairing temperature T_{ph} (yellow line) determined from the BCS gap equation for holons, spinon pairing temperature T_{ps} (red line) determined from the spinon gap equation, crossover PG-SM (dashed line) determined above T_{ps} from the inflection point of resistivity and below T_{ps} from matching the contour lines of the values of the spinon order parameter, T_c (green line) determined from the transition temperature of the XY model of spinons. The Nèel temperature (dot-dashed line) is qualitative from experiments not derived theoretically. (b) Experimental data for T_c (yellow squares) and onset of Nernst signal (green diamonds) in LSCO from Ref. 16, "low pseudogap" (red triangles) in LSCO from Ref. 8 and "high pseudogap" (blue circles) in YBCO from Ref. 9.

– In-plane resistivity $\rho_\parallel$. Since the hole is composite, due to the slave-particle gauge "string" between holon and spinon the in-plane resistivity is dominated by the slower of the two excitations (Ioffe-Larkin rule[42]). The antiferromagnetic spin vortices generate a spinon gap, therefore the resistivity is dominated by the spinons. In an eikonal treatment the interaction of the gauge field a in PG originates a saddle point producing a scattering rate for the spinon, Γ, behaving like $\Im\sqrt{m_s^2 + icT/k_F}$ with k_F the Fermi momentum of the small holon FS[28]. This in turn produces an inflection point of the resistivity at $T^* \sim m_s^2/k_F \sim |\log \delta|$. At a lower temperature, T_{MIC}, one finds also a metal-insulator crossover, intrinsic in this approach and not due to disorder[43]. It occurs when the insulating behaviour due to the spinon gap dominates, the time scale being set by spinon diffusion increasing with T, over the dissipative behaviour due to the gauge fluctuations. Furthermore, due to spinon dominance, the normalized resistivity defined by

$$\rho_{\parallel n}(T) = \frac{\rho_\parallel(T) - \rho_\parallel(T_{MIC})}{\rho_\parallel(T^*) - \rho_\parallel(T_{MIC})} \tag{25}$$

exhibits an (approximate) universal behaviour[28][45] when expressed as a function of the normalized temperature T/T^*, in agreement with experiments as discussed in Refs. 44. A comparison between experiments and theory is shown in Fig. 2(a). T^* appears also as an inflection point in the spin-lattice relaxation rate of the Cu $^63(1/T_1T)$ (see e.g. Ref. 46) in PG and is reproduced in the above approach, see

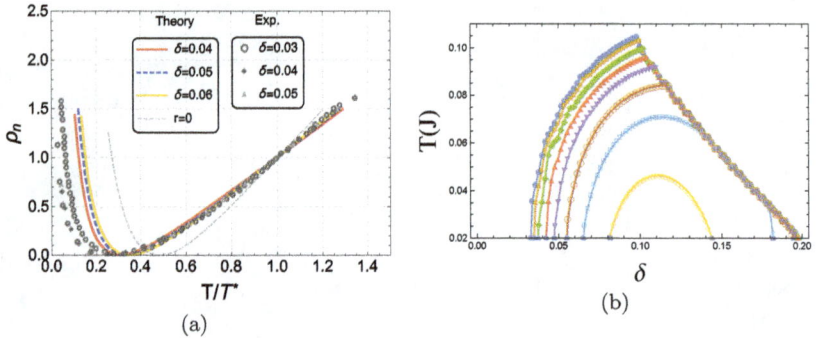

Fig. 2. (a) The normalized in-plane resistivity $\rho_{\|n}$ theoretically calculated, including a holon contribution with relative weight r w.r.t. the spinon contribution, from Ref. 45, compared with experimental data from Ref. 52. The $r = 0$ curve corresponds to the universal, pure spinon contribution [28]. The discrepancy at low T might be due to the missing account of holon and spinon pair formation in the above treatment.(b) Equi-level lines of Δ_s from 0.19 to 0.26 in units of J with T in units of J; compare with experimental data in Ref. 16.

Ref. 27. In SM there is no saddle point in Γ and one recovers [34] the experimental resistivity linear in T and the almost constant $^6 3(1/T_1)$.

– Nernst signal. One expects that the amplitude of the spin-pair order parameter Δ_s is roughly proportional to the intensity of the Nernst signal and a comparison of the equi-level curves for Δ_s obtained solving the gap equation (22) reasonably well compare with the intensity of the Nernst signal [16], as shown in Fig. 2(b).

– Superfluid density $\rho^{(s)}$. For the same reason given for in-plane resistivity one finds that the superfluid density satisfies Ioffe-Larkin rule and in the underdoped region it is dominated by the spinons. From (23), neglecting a^μ that is suppressed in the superconducting phase, one obtains a 3DXY model with effective temperature $\Theta(T) \sim M_s(T)/|\Delta_s|^2(T)$. Hence the superfluid density exhibits a 3D XY behaviour, with critical exponent 2/3, so definitely non-BCS, as already advocated for the experimental data in Ref. 48. Furthermore $\rho^{(s)}$ normalized as

$$\rho_n^{(s)}(T/T_c) = \frac{\rho^{(s)}(T/T_c)}{\rho^{(s)}(T=0)} \qquad (26)$$

shows an universal behaviour [50], shown in Fig. 3(a), agreeing with this phenomenology noticed for a variety of materials in Ref. 49. Because T_c is determined by the XY transition for the effective temperature $\Theta(T)$ and $\Theta(0) = 0$, expanding to the first order in T one finds $T_c^{XY} = \Theta(T_c) \approx (d\Theta/dT)(0)T_c$, where T_c^{XY} denotes the critical temperature of the XY model. But from the theory of the XY model we know that $\rho^{(s)}(0) \approx [(d\Theta/dT)(0)]^{-1}$, hence we derive an approximate Uemura relation [51] $\rho^{(s)}(0) \sim T_c$.

– Spectral weight. The dependence on the FS angle of the calculated symmetrized spectral weight of the hole is shown in Fig. 3(b), it exhibits the Fermi arc phenomenology discussed in Sec. 9 and it is compared with experimental data.

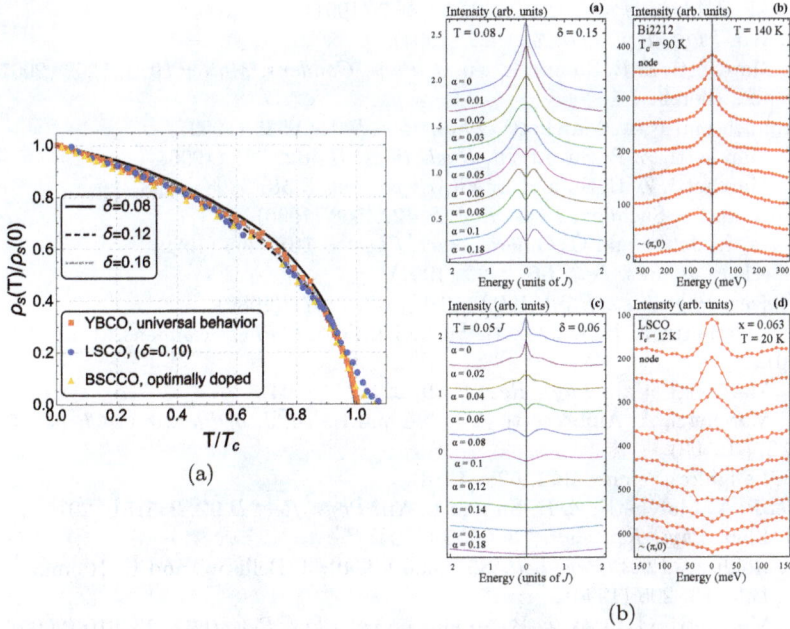

Fig. 3. (a) The normalized superfluid density $\rho_n^{(s)}$ from Ref. 45, compared with experimental data for underdoped (LSCO, YBCO) and optimally doped (BSCCO, YBCO) samples, from Refs. 49 53. (b) Panel (a, c) shows the symmetrized spectral weight of the hole as a function of the FS angle α from the nodal direction in SM and in PG, respectively, from Ref. 37. Panel (b) shows the experimental data for Bi2212 in Ref. 47 and panel (d) for LSCO in Ref. 54.

References

1. J. G. Bednorz and K. A. Müller, *Z. Phys. B* **64**, 189 (1986).
2. D. Manske, *Theory of Unconventional Superconductors. Cooper-Pairing Mediated by Spin Excitations*, Springer Tracts in Modern Physics 202 (Springer-Verlag Berlin Heidelberg 2004) and references therein.
3. S. Uchida, *High Temperature Superconductivity. The Road to Higher Critical Temperature*, Springer Series in Materials Science 213 (Springer Japan 2015).
4. J. M. Bok et al., *Sci. Adv.* **2**, 1501329 (2016).
5. F. C. Zhang and T. M. Rice, *Phys. Rev. B* **37**, 3759(R) (1988).
6. N. Hussey, *Nat. Phys.* **12**, 290 (2016).
7. T. Yoshida et al., *J. Phys. Soc. Jpn.* **81**, 011006 (2012).
8. T. Honma et al., *Phys. Rev. B* **70**, 214517 (2004).
9. N. Barisic et al., *PNAS* **110**, 12235 (2013).
10. L. Li et al., *Phys. Rev. B* **81**, 054510 (2010).
11. S. Badoux et al., *Nature* **531**, 210 (2016).
12. G. Baskaran, Z. Zou and P. W. Anderson, *Solid State Commun.* **63**, 973 (1987).
13. S. A. Kivelson, D. S. Rokhsar and J. P. Sethna, *Phys. Rev. B* **35**, 8865 (1987).
14. T. Giamarchi, *Quantum Physics in One Dimension* (Clarendon Press Oxford 2003).
15. F. Wilczek, *Fractional Statistics and Anyon Superconductivity* (World Scientific Singapore 1990).
16. Y. Wang, L. Li and N. P. Ong, *Phys. Rev. B* **73**, 024510 (2006); L. Li et al., *Phys. Rev. B* **81**, 054510 (2010).

17. F. D. M. Haldane, *Phys. Rev. Lett.* **67**, 937 (1991).
18. Y. S. Wu, *Phys. Rev. Lett.* **73**, 922 (1994).
19. P. A. Marchetti, Z. B. Su and L. Yu, *J. Phys. Condens. Matter* **19**, 125209 (2007) and references therein.
20. J. Frohlich and P. A. Marchetti, *Phys. Rev. B* **46**, 6535 (1992).
21. P. A. Marchetti, Z. B. Su, L. Yu, *Nucl. Phys. B* **482**, 731 (1996).
22. P. A. Marchetti, Z. B. Su and L. Yu, *Phys. Rev. B* **58**, 5808 (1998).
23. N. Read and S. Sachdev, *Phys. Rev. B* **42**, 4568 (1990).
24. E. Gava, R. Jengo and C. Omero, *Nucl. Phys. B* **168**, 465 (1980).
25. I. Affleck *Phys. Rev. Lett.* **66**, 2429 (1991).
26. M. Reizer, *Phys. Rev. B* **39**, 1602 (1989); **40**, 11571 (1989).
27. P. A. Marchetti, J. H. Dai, Z. B. Su and L. Yu, *J. Phys. Condens. Matt.* **12**, L329 (2000).
28. P. A. Marchetti et al., *Phys. Rev. B* **69**, 024527 (2004).
29. P. A. Marchetti, A. Ambrosetti, Z. B. Su and L. Yu, *J. of Phys. and Chem. of Solids* **69**, 3277 (2008).
30. R. B. Laughlin, *Science* **242**, 525 (1988).
31. F. Ye, P. A. Marchetti, Z. B. Su and L. Yu, *Phys. Rev. B* **92**, 235151 (2015).
32. E. H. Lieb, *Phys. Rev. Lett.* **73**, 2158 (1994).
33. D. S. Rokhsar, *Phys. Rev. Lett.* **65**, 1506 (1990); J. Bellisard and R. Rammal, *Europhys. Lett.* **13**, 205 (1990).
34. P. A. Marchetti, G. Orso, Z. B. Su and L. Yu, *Phys. Rev. B* **71**, 134510 (2005).
35. P. A. Lee, N. Nagaosa and X. G. Wen, *Rev. Mod. Phys.* **78**, 17 (2006).
36. P. A. Marchetti, F. Ye, Z. B. Su and L. Yu, *Phys. Rev. B* **84**, 214525 (2011).
37. P. A. Marchetti and M. Gambaccini, *J. Phys. Condens. Matt.* **24**, 475601 (2012).
38. P. Nozieres and S. Schmitt-Rink, *J. Low Temp. Phys.* **59**, 195 (1985).
39. Y. J. Uemura, *Physica* **282**, 194 (1997); Q. Chen, I. Kostin, B. Janko and K. Levin, *Phys. Rev. Lett.* **81**, 4708 (1998).
40. H. A. Weldon, *Phys. Rev. D* **40**, 2410 (1989).
41. H. Harrison and S. E. Sebastian, *Phys. Rev. B* **92**, 224505 (2015).
42. L. Ioffe and A. Larkin, *Phys. Rev. B* **39**, 8988 (1989).
43. P. A. Marchetti, Z. B. Su and L. Yu, *Phys. Rev. Lett.* **86**, 3831 (2001).
44. B. Wuyts, V. V. Moshchalkov, and Y. Bruynseraede, *Phys. Rev. B* **53**, 9418 (1996); L. Trappeniers et al., *J. Low Temp. Phys.* **117**, 681 (1999); H. G. Luo, Y. H. Su, and T. Xiang, *Phys. Rev. B* **77**, 014529 (2008).
45. P. A. Marchetti and G.Bighin, *J. Low Temp. Phys.* **185**, 87 (2016).
46. C. Berthier et al., *Physica C* **235-240**, 67 (1994).
47. T, Kondo et al., *Nature Phys.* **7**, 21 (2010).
48. E. Carlson, S. Kivelson, V. Emery and E. Manousakis, *Phys. Rev. Lett.* **83**, 612 (1999).
49. W. N. Hardy, S. Kamal, and D. Bonn, in *The Gap Symmetry and Fluctuations in High-T_c Superconductors* (Plenum Press, New York 1998).
50. P. A. Marchetti and G. Bighin, *Europhys. Lett.* **110**, 37001 (2015).
51. Y. J. Uemura, *Phys. Rev. Lett.* **62**, 2317 (1989).
52. Y. Ando et al., *Phys. Rev. Lett.* **93**, 267001 (2004).
53. T. Jacobs et al., *Phys. Rev. Lett.* **75**, 4516 (1995); C. Panagopoulos et al., *Phys. Rev. B* **60**, 14617 (1999).
54. T. Yoshida et al., *J. Phys. Condens. Matt.* **19**, 125209 (2007).

Some topological phases for sound

Baile Zhang

*Division of Physics and Applied Physics, School of Physical and Mathematical
Sciences, Nanyang Technological University, Singapore, 637371, Singapore
E-mail: blzhang@ntu.edu.sg
www.ntu.edu.sg/home/blzhang*

Topological acoustics is a subfield of topological physics that studies the construction
and behavior of topological phases for sound propagation. Different from electrons and
photons, sound waves travelling through air are essentially longitudinal waves that carry
no intrinsic spin and do not respond to magnetic fields, thus lacking sufficient degrees
of freedom. Here we review the recent progress in our group in constructing topological
phases for sound. The emphasis is laid on the underlying insights as inspired from the
corresponding development in condensed matter and photonic systems. The first example is the acoustic quantum Hall effect in a phononic crystal that consists of spinning
rods. This design follows Haldane and Raghu's work in constructing the first topological photonic crystal. The second example is the acoustic Weyl nodes by stacking
one-dimensional dimerized chains of acoustic resonators. It can be identified that its
topological phase corresponds to the recently studied type-II Weyl semimetal phase.
The last example is the acoustic Landau levels by introducing strain engineering to a
two-dimensional aperiodic acoustic lattice. This offers a path to previously inaccessible
magnetic-like effects in traditional periodic acoustic structures.

Keywords: Topological acoustics; gauge fields; type-II Weyl nodes; Landau levels.

1. Introduction

The concept of topology has transformed various branches of physics, from quantum
matter such as quantum Hall effects and topological insulators, to classical systems
such as topological photonics [1] and topological mechanics [2]. A very special kind
of classical waves is the air-borne acoustic waves. Because of their longitudinal properties, they have no intrinsic spin, and do not respond to magnetic fields. However,
by constructing effective gauge fields, it is possible to explore magnetic-like effects
(e.g., the quantum Hall effect) for sound. This is the underlying mechanism supporting the emerging field of topological acoustics [3–13]. The macroscopic scales
of acoustic structures make it convenient to study topological physics at room temperature without magnetic field. The "wavefunction" information that is elusive at
a microscale in quantum systems can be directly measured in an analogous acoustic
system for classical acoustic waves.

Modern acoustics is generally developed in tandem with photonics. It is Haldane and Raghu who firstly transferred topological physics into photonics by de-
signing a topological photonic crystal exhibiting features of Haldane's model [14].
They predicted that such a quantum Hall (without Landau levels [15]) phenomenon
can arise in the context of classical electromagnetism, which was subsequently

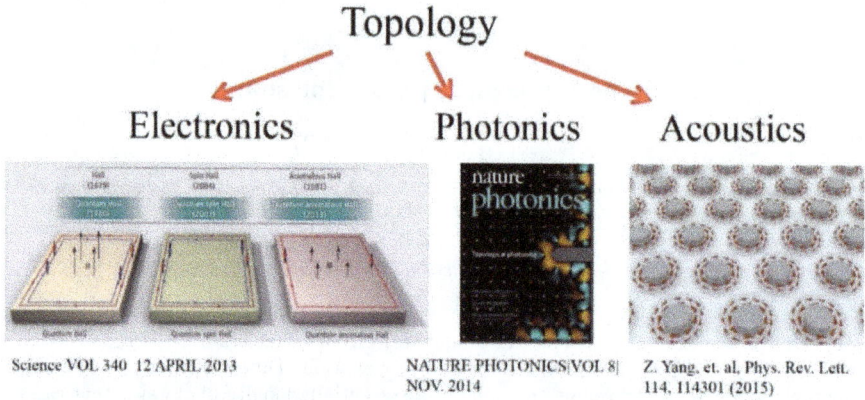

Fig. 1. The concept of topology applied to electronics, photonics and acoustics. References of corresponding figures have been indicated respectively.

demonstrated in a microwave-scale magneto-optic photonic crystal [16]. Various photonic devices have been proposed in recent years to emulate different topological phases.

Here we focus on the development of topological acoustics, following the development in condensed matter electronic systems and topological photonics. Figure 1 shows the relation among topological electronics, topological photonics, and topological acoustics.

2. Quantum Hall effect of sound

In this section we discuss the quantum Hall effect of sound, following Haldane and Raghu's work in constructing the first topological photonic crystal [14]. Most of the content has been published in Ref. [3].

2.1. Nonreciprocal acoustic crystal

In order to realize topological band theory in acoustics, we begin with a spatially periodic acoustic medium, sometimes called a "phononic crystal", and introduce a mechanism that breaks time-reversal (T) symmetry, similar to Haldane and Raghu's work [14]. The proposed acoustic structure is shown in Fig. 2(a). It is a triangular lattice of lattice constant a. Each unit cell consists of a rigid solid cylinder (e.g. a metal cylinder) with radius r_1, surrounded by a cylindrical circulating-fluid-filled region of radius r_2. Figure 2(b) shows the unit cell with circulating flow. The rest of the unit cell ($r > r_2$) consists of a stationary fluid, separated from the fluid in the cylindrical region by a thin impedance-matched layer at radius r_2 (this layer can be achieved using a thin sheet of solid material that is permeable to sound). The central cylinder spins along its axis with angular speed Ω, which produces a circulatory flow in the surrounding fluid. We will not consider the possibility of

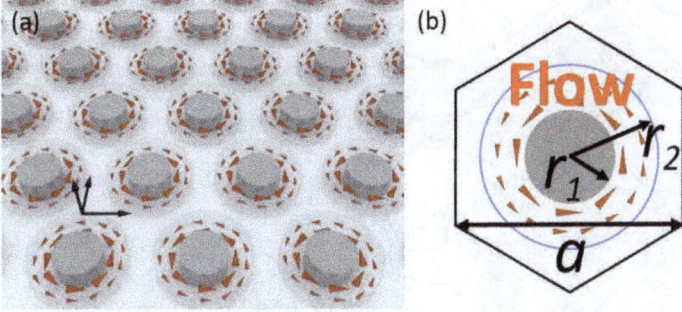

Fig. 2. (a) Triangular acoustic lattice with lattice constant a. (b) The unit cell containing a central metal rod of radius $r_1 = 0.2a$, surrounded by an anticlockwise circulating fluid flow (flow direction indicated by red arrows) in a cylindrical region of radius $r_2 = 0.4a$. Figure adopted from Ref. [3].

Taylor vortex because we here focus on two-dimensional (2D) model and Taylor vortex does not contribute an effective flux through x–y plane. The fluid velocity is considered much slower than the speed of sound (i.e., the Mach number, which is defined as v/c, is less than 0.3). The motion of the fluid can be described by a circulating "Couette flow" distribution; the velocity field points in the azimuthal direction, with component $v_\theta = -\frac{\Omega r_1^2}{r_2^2 - r_1^2}r + \frac{\Omega r_1^2 r_2^2}{r_2^2 - r_1^2}\frac{1}{r}$, where r is measured from the origin at the axis of the cylinder. This angular velocity is equal to Ω at radius $r = r_1$, and zero at radius $r = r_2$.

2.2. Governing equation

Assuming that the viscosity and heat flow are negligible, we can start from three independent equations — Euler, continuity and state equations in terms of acoustic disturbances of p, ρ and v: $\rho_0 \partial_t v + \nabla p = 0$, $\partial_t \rho + \rho_0 \nabla \cdot v = 0$ and $p = c_0^2 \rho$. From the above three equations, we can arrive at the linearized approximate sound master equation:

$$\frac{1}{\rho}\nabla \cdot \rho\nabla\phi - (\partial_t + \vec{v}_0 \cdot \nabla)\frac{1}{c^2}(\partial_t + \vec{v}_0 \cdot \nabla)\phi = 0 \qquad (1)$$

where ρ is the fluid density, c is the speed of sound, and $\vec{v}_0$ is the background fluid velocity. The relation between velocity potential ϕ and sound pressure p is $p = \rho(\partial_t + \vec{v}_0 \cdot \nabla)\phi$. We take the surface of each cylinder as an impenetrable hard boundary by setting $\vec{n} \cdot \nabla\phi = 0$ where $\vec{n}$ is the surface normal vector.

We neglect second order terms as $|\vec{v}_0/c|^2 \ll 1$. With a change of variables $\Psi = \sqrt{\rho}\phi$ the master equation can be rewritten as

$$[(\nabla - i\vec{A}_{eff})^2 + V(x, y)]\Psi = 0 \qquad (2)$$

where the effective vector and scalar potentials are

$$\vec{A}_{eff} = -\frac{\omega\vec{v}_0(x, y)}{c^2} \qquad (3)$$

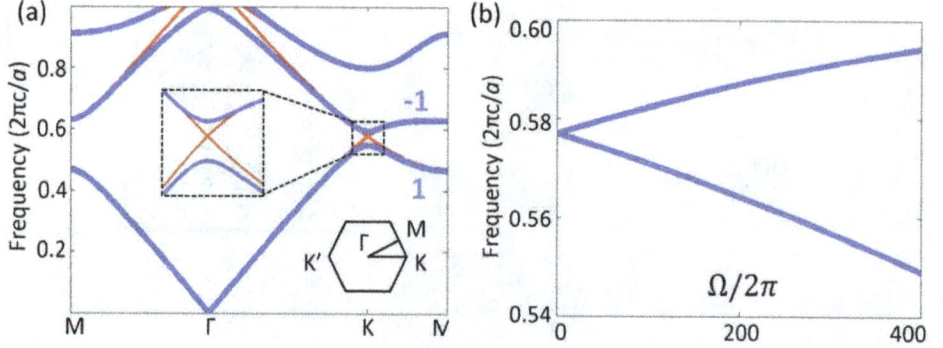

Fig. 3. (a) Band structures of the acoustic lattice without the circulating fluid flow (red curves; $\Omega = 2\pi \times 0$ rad/s) and with fluid flow (blue curves; $\Omega = 2\pi \times 400$ rad/s). In the gapped band structure, the bands have Chern number ± 1 (blue labels). (b) Frequency splitting as a function of the angular velocity of the cylinder in each unit cell. Figure adopted from Ref. [3].

$$V(x, y) = -\frac{1}{4}|\nabla \ln \rho|^2 - \frac{1}{2}\nabla^2 \ln \rho + \frac{\omega^2}{c^2} \tag{4}$$

Evidently Eq. (2) maps onto the Schrodinger equation for a spinless charged quantum particle in non-uniform vector and scalar potentials in Eqs. (3–4). For non-zero Ω, the inner boundary of the Couette flow contributes positive effective magnetic flux, and the rest of the Couette flow contributes negative effective magnetic flux; the net magnetic flux, integrated over the entire unit cell, is zero. The acoustic system thus behaves like a "zero field" quantum Hall system.

If we focus on only a single unit cell, it should be pointed out that a similar approach to construct an effective magnetic vector potential for fluid wave propagation has been discussed by Berry and colleagues [17]. They showed that an irrotational ("bathtub") fluid votex exhibits a classical wavefront dislocation effect, analogous to the Aharanov-Bohm effect. Here instead we study a periodic acoustic crystal, so that the effective magnetic vector potential gives rise to a topologically nontrivial acoustic band structure.

2.3. Topological band structure

We can calculate the acoustic band structures by using the finite-element commercial software COMSOL Multiphysics. For simplicity, we assume the background fluid involved is air. The results, with $\Omega = 0$ and $\Omega \neq 0$, are shown in Fig. 3(a). For $\Omega = 0$ [red curves in Fig. 3(a)], the acoustic band structure exhibits a pair of Dirac points at the corner of the hexagonal Brillouin zone, at frequency $\omega_0 = 0.577 \times 2\pi c_a/a$ (992 Hz when $a = 0.2$ m), where c_a is the sound velocity in air.

For $\Omega \neq 0$ the circulating air flow lifts the Dirac point degeneracy, producing a finite complete bandgap (we set $\Omega = 2\pi \times 400$ rad/s). The frequency splitting at the

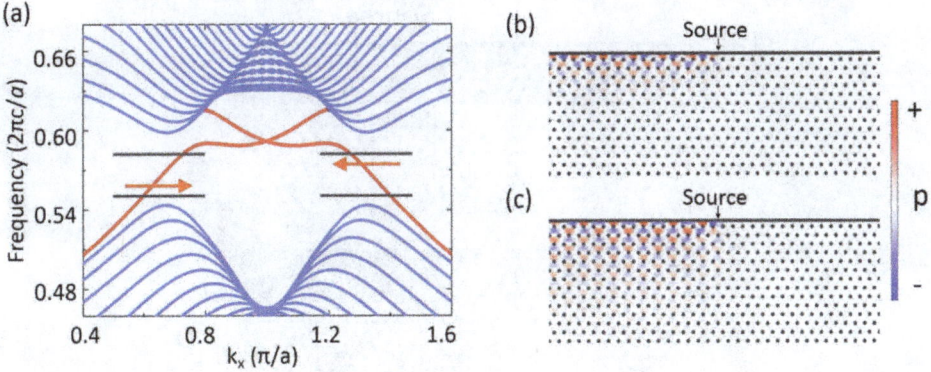

Fig. 4. (a) Dispersion of the one-way acoustic edge states (red curves) occurring in a finite strip of the acoustic lattice, for $\Omega = 2\pi \times 400$ rad/s. (b–c) The normalized acoustic pressure p for a left-propagating acoustic edge state at a frequency $\omega_0 = 0.577 \times 2\pi c_a/a$ (992 Hz) for $\Omega = 2\pi \times 400$ rad/s (b) and $\Omega = 2\pi \times 200$ rad/s (c). Figure adopted from Ref. [3].

zone corners as a function of Ω, is plotted in Fig. 3(b). The ratio of the operating frequency to the bandgap, which is an estimation for the penetration depth of the topological edge states in units of the lattice constant, is on the order of $\omega/\delta\omega \approx 10$ for the range of angular velocities plotted here. The Chern number of the nth acoustic band can be defined as $C_n = \frac{1}{2\pi} \oiint ds \cdot \nabla_k \times \mathcal{A}$, where the Berry connection $\mathcal{A} = i\langle u_{n,k} | \nabla_k | u_{n,k} \rangle$. The function $u_{n,k}(r)$ obeys the orthonormality condition and is cell-periodic eigen-function of the Bloch Hamiltonian. By exporting the Bloch wave functions calculated from finite-element method, we can numerically verify that the lowest two bands in Fig. 3(a) have Chern numbers of ± 1.

2.4. One-way edge modes

The principle of bulk-edge correspondence then predicts the existence of topologically-protected acoustic edge states. We numerically calculate the band structure for a 20×1 super-cell (a ribbon that is 20-unit-cell wide in y direction and infinite along x direction). As shown in Fig. 4(a), for $\Omega = 2\pi \times 400$ rad/s, the bandgap contains two sets of edge states confined to opposite edges of the ribbon with opposite group velocities.

Figures 4(b–c) show the propagation of these edge states in a finite (34×14) lattice. Here the upper edge of the acoustic crystal is enclosed by a sound-impermeable hard boundary (e.g., a flat metal surface), in order to prevent sound waves from leaking into the upper half space. A point sound source with mid-gap frequency ω_0 is placed near the upper boundary. For $\Omega = 2\pi \times 400$ rad/s, this excites a unidirectional edge state which propagates to the left along the interface [Fig. 4(b)]. The field distribution for a reduced angular velocity $\Omega = 2\pi \times 200$ rad/s [Fig. 4(c)] shows an edge state with a longer penetration depth because of a narrower bandgap. Note

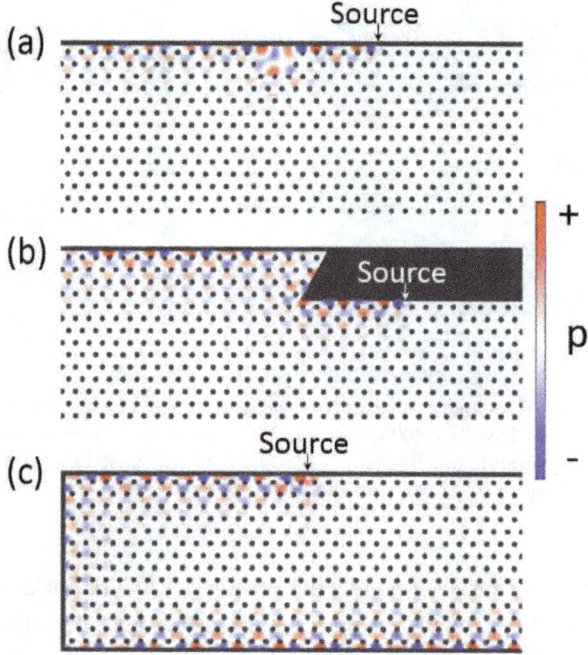

Fig. 5. Demonstration of the robustness of acoustic one-way edge states against disorder. Topological protection requires the acoustic waves to be fully transmitted through an acoustic cavity (a), a Z-shape bend along the interface (b) and a 180-degree bend (c). The operating frequency is $\omega_0 = 0.577 \times 2\pi c_a/a$ (992 Hz) and $\Omega = 2\pi \times 400$ rad/s. Figure adopted from Ref. [3].

that the group velocity flips sign within a very small frequency range near the bulk states. This behavior is dependent on boundary details and does not violate the bulk-edge correspondence principle.

We then demonstrate the topological protection in the propagation of these edge states. Figure 5(a) shows an acoustic cavity located along the interface; the incident wave flows through the cavity, and excites localized resonances within the cavity, but does not backscatter. Figure 5(b) shows a Z-shape step connecting two parallel surfaces at different heights; again, no backscattering occurs. Finally, Fig. 5(c) shows a 180-degree bend which allows acoustic edge states to be guided from the top to the bottom of the structure. Note that the top and bottom boundaries are called "zigzag" boundaries. Whereas the left boundary is an "armchair" boundary, which supports one-way edge states with different dispersion relations.

In addition, we numerically calculate the dispersion relation of a 20×1 super-cell (a ribbon that is 20-unit-cell wide in x direction and infinite along y direction). In this case, the armchair shape of the edges can still support the topologically protected one-way surface modes, as shown in Fig. 6(a). For $\Omega = 2\pi \times 400$ rad/s, within the bandgap there are two edge states corresponding to right and left edges of the ribbon, which have positive and negative group velocities, respectively.

(a)

(b)

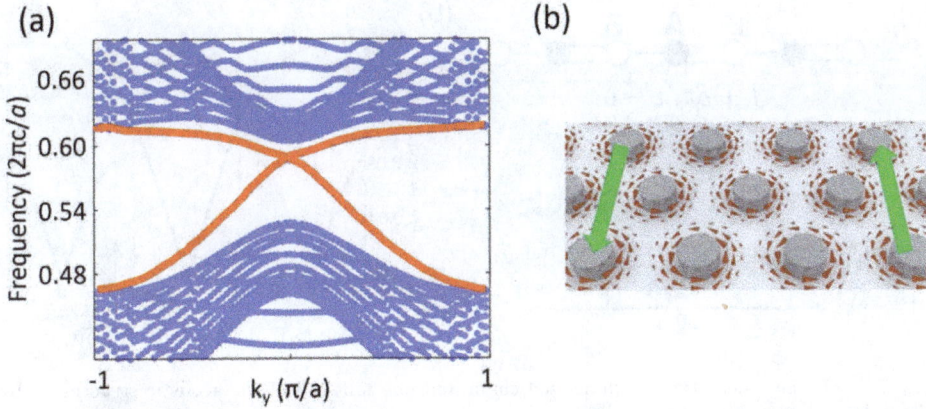

Fig. 6.   (a) The dispersion of the edge states projected along x direction. There are two edge states corresponding to opposite edges. (b) The schematic shows the propagation directions of the edge states shown in panel (a).

3. Type-II Weyl nodes of sound

Weyl nodes are three-dimensional (3D) extension of 2D Dirac nodes. Their dispersions are similar to those of Weyl fermions predicted by Weyl in 1929 in the development of quantum field theory, but Weyl fermions have not been found as realistic particles. Recently, it has been found that the electronic low-energy excitations as quasiparticles in some novel Weyl semimetals can support the Weyl fermion Hamiltonian. What is beyond the context of traditional Weyl fermions is that, on the platform of condensed matter system, a new type of Weyl fermions, nowadays classified as "type-II" Weyl fermions, can emerge at the contact of the electron and hole pockets. This new type of type-II Weyl fermions violate the fundamental Lorentz symmetry in quantum field theory, and thus were actually missed by Weyl in his original studies. The Weyl nodes in acoustic systems have been discussed in Ref. [6], in the context of type-I Weyl Hamiltonian. Here we discuss how to construct explicitly type-II Weyl nodes for sound propagation and discuss the difference between type-I and type-II Weyl systems. Most content has been published in Ref. [8].

3.1. *Acoustic dimerized chain*

3.1.1. *Reducing the resonator model to tight-binding Hamiltonian*

We start with constructing a simple one-dimensional (1D) tight-binding model which consists of acoustic resonators that are only coupled to nearest-neighbor resonators through a coupling waveguide. This method can be extended to 2D and 3D acoustic resonator systems, which will be applied in the next section.

Assuming there is an infinitely long 1D resonator chain as shown schematically in the upper panel of Fig. 7(a) – two resonators per unit cell. The filled (open) circle

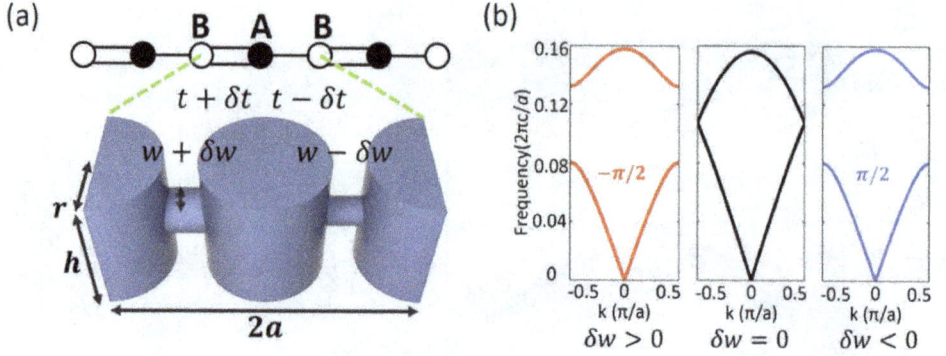

Fig. 7. (a) The schematic of a dimerized chain and one unit-cell of the acoustic structure. (b) Three band structures with $\delta w = 0.3w$, $\delta w = 0$ and $\delta w = -0.3w$. The band gap closes at the middle panel, which indicates the existence of the topological transition. Figure adopted from Ref. [8].

indicates A (B) type resonator. The lth resonator satisfies the following coupled-mode equations:

$$i\partial_t a_l = \sum_{\langle m \rangle} \kappa_1 b_m \quad \text{and} \quad i\partial_t b_l = \sum_{\langle m \rangle} \kappa_2 a_m \tag{5}$$

where a (b) is the amplitude of resonator mode for type A (B) resonator. The summation $\langle m \rangle$ is taken over the nearest-neighbor resonators. The hopping strength can be obtained from the above coupled-mode equations as $\kappa_1 = -\omega_n \int \langle b_l | a_l \rangle dV$ and $\kappa_2 = -\omega_n \int \langle a_{l+1} | b_l \rangle dV$, where the integration is taken over the volume of the coupling waveguide. The above coupled-mode equations can be viewed as a tight-binding eigenvalue problem $H\psi = E\psi$.

In particular, for the two-band model of 1D acoustic dimerized chain, the construction is shown in the lower panel of Fig. 7(a). The left and right nearest-neighbor hopping strengths of A type resonator are $t + \delta t$ and $t - \delta t$, respectively. The dispersion from coupled-mode equations is the same as the Hamiltonian $H = \sum_l [(t + \delta t)a_l^\dagger b_l + (t - \delta t)b_l^\dagger a_{l+1} + h.c.]$, where a (b) and $a^\dagger$ ($b^\dagger$) are the annihilation and creation operators on the sub-lattice sites, t and δt are the nearest hopping and the tuning strength. Furthermore we transform the Hamiltonian into k-space and set zero energy offset between two sites. The Bloch Hamiltonian $H(k)$ for the 1D acoustic resonator system can be obtained as follows:

$$H_1(k) = 2t\cos(k_x a)\sigma_x - 2\delta t \sin(k_x a)\sigma_y \tag{6}$$

3.1.2. Acoustic dimerized chain consisting of resonators

Now, we step into the details of the acoustic structure and characterize its physical properties, since the Hamiltonian Eq. (6) can be implemented in an acoustic dimerized chain. One unit cell of the dimerized chain consists of two resonators,

(a)

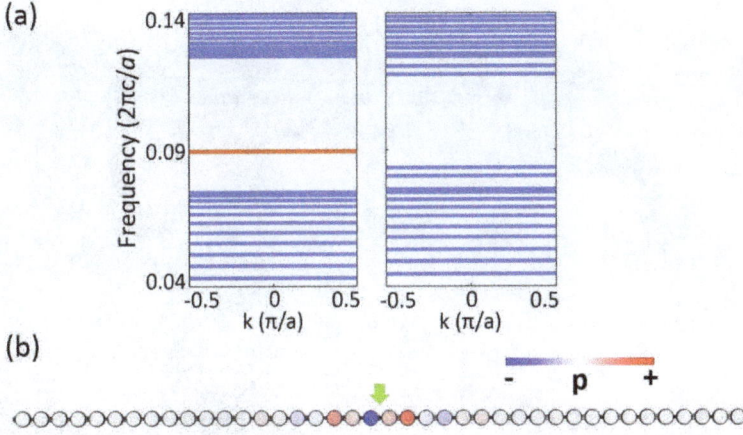

(b)

Fig. 8. (a) Topologically non-trivial and trivial band structures for the interface between two connected chains. The red line indicates the topological interface state. (b) The acoustic field of the interface state. The green arrow points to the interface. Figure adopted from Ref. [8].

connected by two coupling waveguides with different radii, as shown in the lower part of Fig. 7(a). The periodic boundary condition is applied to the left and right surfaces. Other surfaces (blue color) of the unit cell are treated as hard boundaries for sound. The distance between two nearest resonators is $a = 0.1$ m. The radius and height of the cylinder (resonator) is $r = 0.4a$ and $h = 0.8a$. For dimerization, we apply modulation of $\delta w = 0.3w$ to the original radius of coupling waveguide $w = 0.26r$. We thus have $w + \delta w$ for one coupling waveguide, and $w - \delta w$ for the other, as shown in the lower panel of Fig. 7(a).

By choosing three values of modulation $\delta w = 0.3w$, 0, $-0.3w$, we arrive at three band structures by solving the acoustic wave equation, as shown in Fig. 7(b). The bandgap closes at $\delta w = 0$ (black curve), implying a topological phase transition. The topological properties of the lower bands can be characterized by calculating the topological invariant — Zak phase,

$$\varphi_{Zak} = -\int_{-\pi/2a}^{\pi/2a} u_{n,k} |\nabla_k| u_{n,k} dk \qquad (7)$$

The results are $-\pi/2$ and $\pi/2$ for $\delta w > 0$ and $\delta w < 0$, respectively. Note that the Zak phase of each dimerization is a gauge dependent value, but the difference between the Zak phases of two dimerized configurations with $\delta w > 0$ and $\delta w < 0$, which is $\Delta\varphi_{Zak} = \varphi_{Zak2} - \varphi_{Zak1} = \pi$ in our acoustic model, is topologically defined.

The above topologically nontrivial phases in acoustic resonators ensure the existence of interface states between two configurations of dimerized lattices with different Zak phases. Hereafter we cut and connect the two chains through their mirror centers for physical reasons. The interface dispersions demonstrated in Fig. 8(a) are the results from numerical simulation. For the left panel, we apply $\delta w = 0.3w$ and $\delta w = -0.3w$ on two sides of an interface. For the right panel, $\delta w = 0.3w$ and

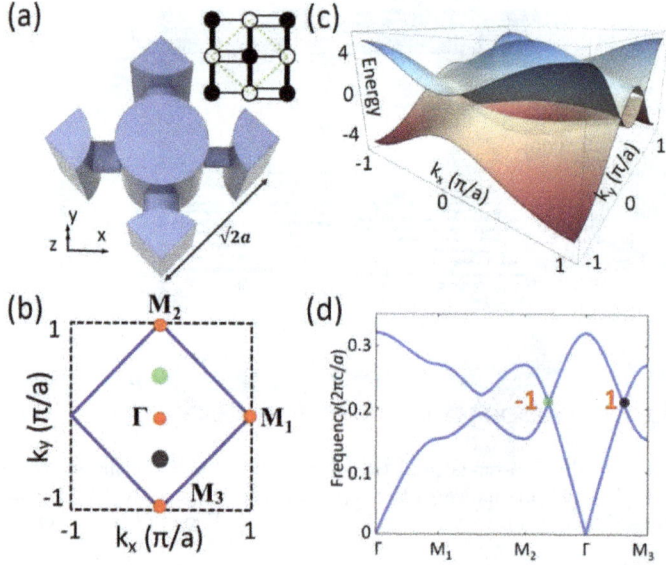

Fig. 9. (a) The schematic of the 2D dimerized lattice and one unit-cell of the acoustic structure. (b) The first Brillouin zone enclosed by the blue lines. Green and black dots are locations of the Dirac nodes. (c) The band structure calculated from the tight-binding model shows the existence of the accidental degeneracy. (d) The band structure of the acoustic lattice from numerical simulation. Red numbers indicate the chirality. Figure adopted from Ref. [8].

$\delta w = 0.2w$ are applied. There is an interface state, as predicated, locating inside the bandgap in the left panel of Fig. 8(a), as highlighted by the red line. The acoustic pressure field pattern of the interface state is shown in Fig. 8(b).

3.2. Dirac nodes in a dimerized square lattice

Utilizing these 1D dimerized chains shown in last section as building blocks, we can construct 2D Dirac nodes by stacking these 1D dimerized chains in a staggered manner. We find that the 2D Dirac nodes can be constructed by the Bloch Hamiltonian of a 2D dimerized acoustic lattice shown below:

$$H_2(k) = [2t_x \cos(k_x a) + 2t_y \cos(k_y a)]\sigma_x - 2\delta t_x \sin(k_x a)\sigma_y \qquad (8)$$

where t_x (t_y) is the hopping strength along x (y) direction, and δt_x is the modulation of the hopping strength along x direction. Following the above tight-binding model, we set the unit cell of the acoustic lattice as shown in Fig. 9(a). The right inset is the schematic of 2D lattice whose unit cell is enclosed by green dashed lines. Through tuning the coupling strength, we choose $t_x = -1$, $t_y = -2$ and $\delta t_x = -0.5$. The band structure in the 2D momentum space ($k_x k_y$), as shown in Fig. 9(b), can be calculated from the Hamiltonian Eq. (8), as shown in Fig. 9(c). There are two isolated degenerate points located at $(0, \pm 2/3)$ in units of π/a in the first 2D Brillouin zone.

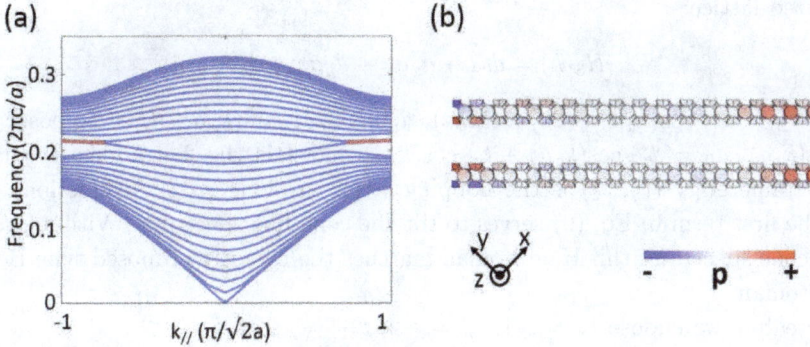

Fig. 10. (a) The band structure for a finite acoustic lattice. The red curves indicate the flat edge states, which means the nearly zero group velocity. (b) The acoustic fields of the edge states. Acoustic waves are localized at the edges of the sample. Figure adopted from Ref. [8].

For the acoustic structure, the lattice constant and parameters of the resonator (radius and height) are the same with those in 1D dimerized chain. The modulation $\delta w_x = 0.3 w_x$, where $w_x = 0.26 r$, is applied to the coupling waveguides along x direction, whose radii are $w_x \pm \delta w_x$, respectively. Coupling waveguides along y direction with radius $w_y = 2 w_x$ connect these 1D dimerized chains. For this real acoustic structure, the band structure along high symmetry lines in the first Brillouin zone is shown in Fig. 9(d). It can be seen that there are two degenerate points (2D Dirac nodes) at frequency $0.21 \times 2\pi c/a$ (718.05 Hz), where c is the speed of sound, located at $(k_x, k_y) = (0, +0.61)$ and $(k_x, k_y) = (0, -0.61)$ in units of π/a within high symmetry lines $M_2 \Gamma$ and ΓM_3. After calculating the winding number [18], we find $w = -1$ ($w = 1$) for degenerate point on $M_2 \Gamma$ (ΓM_3), as indicated in Fig. 9(d).

Here the 2D acoustic lattice has flat edge states similar to those in graphene. We investigate the 2D acoustic structure that is finite in the x–y direction having a width of 15.5 unit cells and is infinite in the $x+y$ direction, as plotted in Fig. 10(b). Figure 10(a) shows the band structure of the finite acoustic system. The 2D Dirac nodes are projected onto the $k_x - k_y$ direction with good quantum number $k_{//} = (k_x + k_y)/\sqrt{2}$. The red curves with degeneracy of two indicate the nearly flat edge states connecting two projected Dirac nodes with opposite chirality. Note that the little derivation from a perfectly flat dispersion is a result of the real acoustic structure, mesh settings and boundary conditions adopted. The acoustic fields are shown in Fig. 10(b). The acoustic waves of degenerate edge states almost do not propagate due to nearly zero group velocity.

3.3. Type-II Weyl nodes in a dimerized cubic lattice

By stacking the 2D dimerized lattice along the z direction with periodicity a and tuning the coupling strength, we can construct a Bloch Hamiltonian for the 3D dimerized lattice:

$$H_3(k) = d_0 I + d_x \sigma_x + d_y \sigma_y + d_z \sigma_z \tag{9}$$

where $d_x = 2t_x \cos(k_x a) + 2t_y \cos(k_y a)$, $d_y = -2\delta t_x \sin(k_x a)$, $d_z = t_{z1} \cos(k_z a) - t_{z2} \cos(k_z a)$, $d_0 = t_{z1} \cos(k_z a) + t_{z2} \cos(k_z a)$ and I is the 2×2 identity matrix. The parameter t_x (t_y, t_z) is the hopping strength along x (y, z) direction. Note that the first term in Eq. (9) serves to tilt the cone-like spectrum. With a strongly tilted cone spectrum, this Hamiltonian matches the recently proposed type-II Weyl Hamiltonian.

Hereafter we choose $t_x = -1$, $t_y = -2$, $t_{z1} = -1$, $t_{z2} = -2$ and $\delta t_x = -0.5$, as schematically shown in Fig. 11(a). We calculate the band structure as shown in Fig. 11(b) with $k_z = 0.5$ in the $k_x - k_y$ Brillouin zone plane. Four linear degenerate points locate at $(0, \pm 2/3, \pm 1/2)$ in units of π/a in the 3D first Brillouin zone. Typically, we plot in Fig. 11(c) the cone spectrum in the vicinity of the degenerate point $(0, 2/3, 1/2)$ in $k_x - k_y$ Brillouin zone plane. It can be seen that the cone spectrum has been strongly tilted. Since the group velocities along z direction near the degenerate point are $2t_{z1} \sin(k_{z0})$ and $2t_{z2} \sin(k_{z0})$ where k_{z0} is the location of degenerate point, the two bands acquire the same sign of group velocity.

Figure 11(a) shows one unit-cell of the 3D acoustic structure. The inset presents the schematic of the 3D dimerized lattice. The radii of the coupling waveguides along the z direction are $w_{z1} = w_x + 2\delta w_x$ for the A resonator and $w_{z2} = w_x$ for the B resonator. The other parameters are the same as in previous 1D and 2D acoustic structures. The band structures in Brillouin zone planes $(k_x, k_y, 0)$ and $(k_x, k_y, 0.51)$ are shown in Fig. 11(d–e), which reveal a band gap at $k_z = 0$, and two degenerate points with frequency $0.26 \times 2\pi c/a$ (900.10 Hz) at $k_z = 0.51$. In the left part of Fig. 11(f), by sweeping k_z at $(k_x, k_y) = (0, 0.61)$, we get the band structure with a degenerate point at $k_z = 0.51$. It can be seen that the two bands have the same sign of group velocity. Therefore, there are four acoustic type-II Weyl nodes at $(0, \pm 0.61, \pm 0.51)$.

One significant distinction between the type-I and type-II Weyl nodes appears in the density of states. As shown in Fig. 11(f), the density of states acquires finite values for type-II Weyl nodes due to the presence of unbounded two-band pockets. This will be further discussed later.

Because the Chern number is not changed by the first term of the Hamiltonian, we expand the Pauli-matrix components around the Weyl node by substituting $k_x = k_{x0}/a + \delta k_x/a$, $k_y = k_{y0}/a + \delta k_y/a$ and $k_z = k_{z0}/a + \delta k_z/a$ around the Weyl nodes (k_{x0}, k_{y0}, k_{z0}) and keeping the first-order terms and Pauli-matrix terms, we have:

$$h_3(\delta k) = -2t_y \sin(k_{y0}) \delta k_y \sigma_x - 2\delta t_x \delta k_x \sigma_y - (t_{z1} \sin(k_{z0}) \delta k_z - t_{z2} \sin(k_{z0}) \delta k_z) \sigma_z \tag{10}$$

In the 3D momentum space, the Weyl nodes are topological monopoles of quantized Berry flux characterized by the chirality $c = \text{sgn}[\det(v_{ij})]$, or the topological

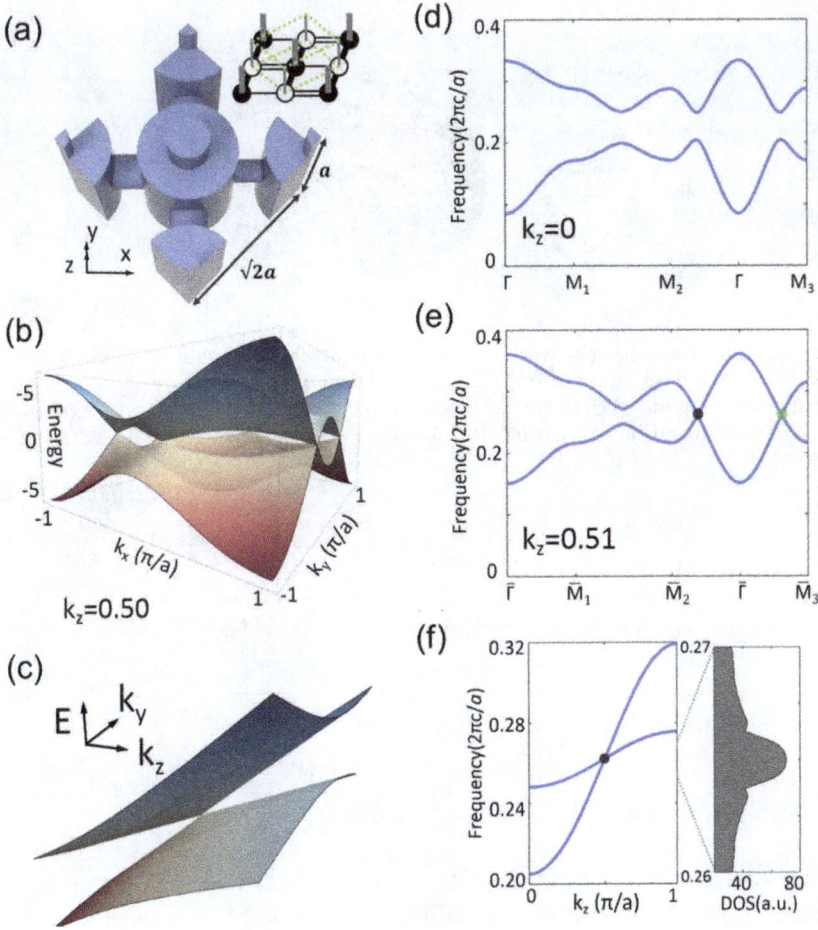

Fig. 11. (a) A unit-cell of the 3D acoustic structure. Inset is the schematic of the 3D dimerized lattice. (b–c) The band structures calculated from the tight-binding model in the first Brillouin zone plane with $k_z = \pi/2a$ (b) and around the Weyl node with $k_x = 0$ (c). (d–e) The band structures along the high symmetry lines in 2D Brillouin zone planes. The black (green) dot indicates the Weyl node with positive (negative) chirality. (f) Left: the band structure as a function of k_z with fixed $(k_x, k_y) = (0, 0.61)$ in units of π/a. Right: non-vanishing density of states. Figure adopted from Ref. [8].

invariant — Chern number. Here the group velocity matrix is:

$$v_{ij} = \begin{pmatrix} 0 & -2\delta t_x & 0 \\ -2t_y \sin(k_{y0}) & 0 & 0 \\ 0 & 0 & -t_{z1}\sin(k_{z0}) + t_{z2}\sin(k_{z0}) \end{pmatrix} \qquad (11)$$

Therefore the chirality is $c = 1$ ($c = -1$) for $(0, 2/3, 1/2)$ and $(0, -2/3, -1/2)$ $[(0, 2/3, -1/2)$ and $(0, -2/3, 1/2)]$, which are indicated by black "+" (green "−") in the momentum plane $(0, k_y, k_z)$ as shown in Fig. 12(a). From above equations, an analytical expression for Berry curvature around the Weyl point $(0, 2/3, -1/2)$

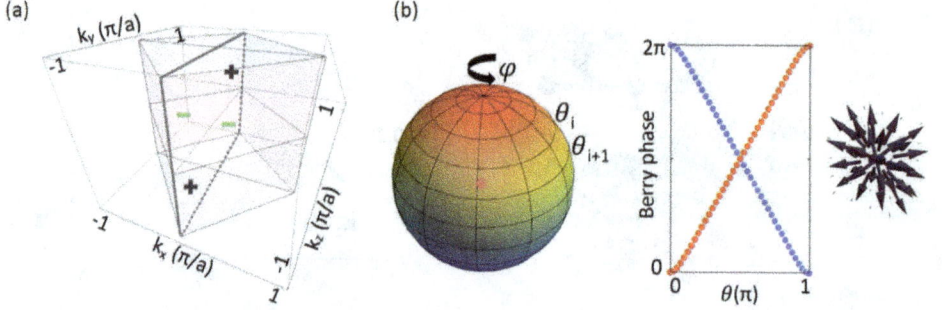

Fig. 12. (a) The distribution of type-II Weyl nodes in 3D first Brillouin zone. Black "+" (green "−") indicates the positive (negative) chirality. (b) Left panel: A sphere in momentum space enclosing one Weyl point. The radius of the sphere is $0.05 \times 2\pi/a$. Right panel: Berry phase and Berry curvature around the Weyl point $(0, 2/3, 1/2)$. Figure adopted from Ref. [8].

is

$$\left[k_x \frac{2\sqrt{3}\delta t_x t_y (t_{z2} - t_{z1})}{(4(\delta t_x k_x)^2 + 3(t_y k_y)^2 + 4(k_z(t_{z1} - t_{z2}))^2)^{\frac{3}{2}}}, \right.$$

$$k_y \frac{2\sqrt{3}\delta t_x t_y (t_{z2} - t_{z1})}{(4(\delta t_x k_x)^2 + 3(t_y k_y)^2 + 4(k_z(t_{z1} - t_{z2}))^2)^{\frac{3}{2}}}, \tag{12}$$

$$\left. k_z \frac{2\sqrt{3}\delta t_x t_y (t_{z2} - t_{z1})}{(4(\delta t_x k_x)^2 + 3(t_y k_y)^2 + 4(k_z(t_{z1} - t_{z2}))^2)^{3/2}} \right]$$

Also, we adopt the method from Ref. [19] and calculate the Berry phases of the two bands over a closed sphere of radius $r = 0.05 \times 2\pi/a$, presented in the left panel of Fig. 12(b), centered at a 3D degenerate node $(0, 2/3, 1/2)$. We discretize the sphere into closed loops in terms of polar angle θ (0 to π), as indicated by horizontal black circles. For each closed loop θ_i, the Berry phase is numerically calculated by using the discretized formula:

$$\gamma_n = -\mathrm{Im} \ln \prod_j \langle u_{n,k_j} | u_{n,k_{j+1}} \rangle \tag{13}$$

where k_j represents jth point at closed loop θ_i.

After manipulating the eigenvectors calculated theoretically, as shown in the middle of Fig. 12(b), we plot the Berry phases as a function of polar angle θ. The red (blue) dotted line represents the first (second) band. The Berry phase of the lower (upper) band changes from 0 to 2π (2π to 0), which verifies that the crossing point in 3D momentum space is a Weyl node with Chern number of 1. The right inset shows the Berry curvature around the Weyl node. The same calculation can be applied to identify the charges of other Weyl nodes.

(a)

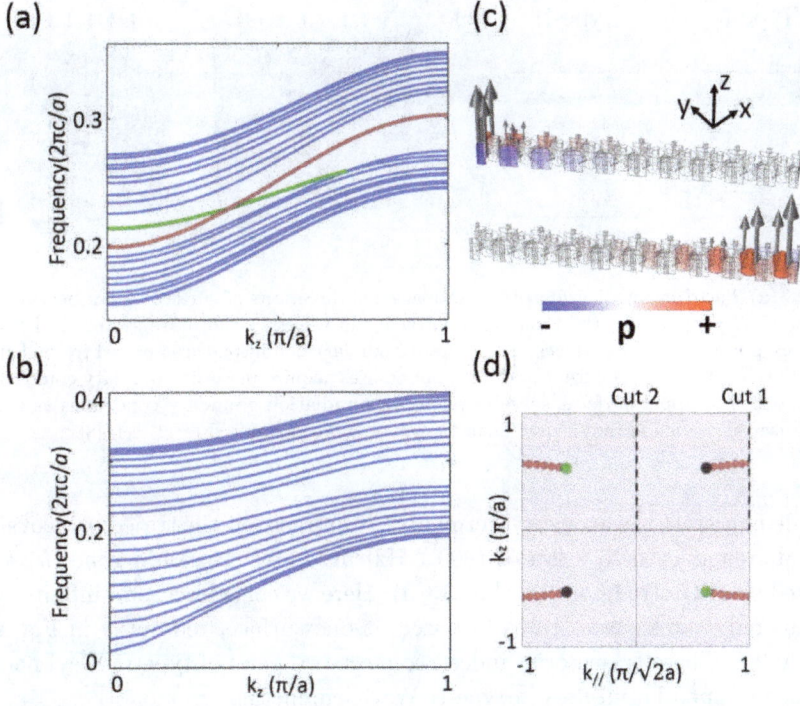

(b)

(c)

(d)

Fig. 13. (a–b) The band structures with $k_{//} = 1/\sqrt{2}$ (a) and $k_{//} = 0$ (b). The red and green curves indicate the surface states. (c) The acoustic fields of the surface states in (a). The grey arrows represent the sound intensity. (d) Red dotted points indicate the trajectories of the "Fermi arc" in the 2D surface Brillouin zone. Black dots (green dots) indicate Weyl points with positive (negative) chirality. Figure adopted from Ref. [8].

3.4. Fermi-arc-like surface states

The nonzero Chern numbers imply the existence of topological surface states. We investigate the 3D acoustic structure that is finite in the x–y direction and infinite in the $x + y$ and z directions. In this case, the Weyl nodes are projected along the $k_x - k_y$ direction, indicated by black and green dotted points in Fig. 13(d), with good quantum numbers $k_{//} = (k_x + k_y)/\sqrt{2}$ and k_z. Figure 13(a–b) show the projected band structures with fixed $k_{//} = 1/\sqrt{2}$ [Fig. 13(a)] and $k_{//} = 0$ [Fig. 13(b)] in units of π/a, as indicated by "Cut 1" and "Cut 2" in Fig. 13(d). In Fig. 13(a), two surface states (red and green curves), corresponding to the two opposite surfaces, are located in an incomplete bandgap, both acquiring positive group velocity. In Fig. 13(b), no surface state shows up. As presented in Fig. 13(c), the upper (lower) acoustic field corresponds to the surface state of green (red) curve in Fig. 13(a). We also plot the sound intensity $\mathbf{I} = p\mathbf{v}$ (p is sound pressure and $\mathbf{v}$ is the velocity) as grey arrows, whose length represents the amplitude of sound intensity. Both the two surface states propagate along the z direction, which is consistent with the positive group velocity in Fig. 13(a).

40

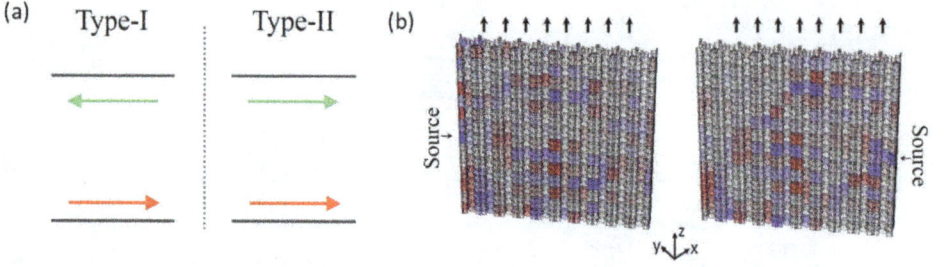

Fig. 14. (a) The distinct diagram of the propagating directions of surface states between type-I and type-II Weyl systems. (b) Excitations of acoustic waves in the finite systems. Two sound sources are put at opposite surfaces. Sound hard boundary conditions are applied to the left, right and bottom boundaries. The front and back boundaries acquire periodic boundary condition with $k_{//} = 1/\sqrt{2}$. The upper surfaces serve as plane wave radiation boundary conditions as indicated by black arrows, which do not reflect acoustic waves. Figure adopted from Ref. [8].

To demonstrate the acoustic "Fermi arcs," we trace out the trajectories of surface states at frequency $0.26 \times 2\pi c/a$ (900.1 Hz) in the 2D Brillouin zone $(k_{//}k_z)$, as indicated by red dotted points in Fig. 13(d). Here we consider a semi-infinite system and thus only the surface states localized at one surface [red curve in Fig. 13(a)] are included. These trajectories indeed connect two pairs of type-II Weyl nodes, as an analog of open Fermi arcs in type-II Weyl semimetals.

3.5. Distinct features between type-I and type-II Weyl systems

Firstly, one significant distinction between the type-I and type-II Weyl nodes appears in the density of states. The equation of density of states is expressed as $g(\omega) = \frac{V}{(2\pi)^3} \int \delta(\omega - \omega(k)) d^3 k$, where V is the volume of contributed momentum box in the vicinity of the Weyl node. For type-I Weyl nodes, the density of states vanishes at the frequency of Weyl nodes. However, the density of states acquires finite values for type-II Weyl nodes due to the presence of unbounded two-band pockets. We retrieve the parameters from fitting the data of bands with the Eq. (9) and plot the density of states that arises due to the type-II Weyl node in the right part of Fig. 11(f). The contribution of the other iso-frequency surface to the density of states is not included. The peak indicates the location of the type-II Weyl node.

Secondly, as presented in Fig. 13(a), the positive group velocity of the surface states is determined by the strongly tilted cone spectrum of type-II Weyl nodes. For $k_z > 0$ with fixed $k_{//} = 1/\sqrt{2}$, the two surface states localized at the opposite surfaces of the system both acquire group velocities of the same positive sign. In contrast, in the previously demonstrated topological surface states of acoustic type-I Weyl system [6], their propagation direction is surface-dependent: if the surface states on one surface propagate in one direction, those on the opposite surface should propagate in the opposite direction. This distinction is schematically illustrated in Fig. 14(a).

Thirdly, another distinction is that the surface states of type-II Weyl nodes stay in an incomplete bandgap. By putting sound sources with frequency $0.26 \times 2\pi c/a$ (900.1 Hz) at the surface of the acoustic type-II Weyl system, we can study the transport features of the surface states. As shown in Fig. 14(b), we consider a structure that is finite in the x–y direction. Under the condition of single-frequency excitation, the surface states as well as the bulk states can both be excited by the sound source, as a consequence of the incomplete bandgap associated with type-II Weyl node. At the two bottom corners, the surface states will get scattered into the bulk because of the existence of backscattering modes. Therefore, they do not have the same robustness (scattering-immunity against defects) as demonstrated in Ref. [6], but the existence of open "Fermi arcs" connecting Weyl nodes in the momentum space is robust.

4. Landau levels of sound

Haldane's model for the quantum Hall effect is "without Landau levels" [15]. It is thus interesting to discuss the possibility of acoustic Landau levels. In a magnetic-free circumstance, it has been reported that strained graphene can form gauge field with effective magnetic field reaching 300 tesla [20], opening the door to the strain engineering of graphene electronics. Shortly, this idea of strain engineering has been introduced into photonics [21], where a strained honeycomb photonic lattice, being invariant along the z direction, can form photonic Landau levels. However, these photonic Landau levels refer to the quantization of k_z momentum, rather than the energy (frequency), being fundamentally difficult to be implemented in a 2D geometry.

Here we introduce the strain-induced gauge fields into a 2D acoustic structure hosting air-borne acoustic waves. The strain effect makes the acoustic lattice aperiodic, which can be realized by simply displacing the lattice sites away from their original positions. The strain-induced gauge field corresponds to a strong uniform magnetic field, enabling the emergence of discretized Landau levels separated by significant band gaps. Comparing the previously demonstrated photonic Landau levels [21] and the acoustic Landau levels studied in the current work, one can find the following fundamental distinctions. First, the photonic Landau levels are discrete momentum levels of wavevector k_z, while the current acoustic Landau levels are energy levels. Second, the previous photonic Landau levels are demonstrated in a 3D structure whose z dimension should be infinitely long, while the current acoustic Landau levels are designed in a 2D geometry. Most content in this section has been published in Ref. [12].

4.1. *Acoustic honeycomb lattice with Dirac cones*

We start with an acoustic honeycomb lattice. The unit cell of the lattice consists of two identical acoustic resonators (cylinders A and B) connected with a thin

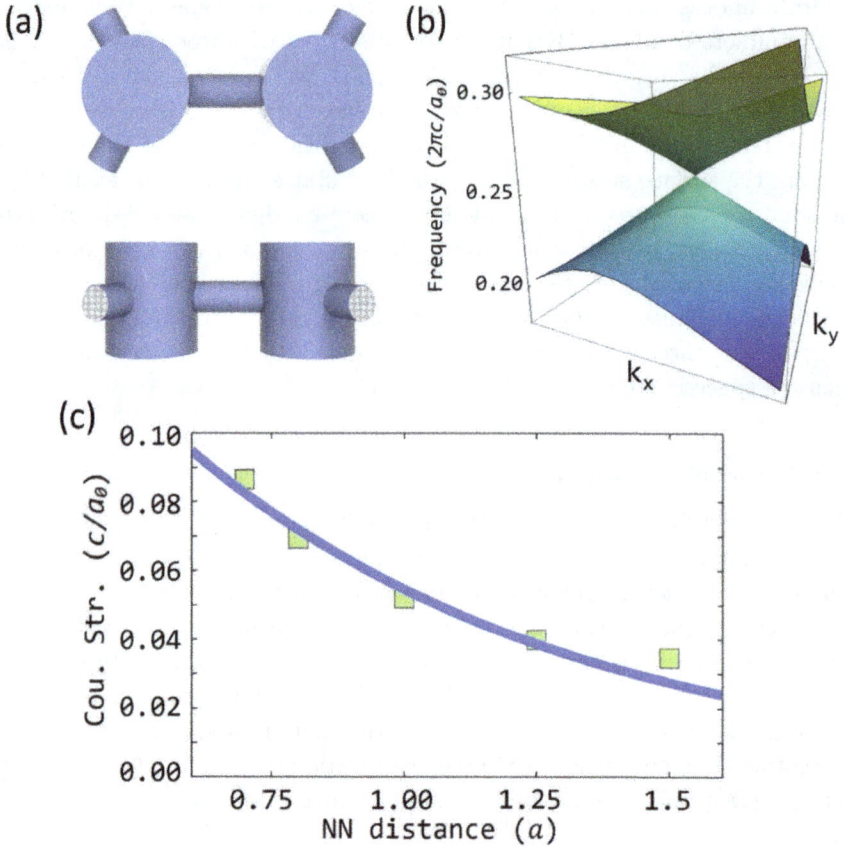

Fig. 15.   (a) Top and side views of a unit cell of the acoustic honeycomb lattice. The unit cell consists of two identical resonators at inequivalent sites. Each pair of resonators are connected with a thin coupling waveguide. (b) The Dirac cone spectrum in a square area with side length $0.4\pi/a$ locates at a corner of the Brillouin zone. (c) Coupling strength and the fitted curve from fitting model $\kappa = \kappa_0 \exp[-\beta(r/a - 1)]$. Figure adopted from Ref. [12].

cylindrical coupling waveguide, as shown in Fig. 15(a). The system is filled with air and the blue surfaces in Fig. 15(a) are treated as acoustic hard boundaries. The lattice constant and nearest-neighbor distance between two sites are $a_0 = \sqrt{3}a$ and a, respectively. This lattice of acoustic resonators can be described by the coupled-mode equation, $i\partial_t p_n = \sum_{\langle m \rangle} \kappa(|\mathbf{r}_{n,m}|)p_m$, where p_n is the amplitude of the mode for the nth resonator, κ is the coupling strength between two resonators, and the summation is taken over the nearest-neighbour resonators. By considering only the lowest two acoustic eigenmodes, the solution to the coupled-mode equation can be equivalent to the one to the tight-binding Hamiltonian of graphene. As a result, at each of the six corners of Brillouin zone, the energy bands will be expected to cross at the Dirac point. In the following, we choose the radius and height of each resonator as $r = 0.3a$ and $h = 2a/3$. The coupling waveguide connecting resonators

A and B has the radius of $r_c = 0.1a$. We numerically calculate the band structure of the acoustic honeycomb lattice with finite-element method, in a square area (side length $0.4\pi/a$) centered at a corner of the first Brillouin zone, as shown in Fig. 15(b). It can be seen that the lowest two bands exhibit the Dirac cone spectrum in the valley K [near $\pi/a_0(4/3,0)$]. The frequency at the Dirac point is $0.255 \times 2\pi c/a_0$ (or 5041 Hz for $a = 0.01$ m), where c is the speed of sound in air.

The coupling strength κ, which is physically realized by the coupling waveguide, decays almost exponentially as the nearest-neighbour distance increases. Here we assume the coupling strength takes the form of function $\kappa = \kappa_0 \exp[-\beta(r/a - 1)]$ as it is in graphene, where κ_0 is the coupling strength for $r = a$, and β describes the decay rate of coupling strength. The parameters of $\kappa_0 = 0.056 \times c/a_0$ and $\beta = 1.271$ are retrieved by fitting the data from finite-element simulation, as shown in Fig. 15(c).

4.2. *Triaxial strain, uniform magnetic field and Landau levels*

Strain engineering in graphene has shown that a 2D strain field can induce a gauge field with the vector potential $A(r) = \frac{\beta}{a_0}(u_{xx} - u_{yy}, -2u_{xy})$, where u_{xx}, u_{yy} and u_{xy} are the elements of the strain tensor. The magnetic field then can be calculated as $B(r) = \nabla \times A$ (given units $\hbar/e = 1$). In the following, we apply the triaxial strain on the acoustic honeycomb lattice. The triaxial strain determines the space-dependent displacement for each resonator as $(d_x, d_y) = q(2xyx^2 - y^2)$, where x, y are the original location of the resonator and q describes the strength of strain. This displacement will construct the vector potential $A(r) = \frac{4\beta q}{a_0}(y, -x)$, which leads to a uniform magnetic field $B = 8\beta q/a_0$.

Once the pseudo-magnetic field is formed, Landau levels can emerge in the Dirac cone region. To demonstrate the existence of acoustic Landau levels, we first calculate the eigenvalues of a large lattice in the tight-binding limit. We adopt a honeycomb disc with radius $40a_0$, which contains 11600 resonators, and calculate the eigenvalues in the tight-binding limit. A triaxial strain with strength $q = 0.004/a$ is applied. Fig. 16 shows, in ascending order of the state number, the calculated 800 eigenvalues near the Dirac cone region for the strained lattice. It can be clearly seen that highly degenerate states and flat levels emerge in the energy spectrum. The energy gaps of Landau levels are proportional to $\sqrt{BN}$ (N is the level number), being consistent with the behavior of Dirac fermions in a strong magnetic field. The states connecting the discretized Landau levels are edge states that contribute to the Hall conductivity in the quantum Hall effect.

To facilitate potential experiments, we now consider a smaller disc with radius $6a_0$, which contains only 262 resonators. The triaxial strain with different strengths of $q = 0.000/a$, $q = 0.015/a$ and $q = 0.030/a$ are applied to the disc, as shown in Fig. 17(a–c), respectively. Note that the strain strength $q = 0.030/a$, if a piece of graphene were strained to this extent, would yield a pseudo-magnetic field of 8590 tesla. The eigenvalues of the states calculated with finite-element method are shown

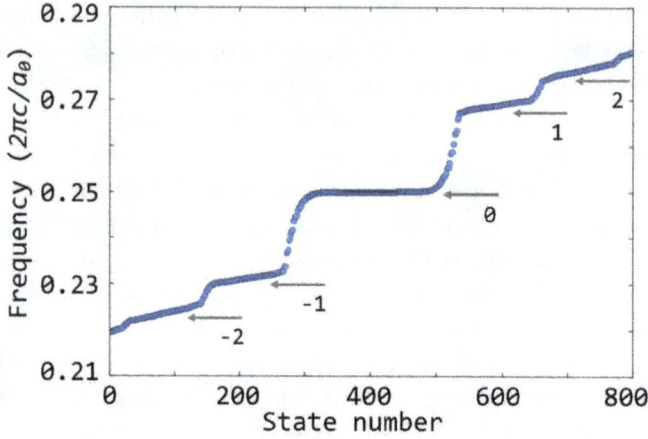

Fig. 16. Triaxial-strain induced acoustic Landau levels near the Dirac region. A disc with radius $40a_0$ which consists of 11600 resonators is adopted in the calculation. The strength of the triaxial strain is $q = 0.004/a$. The numbers indicate the orders of Landau levels. Figure adopted from Ref. [12].

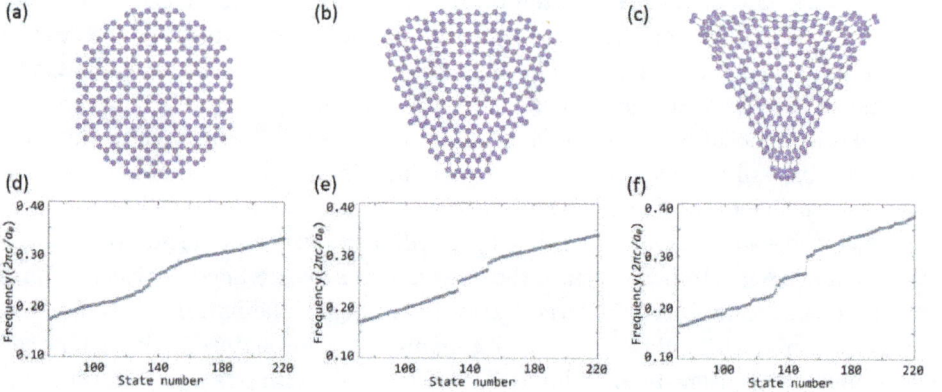

Fig. 17. (a–c) The schematics of the unstrained and strained acoustic lattices with increasing strength of triaxial strain $q = 0.000/a$ (a), $q = 0.015/a$ (b) and $q = 0.030/a$ (c). A disc with radius $6a_0$ which consists of 262 resonators is adopted in the calculation. (d–f) The eigenvalues of the acoustic lattices are plotted in terms of the state number in the ascending order. (d) Without strain, the frequency spectrum shows a continuous behavior. (e-f) Whereas under nonzero strain, the band gaps emerge and widen with increasing strain. Figure adopted from Ref. [12].

in Fig. 17(d–f) for the three cases ordered with increasing strain. In Fig. 17(d), the unstrained lattice shows continuous eigen-frequency spectrum near the Dirac cone region. Whereas for the triaxially strained lattice in Fig. 17(e) with strain strength $q = 0.015/a$, the Landau energy quantization emerges and results in band gaps between the 0th-order and $\pm 1^{st}$-order Landau levels in the frequency ranges of $(0.231, 0.252) \times 2\pi c/a_0$ and $(0.267, 0.286) \times 2\pi c/a_0$. With the increased strain of

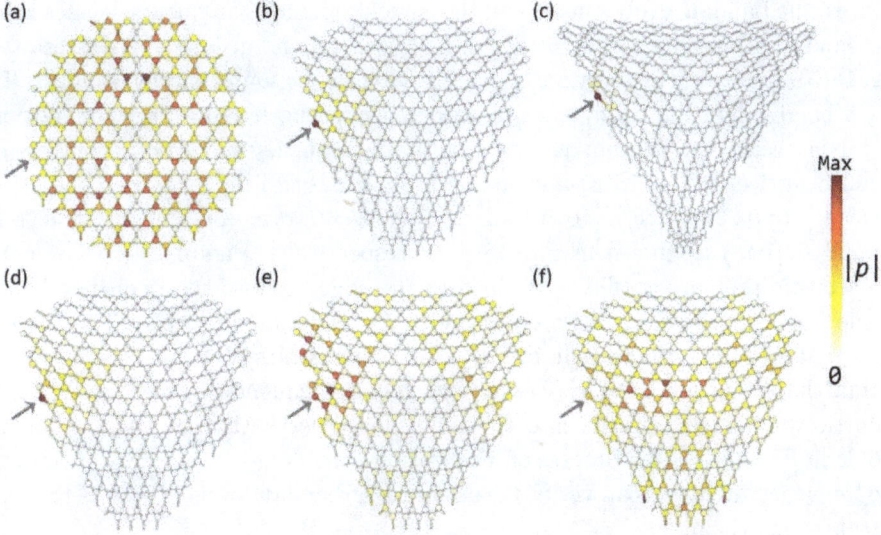

Fig. 18. (a–c) The excitation of acoustic waves at the edges of the unstrained and strained lattices. The operating frequency of the source is $0.281 \times 2\pi c/a_0$ (5565 Hz). The black arrow points to the location of the acoustic source. The extended state is excited in panel (a). Highly confined states are excited in panels (b) and (c), because the frequency locates in the band gap between 0th-order and 1st-order Landau levels. (d–f) The frequency of the source is $0.237 \times \frac{2\pi c}{a_0}$ (d), $0.261 \times \frac{2\pi c}{a_0}$ (e), $0.293 \times \frac{2\pi c}{a_0}$ (f), respectively. The intensity profiles show the transition from localization (d) – spreading (e) – localization (b) – spreading (f). The thermal color indicates the amplitude of acoustic pressure. Figure adopted from Ref. [12].

$q = 0.030/a$ in Fig. 17(f), as the effective magnetic field doubles, these band gaps expand to frequency ranges of $(0.223, 0.252) \times 2\pi c/a_0$ and $(0.266, 0.298) \times 2\pi c/a_0$. In other words, the gap sizes in Fig. 17(f) are roughly 1.5 times the gap sizes in Fig. 17(e), being consistent with the energy gaps of Landau levels that increase as $\sim \sqrt{BN}$. With increased strain, the states of the 0th-order Landau level becomes more flat and thus achieves higher degeneracies leading to slower acoustic wave propagation. There can be relatively more edge states connecting the Landau levels, as it is seen in Fig. 17(f) between the -1st-order and 0st-order levels. Note that the Landau levels of higher orders can also be labeled in principle based on the relationship of $\sim \sqrt{BN}$, where the coefficient can be obtained through data fitting.

4.3. A potential scheme to experimental demonstration

We then propose a simple scheme that can be adopted in experiment to demonstrate the acoustic Landau levels. By fixing the operating frequency of $0.281 \times 2\pi c/a_0$ (or 5565 Hz for $a = 0.01$ m), we can put a sound source, as indicated by the black arrow in Fig. 18(a–c), at edges of the unstrained and strained lattices, to excite the same resonator. In Fig. 18(a), the acoustic pressure spreads into the bulk of the unstrained lattice because of the excitation of extended bulk state. For strained lattices, the Landau levels emerge and the current operating frequency locates inside the band gap between the 0th order and 1st order Landau levels. As a result, in Fig. 18(b), the strong localization on one edge makes the sound energy confined. The larger band gap in Fig. 18(c) leads to shorter decay length and tighter confinement. To further verify the Landau quantization in the frequency spectrum, we choose the strained lattice in Fig. 18(b) and simulate the acoustic field by setting the source frequency to $0.237 \times 2\pi c/a_0$, $0.261 \times 2\pi c/a_0$ $0.293 \times 2\pi c/a_0$ (or 4700, 5158, 5808 Hz for $a = 0.01$ m) as shown in Fig. 18(d–f), respectively. Figure 18(d) shows that the acoustic pressure profile is confined to the edge because the frequency locates in the band gap between -1st order and 0th order Landau levels. The acoustic wave that spreads into the bulk in Fig. 18(e) demonstrates one localized eigenstate within the 0th order Landau level. With higher frequency as in Fig. 18(f), the acoustic wave again spreads into the bulk. Together with the acoustic pressure profile in Fig. 18(b), this process of localization-spreading-localization-spreading as the frequency increases can verify the existence of Landau levels as well as the band gaps between them.

5. Summary

Here we have reviewed the recent progress made in our group about topological acoustics. Firstly, we move the first step in constructing acoustic topological phases by following Haldane and Raghu's work of topological photonic crystal. The time-reversal symmetry can be broken in an acoustic crystal by incorporating circulating air flow. For acoustic waves, the system behaves like a zero-field quantum Hall system, similar to Haldane's model. The band below the gap acquires a non-zero Chern number, which ensures the existence of the topologically protected one-way acoustic edge state.

Secondly, we construct the acoustic version of type-II Weyl Hamiltonian by stacking one-dimensional dimerized chains of acoustic resonators. This acoustic type-II Weyl system exhibits distinct features in finite density of states and unique transport properties of Fermi-arc-like surface states. In a certain momentum space direction, the velocity of these surface states are determined by the tilting direction of the type-II Weyl nodes, rather than the chirality dictated by the Chern number. This clarifies the previous proposal of acoustic Weyl nodes by explicitly distinguishing type-I and type-II Weyl systems.

Finally, we construct the acoustic Landau levels by applying strain engineering to a lattice of acoustic resonators. Different from previous photonic Landau levels that are levels of momentum quantization, the acoustic Landau levels we propose are energy (frequency) levels and can be realized in a 2D geometry. This also supplements the above first part where Landau levels are inherently missing because of the Haldane's model.

It has been about two years since the first work in topological acoustics. Although most works are so far limited to the theoretical level, we have noticed considerable rise of experimental efforts from many groups for demonstrating various proposals. We hope this review can help to provide a hindsight on some topological phases for sound developed in our group.

References

1. L. Lu, J. D. Joannopoulos and M. Soljačić, Topological photonics, *Nat. Photon.* **8**, 821 (2014).
2. L. M. Nash, D. Kleckner, A. Read, V. Vitelli, A. M. Turner, and W. T. M. Irvine, Topological mechanics of gyroscopic metamaterials, *Proc. Natl. Acad. Sci.* **112**, 14495 (2015).
3. Z. Yang, F. Gao, X. Shi, X. Lin, Z. Gao, Y. Chong, and B. Zhang, Topological Acoustics, *Phys. Rev. Lett.* **114**, 114301 (2015).
4. X. Ni, C. He, X.-C. Sun, X.-P. Liu, M.-H. Lu, L. Fu, and Y.-F. Chen, Topologically protected one-way edge mode in networks of acoustic resonators with circulating air flow, *New J. Phys.* **17**, 053016 (2015).
5. A. B. Khanikaev, R. Fleury, S. H. Mousavi, and A. Alù, Topologically robust sound propagation in an angular-momentum-biased graphene-like resonator lattice, *Nat. Commun.* **6**, 8260 (2015).
6. M. Xiao, W.-J. Chen, W.-Y. He, and C. T. Chan, Synthetic gauge flux and Weyl points in acoustic systems, *Nat. Phys.* **11**, 920 (2015).
7. M. Xiao, G. Ma, Z. Yang, P. Sheng, Z. Q. Zhang, and C. T. Chan, Geometric phase and band inversion in periodic acoustic systems, *Nat. Phys.* **11**, 240 (2015).
8. Z. Yang and B. Zhang, Acoustic Type-II Weyl Nodes from Stacking Dimerized Chains, *Phys. Rev. Lett.* **117**, 224301 (2016).
9. C. He, X. Ni, H. Ge, X.-C. Sun, Y.-B. Chen, M.-H. Lu, X.-P. Liu, and Y.-F. Chen, Acoustic topological insulator and robust one-way sound transport, *Nat. Phys.* **12**, 1124 (2016).
10. J. Lu, C. Qiu, L. Ye, X. Fan, M. Ke, F. Zhang, and Z. Liu, Observation of topological valley transport of sound in sonic crystals, *Nat. Phys.* **13**, 369 (2017).
11. Y. G. Peng, C. Z. Qin, D. G. Zhao, Y. X. Shen, X. Y. Xu, M. Bao, H. Jia, and X. F. Zhu, Experimental demonstration of anomalous Floquet topological insulator for sound, *Nat. Commun.* **7**, 13368 (2016).
12. Z. Yang, F. Gao, Y. Yang, and B. Zhang, Strain-Induced Gauge Field and Landau Levels in Acoustic Structures, *Phys. Rev. Lett.* **118**, 194301 (2017).
13. F. Li, X. Huang, J. Lu, J. Ma , and Z. Liu, Weyl points and Fermi arcs in a chiral phononic crystal, *Nat. Phys.* (2017) doi:10.1038/nphys4275.
14. F. D. M. Haldane and S. Raghu, Possible Realization of Directional Optical Waveguides in Photonic Crystals with Broken Time-Reversal Symmetry, *Phys. Rev. Lett.* **100**, 013904 (2008).
15. F. D. M. Haldane, Model for a Quantum Hall Effect without Landau Levels: Condensed-Matter Realization of the "Parity Anomaly", *Phys. Rev. Lett.* **61**, 2015 (1988).
16. Z. Wang, Y. Chong, J. Joannopoulos, and M. Soljacic, Observation of unidirectional backscattering-immune topological electromagnetic states, *Nature* **461**, 772 (2009).
17. M. V. Berry, R. G. Chambers, M. D. Large, C. Upstill, and J. C. Walmsley, Wavefront dislocations in the Aharonov-Bohm effect and its water wave analogue, *Eur. J. Phys.* **1**, 154 (1980).
18. K. Sun, W. V. Liu, A. Hemmerich, and S. Das Sarma, Topological semimetal in a fermionic optical lattice, *Nat. Phys.* **8**, 67 (2011).

19. A. A. Soluyanov, D. Gresch, Z. Wang, Q. Wu, M. Troyer, X. Dai, and B. A. Bernevig, Type-II Weyl semimetals, *Nature* **527**, 495 (2015).
20. N. Levy, S. A. Burke, K. L. Meaker, M. Panlasigui, A. Zettl, F. Guinea, A. H. C. Neto, and M. F. Crommie, Strain-Induced Pseudo–Magnetic Fields Greater Than 300 Tesla in Graphene Nanobubbles, *Science* **329**, 544 (2010).
21. M. C. Rechtsman, J. M. Zeuner, A. Tunnermann, S. Nolte, M. Segev, and A. Szameit, Strain-induced pseudomagnetic field and photonic Landau levels in dielectric structures, *Nat. Photon.* **7**, 153 (2013).

Influences of geometry and topology in nuclei

Martin Freer

School of Physics and Astronomy, University of Birmingham,
Birmingham, B15 2TT, UK
E-mail: M.Freer@bham.ac.uk

This contribution describes the latest developments in the field of light nuclei and explores how the tendency of nucleons to cluster into α-particle subunits influences their structure. In particular there is a focus on the structure of the nucleus ^{12}C and in turn its link to nucleosynthesis and the formation of organic life.

Keywords: Nuclear structure, nuclear clustering, geometry and topology.

1. The origins of carbon

The synthesis of nuclear matter was first primordial in the aftermath of the Big Bang with the formation of the lighter elements such as hydrogen and helium, with only traces of heavier elements produced due to both the electrostatic repulsion (Coulomb barrier) of the larger charges of the nuclei and the lack of stable mass 5 elements blocking their production. This resulted in an early Universe dominated by hydrogen and helium. Thereafter, synthesis proceeded within the stellar cycle. First this occurs as hydrogen burning to produce helium and then in the red giant phase, helium burning crosses the mass 5 stability gap to create to the heavier elements; most significant of which is carbon-12.

The synthesis of carbon-12, ^{12}C, in the helium burning phase occurs through what is known as the triple α-process[1,2], where first two α-particles fuse to form ^{8}Be whose instability to α-decay results in an equilibrium concentration of ^{8}Be. The second step is the capture by ^{8}Be to form ^{12}C at an excitation close to 7.5 MeV followed by electromagnetic decay to the ^{12}C ground state (via two gamma-rays, see Fig. 1). Fred Hoyle recognised the need for a $J^{\pi} = 0^{+}$ state close to this energy in order to account for the absolute abundance of ^{12}C and the relative abundance of ^{12}C and ^{16}O[3]. The presence of the Hoyle-state at this energy resonantly boosts the capture process by a factor of close to 10-100 million. Hoyle predicted the existence of a state at 7.68 MeV[3]. Researchers at the Kellogg Laboratory at Caltech searched for the state. They measured the ^{14}N(d,α)^{12}C reaction using a high resolution spectrometer at which point a state at 7.68 MeV was observed[4]. Subsequent measurements refined the energy of the state to 7.653 ± 0.008 MeV and indicated the most probable spin and parity to be 0^{+}[5] (now firmly established).

The connection between the existence of organic life and ultimately human-kind has led to the interpretation[6] that the prediction of the existence of 7.65 MeV state by Hoyle was an example of the anthropic principle, an idea introduced by Carter in 1973[7]. The principle relies on the fact that intelligent life exists, to assert certain properties of the universe must exist, i.e. we exist therefore so must the 7.65 MeV state in ^{12}C. The question as to if Hoyle deployed the anthropic principle or not has been recently reviewed by Kragh[8]. The conclusion was that Hoyle's reliance was on the challenge of understanding the natural abundance of ^{12}C and ^{16}O (the latter produced by capture of an α-particle on ^{12}C) that drove his prediction, rather than a deeper connection to his own existence. Indeed, it is apparent that there was experimental evidence for the existence of the Hoyle-state which predated the measurements of the Caltech group[4].

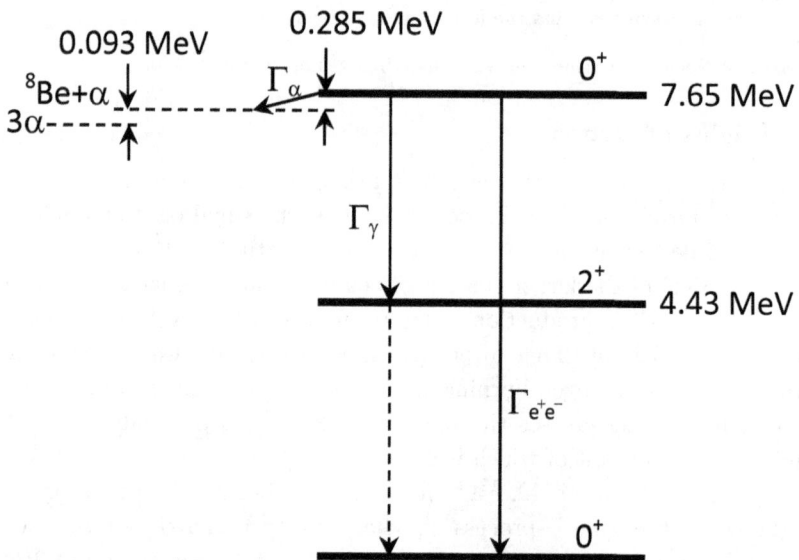

Fig. 1. Schematic of the decay process by which ^{12}C is formed from the radiative ^{8}Be + ^{4}He capture. From [10].

The Hoyle-state is thus intimately linked to the formation of organic life and although it unlikely to have been the anthropic principle that drove the discovery[8], nevertheless, the 7.65 MeV, 0^+, Hoyle-state appears to have an unusual structure, which turns out to be dominated by the presence of α-particle clusters.

2. Clusters and correlations

The structure of nuclear matter is rich and varied. From one perspective, the nucleus may behave like a liquid drop, with its shape and size corresponding to a balance between the long(ish) range attractive and short range repulsive behaviour of the nucleon-nucleon interaction and the charges of the constituent protons. This liquid drop displays collective properties such as vibrations, where vibrational modes distort the nuclear surface. If the nucleus can be encouraged to deform and then can be rotated – as the droplet spins it stretches which provides a mechanism for the determination of the equation-of-state of the fluid. At a critical angular momentum the droplet will fission. Similarly as the mass of a nucleus increases, typically so does the number of protons and hence the charge. The repulsive Coulomb energy should cause the nucleus to spontaneously fission when the number of protons is close to 100. However, it is at this point that another crucial feature contributes which allows nuclei to exist beyond that point - shell effects. Shell structure, which features for light and heavy nuclei alike, is associated with the quantal properties of the nucleus and marks a deviation from the constituent particles to a picture in which the particles are represented by standing waves. The associated quantum states are those of the nuclear shell model and give rise to a sequence of magic numbers which are associated with enhanced stability. A superposition of the macroscopic liquid drop and microscopic shell model-like behaviour is required to describe the stability of nuclei nuclei beyond the point at which the charged liquid drop should explode.

For light nuclei there is a similar interplay between the collective and single-particle nature, but here details of the properties of the interaction between the nucleons becomes increasingly important. Correlations become a dominant feature. The pairing interaction is evident in the nature of the drip-lines, which define the limits of stability on both the proton and neutron-rich side of the chart of nuclides (see Fig. 2). For the helium nuclei, ^{4}He is stable, whereas ^{5}He is not. Similarly, ^{6}He and ^{8}He are stable and ^{7}He is not. The difference being that in addition to the ^{4}He core the stable isotopes have even numbers of neutrons, whereas the unstable ones do not. ^{6}He and ^{8}He are known as Borromean nuclei, as for example in the case of ^{6}He if a neutron is removed then the other two components dissociate; further if the α-particle is extracted then this leaves the unbound 2n system. Such systems are symbolised by the famous 3 interlocking rings.

As an example of potentially exotic structures on the proton-rich side the ^{10}C nucleus sits at the top of a loop around unbound nuclei which include ^{9}B and ^{8}Be. ^{10}C may be thought of being composed of two protons and two α-particles and if any of the components are removed then the other three dissociate. This may be considered as a super-Borromean nucleus, or recognising that Borromean systems belong to a class of mathematical objects called Brunian knots then ^{10}C is a nucleus which is 4th order knot (as illustrated in Fig. 2).

These are rather extreme examples of correlations, but they are rather common-place in light nuclei and have a determining role when it comes to the structure. These correlations can be spatial in addition to energy or momentum and then are referred to as clusters. The most prevalent cluster is the α-particle due to its remarkably high binding and inertness.

Even predating the above first experimental hint of the existence of the 7.65 MeV state in 1940[9], Hafstad and Teller[11] (1938) analysed the binding energy behaviour of the ground states of $N = Z$, N even, nuclei (α-conjugate nuclei), e.g. ^{8}Be, ^{12}C,

^{16}O .... A linear dependence was found between the number of inter α-particle bonds (e.g. ^{8}Be - 1, ^{12}C - 3, ^{16}O - 6, ...) and the binding energies, indicating that there was a fixed energy associated with the α-α interaction and further that the α-clusters featured within the ground-states of these systems.

This view of the ground-states fell from favour as the single-particle description of nuclei became vogue and, in fact, was indeed an over simplification of the structural properties. Some thirty years later Ikeda and co-workers proposed a threshold picture for the appearance of cluster states[12]. The essence is that close to decay thresholds it is possible for nuclei to undergo structural changes. For example, the ground state of ^{8}Be being unbound to decay to two α-particles by 92 keV, might have a well-developed α-cluster structure. The ^{8}Be+α decay threshold, or equivalently 3α decay threshold (7.274 MeV), would permit the internal excitation energy to be transformed into the binding energy of the clusters as displayed in the Ikeda diagram shown in Fig. 3[12].

This then connects the decay thresholds with both nucleosynthesis and also for the formation of clusters. Hence the appearance of nuclear clusters, and in particular α-particles, has been strongly connected with element formation. Once one has established that the α-clusters exist then there are some obvious questions which precipitate, for example how the clusters are arranged and how pervasive is the phenomenon of α-clustering. For the three α-particle system there are at least two options; (i) a triangular geometry and (ii) a linear arrangement. Experiments that have been used to provide an insight into this structure are described later.

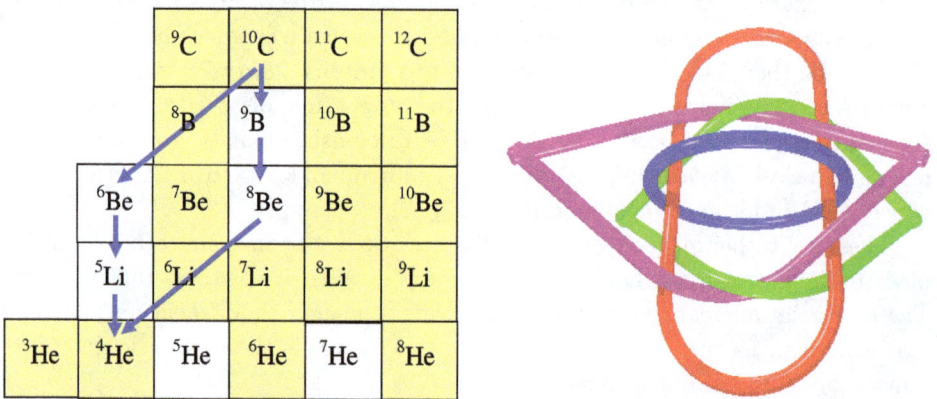

Fig. 2. Light nuclei. Filled squares are either stable or beta-decay, unfilled particle (neutron, proton or α) decay. The arrows show the paths corresponding to the removal of a proton or α-particle from ^{10}C. The diagram on the right hand side illustrates the 4th order Brunian knot.

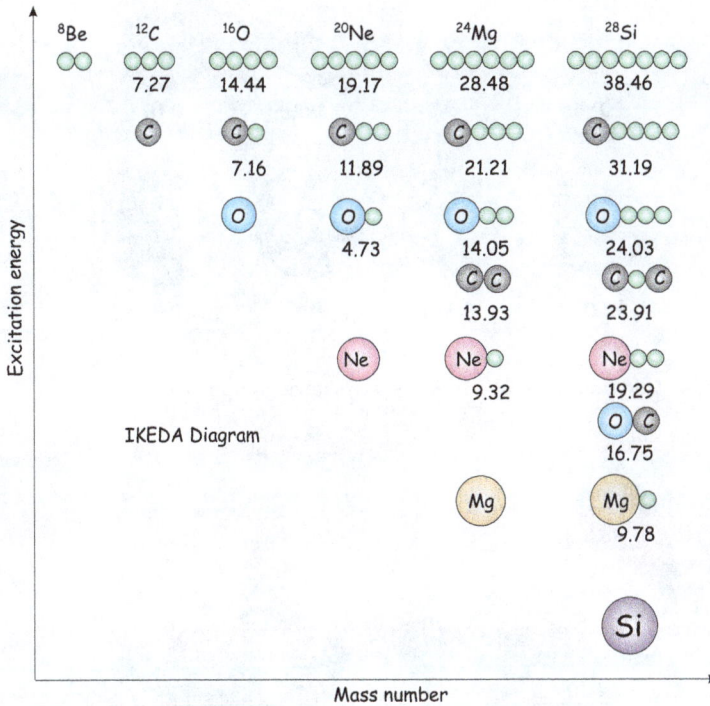

Fig. 3. The Ikeda threshold diagram. This illustrates that for the ^{12}C nucleus above 7.27 MeV it is possible for the system to form into a structure of 3α-particles.

3. Topology

The extension from geometry to topology is perhaps not obvious, but arises from an entirely different approach to the calculation of the properties of nuclei using skyrmions. The skyrmion was developed by Tony Skyrme in 1962[13] to first of all deride the properties of the nucleon. Here the skyrmion was used to describe the topological fields associated with soliton excitations of the pion cloud associated the with strong force. Skyrmions then found favour in solid state physics. More recently, the group of Manton *et al.* have explored how to employ skyrmions in the description of light nuclei[14]. Here the conserved topological charge of the solitons is identified with the baryon number, B, see Ref. 14 for full details. Some of the solutions for baryon number 1-8 are shown in Fig. 4.

Figure 5 shows the solutions corresponding to ^{12}C, B=12, for two configurations where there are three B=4 clusters. Here Lau and Manton[15] have calculated moments of inertia and rotational spectra and made links to the various experimentally known states in ^{12}C. The conclusion of this work was that the ground state has a triangular structure (D_{3h}), whereas the Hoyle state is associated with the linear structure (D_{4h}). Hence, there appears to be an intimate link between the states of

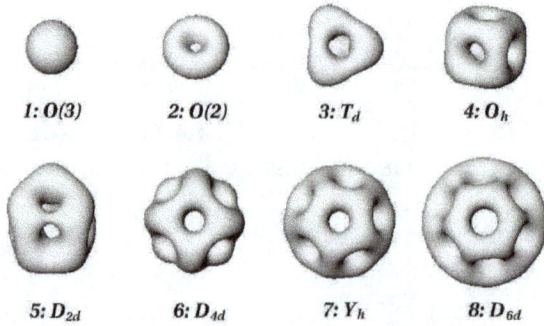

Fig. 4.   Skyrmions for B 1 to 8, the symmetry group is indicated. From[14].

Fig. 5.   Skyrmions for B=12, with D_{3h} (left) and D_{4h} symmetries. From Ref. 15.

^{12}C, geometry and topology. Next we turn to an examination of the experimental evidence, and evidence from more traditional state-of-the-art nuclear theory.

4. Clustering above and below the decay threshold and dynamical symmetries

As described above, it is recognised and accepted that clustering plays a crucial role above the cluster-decay threshold. Here this is well illustrated by the AMD calculations for ^{12}C (Fig. 6). The AMD (Antisymmetrised Molecular Dynamics) approach is based on the calculation of the nuclear properties from a fully anti-symmetrised A nucleon wavefunction with an effective (truncated) nucleon-nucleon interaction. The densities associated with the structures above the 3α decay threshold demonstrate that the nucleus precipitates into 3 clusters. However, below the decay threshold, and even in the ground-state, clustering also appears to have a structural influence. Demonstrating such a structure exists, experimentally, is not simple. The ground-state is compact, though still reveals the 3α structure, which is also strongly influenced by the ^{8}Be+α substructure. The collective excitations of this state are shown in part (b1) and (c1), for the 2^+ and 4^+ states, respectively. Ultimately, measurements of the electromagnetic transition strengths, $B(E2)$, between the states above the α-decay threshold are the litmus test of which states may be strongly structurally linked. However, such measurements are currently too challenging and other approaches are required.

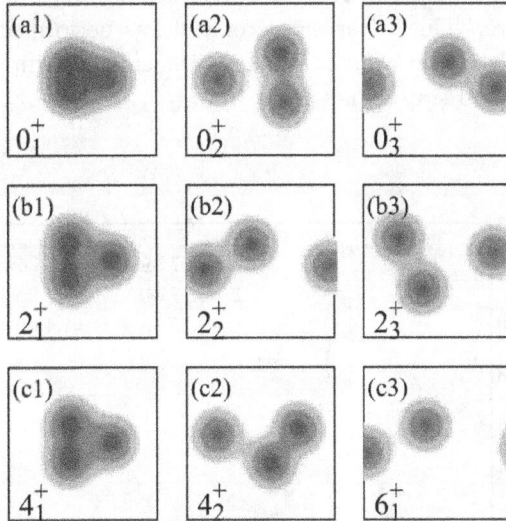

Fig. 6.   AMD densities of the ground state band, 0_1^+: a1, b1 and c1, Hoyle-band, 0_2^+ a2, b2, c2 and 0_3^+ band a3, b3, c3, from Ref. 16.

The dynamical symmetries of a triangular 3α-system correspond to a spinning top with a triangular point symmetry. The rotational properties of these states are given by

$$E_{J,K} = \frac{\hbar^2 J(J+1)}{2\mathcal{I}_{Be}} - \frac{\hbar^2 K^2}{4\mathcal{I}_{Be}} \tag{1}$$

where $\mathcal{I}_{Be}$ is the moment of inertia corresponding to two touching α-particles, which can be determined from the ^{8}Be ground-state rotational band[11]. One would expect that based on this structure there should be a number of rotational bands with different values of K. For $K^\pi=0^+$, the rotations will be around an axis which lies in the plane of the three α-particles, generating a series of states 0^+, 2^+, 4^+ .... These correspond to the rotation of a ^{8}Be nucleus - the rotation axis passing through the centre of the third α-particle. The next set of rotations correspond to the rotation around an axis perpendicular to the plane of the triangle, with each α-particle having one unit on angular momentum - giving $L = 3 \times 1\hbar$; $K^\pi=3^-$. Rotations around this axis and that parallel to the plane combine to give a series of states 3^-, 4^-, 5^-.... Such an arrangement possesses a D_{3h} point group symmetry, as was indicated in Fig. 5.

From the experimental perspective, the ground state rotational band is known for the 2^+ and 4^+ members at 4.44 and 14.08 MeV, as is the 3^- excitation at 9.64 MeV. Recent measurements of the width of the 3^- state have indicated that it has a well-developed alpha-cluster structure[17] and the 4^- member of the band has been identified at 13.2 MeV[18,19]. A key measurement has been the determination of the 5^- member of the K =3 band at 22.5 MeV[20], which has been taken as confirmation of the D_{3h} structure. This latter measurement was performed on the University of Birmingham MC40 cyclotron. These measurements confirm the triangular, D_{3h} structure of the ground state band.

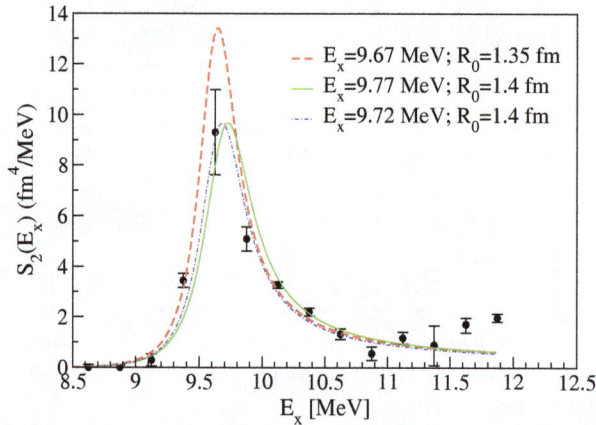

Fig. 7. The spin 2 strength function (S_2(Ex)) extracted from the multipole decomposition analysis[22] (data points) compared with the R-matrix line-shapes associated with different channel radii and excitation energies. Ref. 23.

As early as 1956 it was suggested that the Hoyle state could be highly deformed[21]. However, experimental verification of this conjecture has been slow to materialise. A common analysis of the evidence for a 2^+ resonance from the proton- and α-particle scattering from ^{12}C was reported in Ref. 23, and a discussion of the impact of these measurements is given in Ref. 24. The 2^+ lineshape which is found in the inelastic scattering measurements $^{12}C(\alpha,\alpha')$ and $^{12}C(p,p')$[23] is shown in Fig. 7. The energy of the 2^+ resonance is here determined as $E_x = 9.75(0.15)$ MeV with a width of 750(150) keV.

This existence of the 2^+ resonance seen in the proton- and α-particle scattering experiments was confirmed by a measurement of the $^{12}C(\gamma,3\alpha)$ reaction at the HIγS facility[25]. The angular distributions of the emitted α-particles are shown in Fig. 8

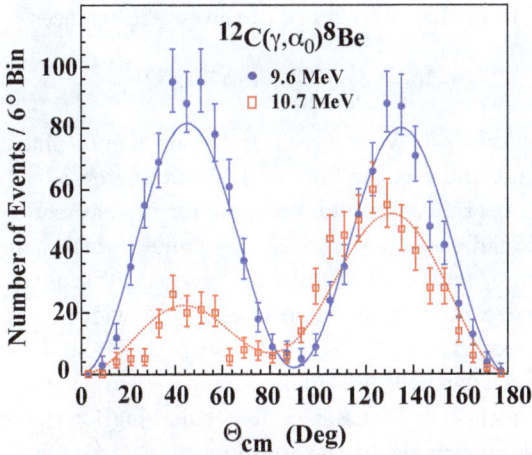

Fig. 8. Angular distribution for $^{12}C(\gamma,\alpha_0)^8Be$ events measured at a beam energy of 9.6 and 10.7 MeV. The solid curve is the fit that included E1 and E2 amplitudes[25].

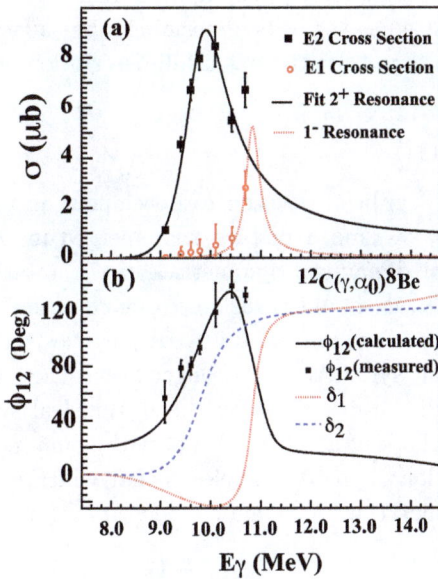

Fig. 9. (a) The measured E1 and E2 cross sections of the $^{12}C(\gamma,\alpha_0)^8Be$ reaction. (b) The measured E1-E2 relative phase angle (ϕ_{12}) together with the phase angle calculated from a two-resonance model[25].

at two different beam energies; 9.6 and 10.7 MeV. These reveal both an E2 term plus an interference with an E1 component. The E2 component is associated with the newly discovered 2^+ resonance and the E1 the known 1^- state at 10.8 MeV. Fig. 9 shows the 2^+ line shape and the phase angle extracted from the data compared with that calculated from the interference of the two resonances

$$\phi_{12} = \delta_2 - \delta_1 + \tan^{-1}(\eta/2) \tag{2}$$

where the nuclear phase shifts δ are given by the resonance phase shift minus the hard sphere phase shift and η is the Sommerfeld parameter[25].

In addition, this experiment provides a value for the electromagnetic transition rate between the ground state and the 2^+ resonance, which can be used to test models of its structure, and is important for assessing the influence of this new resonance on the 3α reaction rate. The energy and width of the 2^+ resonance determined by this experiment is $E_x = 10.13(6)$ MeV and $\Gamma = 2.1(3)$ MeV[26], which indicates a somewhat wider resonance compared to what is determined from the scattering experiments[23]. Evidence is also emerging for the existence of a 4^+ resonance in the neighbourhood of the well-known 4^+ resonance at 14.2 MeV[27]. This is supported by a preliminary report from JYFL[28].

The experimental data for the 0^+ Hoyle state and the 2^+ and 4^+ rotational excitations demonstrate that indeed the deformation of the Hoyle-state is larger than that of the ground state. As yet, though, no definitive understanding of the exact geometry of the state has yet to be determined, but at present it would seem less likely that the linear, D_{4h}, structure is a full description of the state.

5. Other systems: ^{16}O

A very significant effort has been devoted to developing an understanding of the structure of the nucleus ^{12}C and a picture has emerged in which the states can be described in terms of dynamical symmetries of the rotating system. If this description is correct then it should translate to neighboring nuclei. Certainly in ^{8}Be the nucleus can be described in terms of two α-particles, with a set of rotational states consistent with that symmetry. A stronger test is the nucleus ^{16}O.

The 4α system should be described by the tetrahedral symmetry group; T_d. Here the properties are those of a spherical top, with equal moments of inertia. If one assumes the separation of the α-particles is that which is associated with the ^{8}Be ground state, $\mathcal{I}_{Be}$, then the rotational energies are given by

$$E_J = \hbar^2 \frac{J(J+1)}{4\mathcal{I}_{Be}} \tag{3}$$

The rotations of the tetrahedral structure corresponds to the equivalent rotation of two ^{8}Be nuclei around their symmetry axis and hence the $4\mathcal{I}_{Be}$ in the denominator. The symmetry then dictates that all values of J are permitted except $J = 1, 2$ and 5; states with $J = 0, 4$ and 8 have even parity and $J = 3, 7$ and 11 have negative parity. A key feature of this structure would be degenerate 6^+ and 6^- states.

To date, the experimental characterisation of ^{16}O does not match that of ^{8}Be and ^{12}C, and is work for the future.

6. Conclusion

This contribution provides a summary which focused on the structure of light nuclei and some of the influences of topology. Halo nuclei such as ^{6}He and the proton rich system ^{10}C have properties, such as binding, which connect them with mathematical knots of order 3 and 4. These nuclei are strongly influenced by the clustering of nucleons inside the nucleus, particularly into the highly stable α-particles. These α-particles influence the structure of ^{12}C. The properties of ^{12}C have been calculated using an approach based on skymions, topological excitations of the pion field, following the original developments of Tony Skyrme. This approach generates structures which are very close to those produced by more conventional nuclear models and perhaps can be considered a topological influence. Within ^{12}C the Hoyle state is responsible for the formation of carbon in stellar nucleosynthesis. The link of the Hoyle-state to a cluster structure and to the topology beyond, may be argued to point to the role of topology in the origins of life itself, though strictly this may be stretching things.

Acknowledgments

The author would like to acknowledge that the ideas developed here are part of a wider programme of work involving colleagues both from the University of Birmingham and a number of other institutions. The funding support of the UK STFC Research Council is also acknowledged.

References

1. E. Öpik, *Proceedings of the Royal Irish Academy* **A 54**, 49 (1951).
2. E. E. Salpeter, *Astrophysical Journal* **115**, 326 (1952).
3. F. Hoyle, *Astrophysical Journal, Supplement Series* **1**, 12 (1954).
4. D. N. F. Dunbar, R. E. Pixley, W. A. Wenzel and W. Whaling, *Phys. Rev.* **92**, 649 (1953).
5. C. W. Cook, W. A. Fowler, C. C. Lauritsen and T. Lane, *Phys. Rev.* **107**, 508 (1957).
6. J. D. Barrow and F. J. Tipler, *The Anthropic Cosmological Principle* Cambridge University Press, 1986.
7. B. Carter, *Confrontation of Cosmological Theories with Observational Data* Dordecht: Reidel, 291 (1974).
8. H. Kragh, *Archive for History of Exact Sciences* **64**, 721 (2010).
9. M. G. Holloway and B. L. Moore, *Phys. Rev.* **58**, 847 (1940).
10. M. Freer, H. O. U. Fynbo, *Progress in Particle and Nuclear Physics* **78**, 1 (2014).
11. L. R. Hafstad, E. Teller, *Phys. Rev.* **54**, 681 (1938).
12. K. Ikeda *et al.*, *Prog. Theor. Phys. Suppl., Extra Numbers* **464**, 1 (1968).
13. T. Skyrme, *Nuclear Physics* **31**, 556 (1962).
14. N. S. Manton, arXiv:1106.1298.
15. P. H. C. Lau and N. S. Manton, *Phys. Rev. Lett.* **113**, 232503 (2014).
16. Y. Kanada En'yo, *Prog. Theor. Phys.* **117**, 655 (2007).
17. Tz. Kokalova, M. Freer, Z. Buthelezi, J. Carter, R. W. Fearick, S. V. Förtsch, H. Fujita, R. Neveling, P. Papka, F. D. Smit, J. A. Swartz and I. Usman, *Phys. Rev. C* **87**, 057307 (2013).

18. M. Freer *et al.*, *Phys. Rev.* C **76**, 034320 (2007).
19. O. S. Kirsebom *et al.*, *Phys. Rev.* C **81**, 064313 (2010).
20. D. J. Marín-Lambarri *et al.*, *Phys. Rev. Lett.* **113** 012502 (2014).
21. H. Morinaga, *Phys. Rev.* **101**, 254 (1956).
22. M. Itoh *et al.*, *Phys. Rev.* C **84**, 054308 (2011).
23. M. Freer *et al.*, *Phys. Rev.* C **86**, 034320 (2012).
24. H. O. U. Fynbo and M. Freer, Physics, 4, 94 (2011).
25. W. R. Zimmerman *et al.*, *Phys. Rev. Lett.* **110**, 152502 (2013).
26. W. R. Zimmerman, PhD thesis, and private communication.
27. M. Freer *et al.*, *Phys. Rev.* C **83**, 034314 (2011).
28. Jyvaskyla accelerator news, vol 21, March 2013; Oglobin *et al.* EPJ Web of Conferences 66, 02074 (2014).
29. Tz. Kokalova, *EPJ Web of Conferences* **66**, 03046 (2014).

The "glass transition" as a topological defect driven transition in a distribution of crystals and a prediction of a universal viscosity collapse

Z. Nussinov[*] and N. B. Weingartner[†]

Department of Physics and Institute of Materials Science and Engineering, Washington University, St. Louis, Missouri 63130, USA
*E-mail: zohar@wustl.edu,
[†]weingartner.n.b@wustl.edu*

F. S. Nogueira

*Institute for Theoretical Solid State Physics,
IFW Dresden, PF 270116, 01171 Dresden, Germany
E-mail: f.de.souza.nogueira@ifw-dresden.de*

Topological defects are typically quantified relative to ordered backgrounds. The importance of these defects to the understanding of physical phenomena including diverse equilibrium melting transitions from low temperature ordered to higher temperatures disordered systems (and vice versa) can hardly be overstated. Amorphous materials such as glasses seem to constitute a fundamental challenge to this paradigm. A long held dogma is that transitions into and out of an amorphous glassy state are distinctly different from typical equilibrium phase transitions and must call for radically different concepts. In this work, we critique this belief. We examine systems that may be viewed as simultaneous distribution of different ordinary equilibrium structures. In particular, we focus on the analogs of melting (or freezing) transitions in such distributed systems. The theory that we arrive at yields dynamical, structural, and thermodynamic behaviors of glasses and supercooled fluids that, for the properties tested thus far, are in qualitative and quantitative agreement with experiment. We arrive at a prediction for the viscosity and dielectric relaxations that is universally satisfied for all experimentally measured supercooled liquids and glasses over 15 decades.

1. Conventional Melting via the Condensation of Topological Defects

In their landmark studies, Berezinski[1,2] and Kosterlitz and Thouless[3,4] investigated (BKT) transitions in two-dimensional (2D) systems[5]. These systems (as nearly all other 2D theories) do not exhibit usual long range order at low temperatures, e.g.,[6–9]. The vitally important concept of "topological order" that Kosterlitz and Thouless first introduced and coined[3] to aid the description of the transitions in the classical systems that they investigated has evolved over the years and is currently of use also for myriad singular properties of quantum systems at both zero[10] and finite temperatures[11]. Haldane[12], Thouless and his collaborators[13], and other pioneers discovered the novel role of topology in many other (often deeply interrelated) physical arenas including, notably, various quantum systems. It is a pleasure to write in this volume honoring the watershed contributions of Haldane, Kosterlitz, and

Thouless that have been recognized by the 2016 Nobel Prize in Physics. The works of these three laureates ushered very rich applications of topology in condensed matter physics. Apart from providing conceptual breakthroughs in more exotic systems, topology has proven to be very instrumental in understanding everyday behaviors such as usual phase transitions.

Typical melting transitions from solids to liquids proceed by the condensation of topological defects in a crystal (e.g., the appearance of grain boundaries, dislocations and disclinations in ever increasing numbers and volume). Beyond a certain threshold, crystalline order yields to these defects. As they progressively proliferate, the defects ultimately eradicate any trace of crystalline order, rigidity is lost, and the system transitions into a fluid. This process is a simple extension of the condensation of vortices in the 2D models examined by Kosterlitz and Thouless. In expansive books [14,15], Kleinert fleshed out these notions in a detailed way. Kleinert cast the melting of crystals in a field theoretic framework that underscored how melting precisely occurs via the generation of topological defects. This approach has been extended to the quantum arena [16]. In numerous systems (including those studied by Kosterlitz and Thouless), the topological defects that restore the symmetry of the continuum fluid are, in a precise mathematical sense, "dual" to the fields that characterize standard orders. Conventional measures of topological defects often rely on a comparison between a physical configuration vis a vis an idealized one. For instance, dislocations in a crystal, are defined and quantified by their topological charge- the so called "Burgers vector" [14,17]. The Burgers vector is the total sum of the displacements of all atoms that lie on a contour surrounding a given lattice point. The atomic displacements that are measured here are those of the atoms in the deformed solid by comparison to the locations of the very same atoms in a pristine "ideal" crystal that experiences no deformations. In a similar vein, vortices of pertinence to the BKT transitions are defined relative to a uniform background.

While these ideas are extremely alluring, there is, of course, far more to life than such idealized crystalline or uniform backgrounds. Commonplace amorphous low temperature materials such as glasses do not veer towards an idealized regular periodic crystalline array of atoms relative to which displacements may be measured in a meaningful way. Thus, the usual notion of defects seems to be somewhat ill-defined if one were to try to blindly apply it to such materials. This lack of clearly discernible defects is not merely an academic issue. The existence or absence of topological excitations carries very real practical consequences. As is well appreciated, a sufficiently stressed crystalline solid can falter and exhibit "slip" via the "glide" motion of dislocations (the movement of dislocations parallel to their Burgers vector). When an external force is applied to a crystal, dislocations may drift far more readily than the surrounding crystal. The ease with which defects such as dislocations may be made to move as opposed to pushing an entire crystal move *en masse* has a vivid analogy attributed to Orowan [17] (a practical trick that according to legend inspired him to think of glide). This analogy is based on the fact that one may trivially move an extremely heavy carpet (a runner) by repeatedly creating a ruck- a small indentation- in the rug (a "defect") and then merely stepping on this indentation one foot at a time so as to push the ruck from one side of the carpet to the other. Repeatedly marching, rather effortlessly, on the carpet in this way leads to a net displacement of the carpet. In a somewhat similar manner, point like defects may be pushed by an external force and ultimately lead to motions of entire atomic planes. Finding defects may thus enable the delineation of unstable regions in the solid when an external force is applied. There are field theoretic[18] as well as unsupervised[18,19] and supervised[20] machine learning type approaches to define and ascertain defects and structures in amorphous media. Additional measures in amorphous media include the determination of length scales associated with the penetration of external shear[21,22]. However, generally, in glasses, where defects are not recognizable in the usual way, there are no evident "weak spots" that may, so readily, respond to external forces as they do in crystals.

There are numerous metallurgical advantages associated with the lack of usual topological defects in amorphous materials. Indeed, thanks to the absence of sharp topological defects, metallic glasses exhibit extreme strength and hardness, are very fracture resistance, and are exceedingly elastic[23]. These practical benefits, however, come with a prohibitive price tag for theories that aim to explain the transition out of or into the glassy state via conventional melting or freezing. Indeed, a celebrated drawback of the dearth of sharply defined topological defects is the inability to rationalize the transition from rapidly supercooled liquids into the complex (and seemingly largely random) structure of the glass and vice versa by the proliferation of defects that destroy usual order.

2. Freezing or Melting of Simultaneous Equilibrium Phases

In this article, we will consider systems of $N \gg 1$ atoms occupying a volume V of size $\mathcal{O}(N)$. The energy density $\epsilon \equiv E/V$ with E the total system energy. The central question that we focus on is whether glasses may be investigated by a rather trivial extension of the above ideas concerning melting via the proliferation of topological defects. Specifically, we wish to ask what will transpire for an ensemble average of different equilibrium systems each of which may, on its own, "freeze" into an ordered crystal at low temperatures (or energy densities). An illustration of this idea is provided in Fig. 1 (adopted from[25]- an article further building on the notion of "confusion" introduced in[26]). Our concept differs from that of "confusion"[26]. However, the resulting structural picture that our concept gives rise to is similar. The basic premise is that there are multiple equilibrium states of similar energy density. When the system is supercooled, these states of nearly identical energies may appear in unison. A "superposition" of such equilibrium states general can lead to the observed amorphous structures. In Sections 4 and thereafter we will analyze what will occur in the combined system if each of the individual states exhibits its own freezing/melting transition. This "hodge-podge" of overlapping

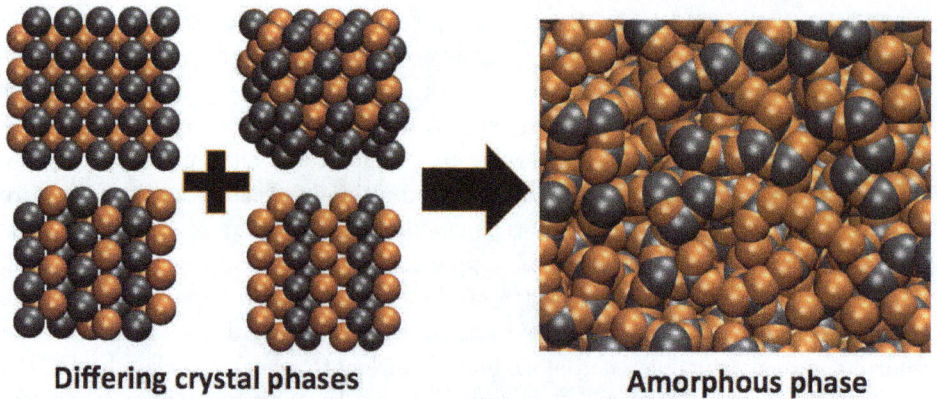

Differing crystal phases   **Amorphous phase**

Fig. 1.   From [25]. A multitude of equilibrium crystal phases of similar energies (left) may "entangle". This thwarts crystallization and leads to complex amorphous structures. process.

states contributes to an internal frustration which greatly influences the dynamics and thermodynamics of the system, chief among them the dynamic viscosity. This analysis leads to a new empirically observed collapse for the viscosities of all glass formers.

3. A Lightning Review of Transitions into the Glassy State

Before proceeding further, we very briefly regress to a discussion concerning the ubiquitous nature of glasses and their striking nature. Glasses are one of the most common states of matter [23,27–34]. Nonetheless, even after millennia of use, they still remain poorly understood. In principle, *any* system may be made glassy by supercooling. "Supercooling" refers to a rapid cooling of a liquid below its melting (or freezing) temperature so that crystallization has insufficient time to occur. It is important to emphasize that these systems are *disorder free*. As the temperature of supercooled liquids is further lowered below melting, they become increasingly sluggish. At low enough temperatures, when the viscosity (η) biomes larger than the (rather arbitrary) threshold value of 10^{12} Pascal $\times$ second, the resulting system is christened a "glass". At temperatures lower than the "glass transition temperature" T_g at which this threshold is obtained, the relaxation times become too long to be easily measured. Glasses exhibit solid-like rigidity on measurable time scales. However, by comparison to ordered equilibrium solids, glasses are amorphous. The "transition" of liquids into the glassy state is unusual. Typical systems display significant changes in all measurable quantities when they transition from one phase to another. This, however, is not the case for glasses. While the dynamics of supercooled liquids typically display significant changes (the viscosity and relaxation times of numerous supercooled liquids increase by many orders of magnitude prior to becoming glasses), [33–35] static properties (such as structure and thermodynamic observables) may exhibit very modest changes.

The most prevalent (Vogel-Fulcher-Tamann-Hesse (VFTH))[36] fit for the viscosities of supercooled liquids is, by now, almost a century old. This fit states that, as a function of the temperature T, the viscosity η of a supercooled liquid is given by $\eta(T) = \eta_0 e^{DT_0/(T-T_0)}$. Here, T_0, D, and η_0 are system specific constants. Taken literally, this fit implies a singular temperature (T_0) at which the viscosity diverges. In tandem with this physical interpretation, T_0 is often called the "ideal glass transition temperature". Since there are very few measurements of the viscosity below T_g, it has been hard to critically assess this putative divergence. All known fits and associated theories of supercooled liquids and glasses argue for various temperature (and other) dependences unrelated to the simple equilibrium melting transition temperature, e.g.,[33,34,38-51]. These often appear as ad hoc additions to existing frameworks. In essence, the existence of the glassy state has always been ascribed to particular processes that are radically different from those in standard *equilibrium melting (or freezing) transitions*.

4. A Universal Collapse of Experimental Viscosity Data Over 15 Decades

The theoretical approach to the glass transition,[52-55] based on the extension of defect melting to an ensemble of *equilibrium states*, that we will review in the upcoming sections motivated us to investigate the viable existence of a universal collapse of the viscosity and other data in which the standard *equilibrium melting transition temperature T_{melt}* (rather specifically, the "liquidus" temperature above which the equilibrium system is its liquid state) is the dominant temperature. Thus, we ask whether the viscosities of supercooled liquids and glasses might be governed by the common equilibrium melting transition temperature. If the answer to this question is affirmative then there might not be a need to introduce singular temperatures (such as T_0), appeal to the possible character of activation processes in complex high dimensional energy landscapes, or invoke other assumptions in order to rationalize the increase of the viscosities of supercooled liquids. With this in mind, we studied[53] 45 supercooled liquids of all known varieties (including silicate, organic, chalcogenide, and metallic glassformers) to discern if the dependence of the viscosity on temperature is governed by the equilibrium melting temperature. Specifically, we asked whether

$$\frac{\eta(T)}{\eta(T_{melt})} = F(\frac{T_{melt} - T}{BT}), \tag{1}$$

with $F(x)$ a universal function and B is a material-dependent (dimensionless) constant. The results of our analysis are displayed in Fig. 2. The observed collapse demonstrates that Eq. (1) is satisfied. The values of B (tabulated towards the end of this article) do not change significantly across the set of all known glassformers. Additional details are found in Refs. [53,55]. The theory of Ref. [52] first predicted

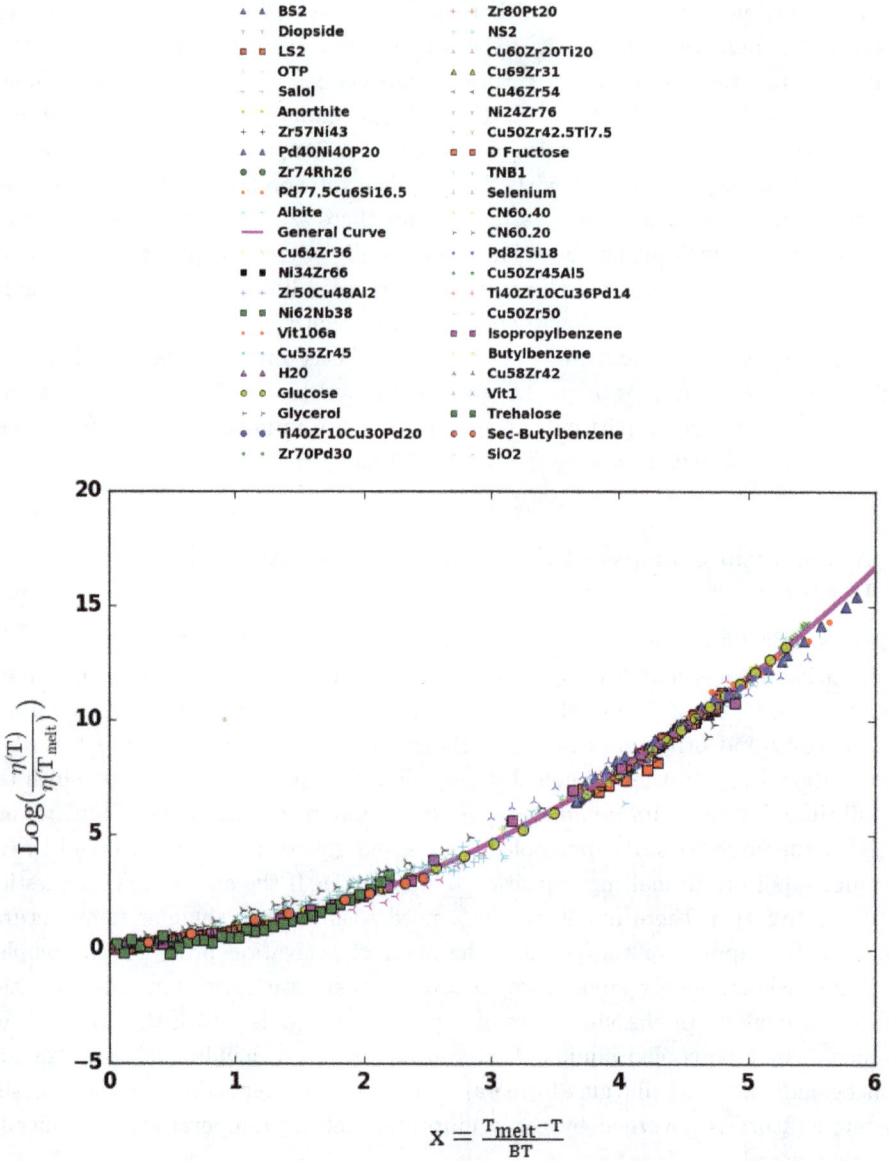

Fig. 2.   The scaling relation of Eq. (1). The "General Curve" is given by the prediction of Eq. (2).

Eq. (1) with

$$F(x) = \frac{1}{\mathrm{erfc}(x)}.$$ (2)

This functional form is consistent with the data[53,55]. Regardless of theoretical prejudice, the raw experimental collapse of Fig. 2 illustrates that, in the broad range of measured $x \equiv \frac{T_{melt}-T}{BT}$ values, Eq. (1) holds very well (with a function F that is either identical or very close to Eq. (2)). Similar to equilibrium transitions,[56] the data collapse that we find hints at an underlying universal description of glass formers. This data collapse suggests that, contrary to a long held belief, the equilibrium melting temperature may play a central role in supercooled liquids and glasses.

5. Spread in Energy Densities Caused by Supercooling: Intuitive Arguments Rationalizing the Universal Viscosity Collapse

In this Section and those that follow, a function that we will focus on is the probability distribution for the energy density of a supercooled liquid (or glass) at temperature T. This function is defined as

$$P_T(\epsilon') = Tr[\delta(\frac{H}{V} - \epsilon')\rho], \tag{3}$$

with $\delta(z)$ a Dirac delta function, H the exact system Hamiltonian (to be explicitly reviewed shortly (Eq. (8)), and ρ the system density matrix just after supercooling. The central feature that we will wish to motivate is that, unlike regular equilibrium systems, in glasses and supercooled liquids, the distribution $P_T(\epsilon')$ may obtain a finite standard deviation σ_ϵ in the thermodynamic limit. We will first turn to an intuitive discussion.

Liquids that are cooled slowly (so that they remain close to equilibrium) may start to crystallize (or veer towards their more general equilibrium solid phase) below a threshold (freezing) transition temperature. By contrast, supercooled liquids have inadequate time for their constituents to crystalize. Thus, a natural viewpoint is that during supercooling multiple low energy structures may start to appear and compete with one another, e.g.,[25,26] (See Fig. 1). Indeed, experiments analyzed via Reverse Monte Carlo techniques[57], attest to a *distribution* of low energy structures, see, e.g.,[58] in various metrics (including so-called Honeycutt-Andersen indices[59]). Taken literally, such a broad distribution not only characterizes the frequency of states of low energies but also those in other tail end- states of elevated energy densities. In a somewhat related vein, supercooled liquids indeed exhibit a broad spectrum of spatially non uniform dynamics[60–64]- some regions of the system are of elevated energy densities by comparison to others. This characteristics is often termed "dynamical heterogeneity"- meaning that the dynamics are, indeed, not spatially uniform. This spatial distribution qualitatively reflects a related aspect of this widening. This spatial distribution qualitatively reflects a related aspect of this widening.

Whenever they appear, states (or modes) of high enough energy densities may support long time flow. That is, just as in equilibrium systems, states of energy densities above those of the system at the onset of its freezing point describe the fluid (in which hydrodynamic motion occurs). Conversely, states having energy densities that are depressed relative to those of the equilibrium system at its melting point, cannot display flow. These melting (or freezing) cutoffs for hydrodynamic flow are associated with the average of the long time flow velocity over *all states* having a well defined energy density. Since the equilibrium average is evaluated over all possible configurations of a given energy density, it may be hardly surprising that precisely the same freezing/melting energy density cutoffs will raise their head for the states of the supercooled liquid. In Section 7, we will sketch a rather simple calculation for the long time average velocity in the fluid and its reciprocal, the viscosity, assuming a Gaussian spread of the energy density (a spread that is linear in temperature as governed by the dimensionless factor B), and a melting (or freezing) energy density cutoff for having a non-vanishing long time velocity readily yields Eqs. (1,2).

When examined from an information theoretic perspective, this may not seem surprising. The long range order that persists in a crystal is associated with a low entropy state. It requires little information to describe a crystal. By contrast, the lack of long range correlations in the equilibrium liquid state implies a considerably higher entropy; much more information is required to elucidate the exact state of the system at any time. The supercooled liquid is observed to fall somewhere in a continuum between these two phases. It is natural to conjecture that the information necessary to describe the state of the supercooled liquid (straddling the line between these two limiting equilibrium states) would require data about the underlying equilibrium crystalline and liquid states. A conditioning of the entropy is natural. On broad grounds, maximizing the Shannon entropy $(-\int d\epsilon' P_T(\epsilon') \log P_T(\epsilon'))$ subject to the constraints of a specified finite standard deviation σ_ϵ that enables the system to straddle different equilibrium states and a given average energy density ϵ leads to a Gaussian distribution $P_T(\epsilon')$. In the absence of such a condition of a finite σ_ϵ, i.e., in the standard equilibrium (equil.) setting of the canonical ensemble, a maximization of the Shannon entropy distribution leads to a Gaussian distribution for the energy density with a standard deviation $\sigma_\epsilon^{\text{equil.}} = \frac{\sqrt{k_B T^2 C_v(T)}}{V} = \mathcal{O}(N^{-1/2})$. Here, k_B is the Boltzmann constant, $C_v = \mathcal{O}(N)$ is the heat capacity. That is, in equilibrium, the distribution $P_T(\epsilon')$ of the energy energy is a Gaussian of a vanishing width in the thermodynamic $(N \to \infty)$ limit. In Section 7, we will invoke precisely such a normal distribution that, as emphasized above, maximizes the Shannon entropy with a constrained standard deviation σ_ϵ that (unlike $\sigma_\epsilon^{\text{equil.}}$) is of a *finite value*.

6. Rigorous Bounds on the Energy Distribution Widths in Systems with a Varying Temperature

At first glance, the results that we next review might seem to be abstract mathematical statements that are conceptually decoupled and have little to do with the very vividly simple considerations of Section 5 and of Fig. 1. These considerations are indeed far afield from simple classical pictures. However, we feel that they provide a glimpse into the generality of the features motivated in Section 5 in very diverse settings. Following Ref.[54], we will now demonstrate that whenever a system is cooled or heated in such a way that its energy density (or measured temperature) varies at a finite rate, then the energy density (even when this density is globally evaluated over the entire macroscopic system) must exhibit a finite width. Naively, such a general statement concerning global averages in a macroscopic systems may appear impossible in systems with local interactions. Indeed, by the central limit theorem, the energy density ϵ (or any other intensive quantity) can only exhibit fluctuations that diminish as $\mathcal{O}(N^{-1/2})$. In particular, in the thermodynamic limit, the standard deviation associated with ϵ may be expected to vanish. Such arguments appealing to the central limit theorem implicitly assume that the system is in equilibrium or that it displays no entanglement.

Although it is not often appreciated, typical thermal states are, in fact, highly entangled. In their ground states $(T = 0)$, most systems display "area law" entanglement entropies. At all other positive temperatures, general systems may exhibit volume law entanglement[65-68]. A trivial yet important point is that although equilibrium states exhibit (by their defining nature) sharp intensive state variables (e.g., energy or number densities), measurements of other generic quantities need not always yield sharp outcomes with a vanishing standard deviation.

As a pedagogical illustration[54], consider a situation in which a cooled or heated subsystem (governed, in the absence of external cooling or heating, by a Hamiltonian H) is part of a larger closed system (defined by the Hamiltonian $\tilde{H}$). When the subsystem is cooled or heated such that the rate of change of its measured temperature dT/dt (or energy density) is finite, the derivative of energy must be extensive, $\frac{dE}{dt} = \langle \frac{dH_H}{dt} \rangle = \mathcal{O}(N)$. Here, $H_H \equiv e^{i\tilde{H}t/\hbar} H e^{-i\tilde{H}t/\hbar}$ is the the Hamiltonian of the subsystem in the Heisenberg representation and the average $\langle - \rangle$ is evaluated with the probability density matrix of the initial $(t = 0)$ state (which we may consider to be an equilibrium state). By Heisenberg's equation, $\langle \frac{dH_H}{dt} \rangle = \frac{i}{\hbar} \langle [\tilde{H}, H_H] \rangle$. We next invoke the generalized uncertainty inequalities to obtain $|\langle [\tilde{H}, H_H] \rangle| \leq 2\sigma_{\tilde{H}} \sigma_{H_H}$. Here, σ_{H_H} and $\sigma_{\tilde{H}}$ respectively denote the widths (or "uncertainties") in the open subsystem and closed global system Hamiltonians. Putting these pieces together,

$$\sigma_{\tilde{H}} \sigma_{H_H} \geq \frac{\hbar}{2} \left| \frac{dE}{dt} \right|. \tag{4}$$

Now if, regardless of the special initial state that we prepare associated with the specific cooling/heating protocol of the subsystem, the closed system exhibits ergodicity at long times such that long time averages become equilibrium averages with $\tilde{H}$ then the uncertainty $\sigma_{\tilde{H}} = \mathcal{O}(1)$ is system size independent. That is, closed systems at equilibrium may be analyzed via the micro-canonical ensemble (in which $\sigma_{\tilde{H}} = \mathcal{O}(1)$). Since the uncertainty in the energy in a closed system is time independent, for systems that equilibrate at long times, $\sigma_{\tilde{H}} = \mathcal{O}(1)$ at all times. From Eq. (4), we then have that the uncertainty in the energy density[54],

$$\sigma_\epsilon = \frac{\sigma_{H_H}}{V} = \mathcal{O}(1). \tag{5}$$

Thus, having a finite width of the energy density is, essentially, inescapable. This general conclusion may be made vivid by exact calculations on various model systems such as generic Heisenberg ferromagnets or antiferromagnets on arbitrary lattices that are quenched by altering external magnetic fields[54]. The bounds of Eqs. (4, 5) draw attention to pervasive broad energy density distributions in systems that are forcefully driven away from equilibrium by external cooling or heating. If the initial ($t = 0$) state is an equilibrium state then the uncertainty of the energy density in that initial state, $\frac{\sigma_H}{V} = \mathcal{O}(N^{-1/2}) \to_{N \to \infty} 0$. However, relative to that same initial state, the operator $H_H(t)$ has a large variance at times t during which the system is cooled or heated. For systems with local Hamiltonians expressible as $H_H(t) = \sum_R \mathcal{H}_R(t)$ with $\mathcal{H}_R(t)$ denoting local Hamiltonians associated with a spatial location R (examples of which can be readily written[54]), the standard deviation of the energy density reads

$$\sigma_\epsilon^2 = \frac{1}{N^2} \sum_{R,R'} (\langle \mathcal{H}_R \mathcal{H}_{R'} \rangle - \langle \mathcal{H}_R \rangle \langle \mathcal{H}_{R'} \rangle) \equiv \frac{1}{N^2} \sum_{R,R'} G(R, R'). \tag{6}$$

A finite width σ_ϵ of the energy density then implies that even for R and R' arbitrarily far apart, the average connected correlation function $G(R, R')$ does not vanish. Thus, cooling or heating may trigger measurable long range correlations in an initial, seemingly trivial, thermal state. Once cooling or heating ceases (so that $\frac{d\epsilon}{dt} = 0$), the lower bound on σ_ϵ vanishes. If there are no impediments for the system to return to equilibrium then it may have a vanishing standard deviation of the energy density shortly after cooling/heating stops. However, in systems that "get stuck" and cannot fully equilibrate, a finite standard deviation of the energy density may persist to long times. Strong hints that this is indeed the case are afforded by the deviations of the average values of various observables in supercooled liquids and glasses vis a vis their average values in equilibrium solids and liquids. By its defining nature, the expectation value of any observable $\mathcal{Q}$ in the microcanonical (m.c.) ensemble is the *average* value of $\mathcal{Q}$ over the $\mathcal{N}(E, \Delta)$ eigenstates that lie in the energy window $E \le E_n \le E + \Delta$, i.e,

$$\langle \mathcal{Q} \rangle_{\text{m.c.}} \equiv \frac{1}{\mathcal{N}(E, \Delta)} \sum_{E \le E_n \le E+\Delta} \langle \phi_n | \mathcal{Q} | \phi_n \rangle. \tag{7}$$

The width of the energies over which this sum extends $\Delta = \mathcal{O}(1)$ is system size independent. In the thermodynamic limit, $\Delta/V \to 0$ and the distribution of energy densities appearing in the sum of Eq. (7) becomes a delta-function. Similarly, as we discussed earlier, also for open equilibrated systems (e.g., the canonical ensemble which we most commonly allude to), the spectral width of the probability density matrix $\sigma^{\text{equil.}} = \mathcal{O}(N^{-1/2})$ also tends to zero as $N \to \infty$ (as it must for the shape state variable ϵ).

Now we turn to experimental facts. The disparity between the values of measured observables in supercooled liquids or glasses vis a vis their values (given by Eq. (7)) in truly equilibrated solids implies that $P_T(\epsilon')$ cannot be a distribution of vanishing width σ_ϵ in the $N \to \infty$ limit. This is so since if σ_ϵ vanished, the average values of all observables in supercooled liquids and glasses would be equal to those in truly equilibrated solids and liquids. Thus, for supercooled liquids and glasses, the distribution $P_T(\epsilon')$ of energy densities must, instead, acquire a finite width σ_ϵ.

Indeed, excusing "many body localized" states[69-75], the expectation values of general observables in typical states of a given energy density may be identical to those found in equilibrium. That is, the equilibrium averages of general physical observables at a certain temperature or energy density are equal to expectation values of the same observables in typical states of the same prescribed energy density. Non-equilibrium behaviors can appear only if, somehow, these individual states (or special sets of such states) are extremely special and display expectation values different from the ensemble averages over all states of a given energy density. Thus, the appearance of a finite energy density width constitutes a very general mechanism for exhibiting non-equilibrium behaviors. The central hypothesis of our approach is that this simple broadening is the sole principal feature separating supercooled liquids and glasses from their equilibrium counterparts. Combined with Eq. (5), the intuitive considerations of Section 5, elicit us to contemplate this simple possibility. For completeness, we further mention that broad distributions σ may also appear in "non self averaging" disordered classical systems[76-79] and, rather trivially, in classical systems with interactions that are of unbounded spatial range or strength.

7. Sketch of the Long Time Averages of General Observables and the Viscosity

Armed with the proof of principle rigorous bounds of Section 6, we now return to the pictorial ideas outlined in Section 5 and couch these in a broader context. The basic strategy that we will follow concerns the logical outcome of a rather tautological statement concerning completeness. The same *complete set* of many body atomic states (whether quantum mechanical eigenstates or classical microstates) describes both (1) the supercooled liquids and glasses and (2) the equilibrium systems. The completeness of these states implies that all observable properties of glasses may be related (via a linear transformation) to the values of the very same observables when measured in equilibrium systems. Since the equilibrium systems undergo a transition only at their melting temperature, the equilibrium melting temperature also plays a vital role in supercooled liquids. More precisely, the eigenstates of the same basic many body Hamiltonian,

$$H = -\sum_i \frac{\hbar^2}{2M_i} \nabla^2_{R_i} - \sum_j \frac{\hbar^2}{2m_e} \nabla^2_{r_j} - \sum_{i,j} \frac{Z_i e^2}{|R_i - r_j|}$$

$$+ \frac{1}{2} \sum_{i \neq i'} \frac{Z_i Z_{i'} e^2}{|R_i - R_{i'}|} + \frac{1}{2} \sum_{j \neq j'} \frac{e^2}{|r_j - r_{j'}|}. \tag{8}$$

governing both systems of types (1) and (2) exhibit a "phase transition" and are non-analytic at the energy densities associated with melting. The Hamiltonian H describes these systems on all scales of experimental interest. In Eq. (8), M_i, R_i, and

$Z_i e$ are, correspondingly, the mass, position, and charge of the i−th nucleus, while r_j is the location of the j−th electron (whose mass and charge are m_e and $(-e)$ respectively). The spectral problem posed by H is, of course, nontrivial. We recall however two points. (i) *Empirically*, the equilibrium averages of various observables display singularities only at phase transitions. (ii) These equilibrium expectation values (in, e.g., the microcanonical or canonical ensembles) are averages over all eigenstates having the same energy density. Taken together (i) and (ii) imply that typical eigenstates of Eq. (8) display non-analyticities at the very same energy densities at which phase transitions occur experimentally. Making contact with the discussion in the Introduction, at the energy densities at which equilibrium transitions appear, topological melting occurs. The typical eigenstates at energies above and below the melting transition are of an inherently different character and reflect the characteristics of the equilibrium liquid and the solid. We reiterate that the only the energy densities of pertinence to non-analyticities of the eigenstates of H are those associated with experimental phase transitions. This, then, suggests that the viscosity and relaxation times of all glass formers should collapse onto a universal curve with the only important temperature scale indeed being that of equilibrium melting. We wish to explicitly further expand on outcomes of the logical steps that we have just invoked. Towards this end, we reiterate that the microcanonical ensemble average of Eq. (7) is (for standard systems that equilibrate and obey ensemble equivalence) equal to the measured value of $\mathcal{Q}$ in thermal equilibrium $\langle \mathcal{Q}(\epsilon) \rangle_{eq}$ at the associated energy density ϵ. Given this $\langle \mathcal{Q}(\epsilon) \rangle_{m.c.} = \langle \mathcal{Q}(\epsilon) \rangle_{eq}$ equivalence, we see that an *average* of any observable over eigenstates $\{|\phi_n\rangle\}$ of energy density ϵ assumes its value in a typical (a) state of an equilibrium solid phase of energy density ϵ when $\epsilon < \epsilon_{melt}$, or of an (b) equilibrium state of the liquid (or higher energy density phases) when $\epsilon > \epsilon_{melt}$. This conclusion regarding the average (or typical) eigenstates of general energy densities ϵ reaffirms the intuitive statements made in Section 5. We *do not need* to address the tall order of diagonalizing Eq. (8) is order to ascertain the exact average of $\langle \phi_n | \mathcal{Q} | \phi_n \rangle$ over the eigenstates $|\phi_n\rangle$ of H in a narrow energy window $[E, E + \Delta]$. Our basic premise is that the *experimental results on the equilibrium system defined by H essentially solve the spectral problem for us.* If particular, since when subject to an infinitesimal external stress there is no long time flow in an *equilibrium* solid, we explicitly deduce a particular corollary of (a). Namely, (c) eigenstates with an energy density $\epsilon < \epsilon_{melt}$ (more precisely, eigenstates of energy density lower than the equilibrium "liquidus" value below which solid inclusions appear) do not, on average, support long time hydrodynamic flow for infinitesimal external stress.

A straightforward calculation[52] illustrates that the long time average $\mathcal{Q}_\infty$ of any quantity or operator ($\mathcal{Q}$ whose experimental equilibrium value $\langle \mathcal{Q}(\epsilon) \rangle_{eq}$ is largely temperature (or energy density ϵ') dependent is given by

$$\mathcal{Q}_\infty \equiv \lim_{\tilde{\mathcal{T}} \to \infty} \frac{1}{\tilde{\mathcal{T}}} \int_{t_{final}}^{t_{final} + \tilde{\mathcal{T}}} dt' \, Tr[\rho(t')\mathcal{Q}] = \int d\epsilon' P_T(\epsilon') \langle \mathcal{Q}(\epsilon') \rangle_{eq}. \qquad (9)$$

In Eq. (9), t_{final} is the time at which the influence of external quenching halts and the supercooled liquid evolves as a closed system with the Hamiltonian of Eq. (8). Eq. (9) is derived by a simple sequence of steps. We write the Schrödinger picture density matrix $\rho(t) = e^{-iH(t-t_{final})/\hbar} \rho e^{iH(t-t_{final})/\hbar}$, insert a resolution of the identity with the eigenstates of H (or, more generally, resolving the identity via the common eigenstates of H and Q when the eigenstates of H are degenerate), replace the latter sum over all eigenstates of H (or of common eigenstates of both H and Q) by an integral over ϵ' (with any additional sum over eigenstates of Q if degeneracy exists) by using Eq. (7), invoke the equivalence $\langle Q(\epsilon) \rangle_{m.c.} = \langle Q(\epsilon) \rangle_{eq}$, note that the ratio $(\int_0^{\tilde{T}} dt'\, e^{i\omega_{nm}t'})/\tilde{T}$ tends to zero for large $\tilde{T}$ whenever the difference in energies amongst eigenstates appearing in the above resolution of the identity is non-vanishing (i.e., the above ratio tends to zero when $\omega_{nm} \equiv (E_n - E_m)/\hbar \neq 0$), and lastly insert Eq. (3). We now use our conclusion (c) and Eq. (9) to predict the measured viscosity of supercooled liquids. One of the oldest methods of determining the viscosity is to measure the terminal velocity of a sphere dropped into the fluid. By Stokes' law, the terminal velocity of a sphere of radius R and mass density ρ_{sphere} in a fluid of viscosity η and mass density ρ_{fluid} is

$$v_\infty = \frac{2}{9} \frac{\rho_{sphere} - \rho_{fluid}}{\eta} g R^2. \tag{10}$$

On the other hand, if the state of the supercooled system at a temperature T emulates a distributed average (with a probability distribution $P_T(\epsilon')$) of the equilibrium result (or typical eigenstate) for an energy density ϵ' then, by invoking Eq. (9),

$$v_\infty = \int d\epsilon'\, P_T(\epsilon')\, v_\infty^{eq}(\epsilon'). \tag{11}$$

Here, $v_\infty^{eq}(\epsilon')$ is the terminal velocity of the sphere when the same system is at equilibrium with an energy density ϵ'. Since the terminal velocity of the sphere will vanish in the solid phase (when $\epsilon' < \epsilon_{melt}$), the integral of Eq. (11) may be performed from $\epsilon' = \epsilon_{melt}$ to $\epsilon' = \infty$). If the distribution $P_T(\epsilon')$ has most of its weight at energies below ϵ_{melt} and the equilibrium terminal velocity $v_\infty^{eq}(\epsilon')$ is a weak function of temperature of energy densities above melting (i.e., $v_\infty^{eq}(\epsilon' > \epsilon_{melt}) \sim v_\infty^{eq}(\epsilon_{melt}^+))$ then

$$v_\infty = v_\infty^{eq}(\epsilon_{melt}^+) \int_{\epsilon_{melt}}^{\infty} d\epsilon'\, P_T(\epsilon'). \tag{12}$$

We consider a Gaussian $P_T(\epsilon')$ (a distribution that maximizes the Shannon entropy given a finite standard deviation σ_ϵ), of standard deviation σ_ϵ such that the ratio

$$B \equiv \frac{\sigma_\epsilon (T_{melt} - T)\sqrt{2}}{T(\epsilon_{melt} - \epsilon)} \tag{13}$$

is, approximately, constant in the temperature regime $(T_g \lesssim T < T_{melt})$ of experimental relevance. Combining Eq. (10) (applied to both the viscosity of the supercooled liquid and the viscosity of the equilibrium system at energy density ϵ')

with Eqs. (12, 13) produces Eqs. (1, 2)[52]. Although, in the rather trivial derivation above, we have invoked the eigenstate decomposition of general states, one may provide different verbal rationalizations to the simple relations of Eqs. (12,13) that led us to suggest and find the collapse of Fig. 2. The key point is that a Gaussian distribution of modes that contribute to long time flow will rather robustly yield Eqs. (1, 2) that is indeed rather universally obeyed as we show in Fig. 2 and discussed in depth in[53,55]. The above calculation may be repeated, *mutatis mutandis*, for other relaxation times (such as those associated with dielectric response relaxations) and a multitude of thermodynamic and structural properties. Elsewhere, we will show that dielectric relaxation data collapse on the very same universal curve shown in Fig. 2. In fact, the very same unique probability distribution P_T should universally relate numerous measurable properties of the glass to those of equilibrium systems at various energy densities ϵ'.

To close the circle of ideas that we started to explore at the beginning of this article, we remark that superposing different crystalline (or other) structures of varying energy densities may generate non-uniform amorphous structures (see, e.g., Fig. 1). This is the structural counterpart of decomposing ill-organized configurations into ordered ("crystalline like") periodic Fourier modes.

8. Conclusions

In summary, we outlined a new approach to the glass transition. Our central proposal is that when the *usual topological defect driven transition* from the equilibrium solid to the liquid is merely smeared by a normal probability distribution of a finite width, then the resulting theory will reproduce quite well the phenomenology of glasses and supercooled liquids. That is, the sole difference between our approach and the usual equilibrium transition is that the probability distribution of the energy density governing supercooled liquids and glasses is a normal distribution of a finite width (instead of a Gaussian of width scaling as $\mathcal{O}(N^{-1/2})$ in finite temperature *equilibrium* systems). A prediction of our theory is that the viscosity and dielectric relaxation times collapse on a simple curve (given by Eqs. (1, 2)). This conjecture is indeed satisfied for all glass formers (as seen in Fig. 2). In a similar vein, numerous other experimental observables should be governed by this same distribution function. We underscore that the empirical collapse of Fig. 2 holds universally regardless of the theoretical considerations that led us to it.

Acknowledgments

ZN and NBW gratefully acknowledge support from the National Science Foundation under grant number NSF 1411229. FSN thanks the German Collaborative Research Center SFB 1143. We further wish to thank the organizers of the workshop on "Topological Phase Transitions and New Developments".

Appendix A. Numerical fit values

In the table that follows, the parameter B in the collapse of Eq. (1) and Fig. 2 is provided. Further detail concerning the analysis leading to these values and many further aspects (both empirical and theoretical) appear in [53,55].

Table 1. Values of Relevant Parameters for all liquids studied.

Composition	B	T_{melt} [K]	$\eta(T_{melt})$ [Pa^*s]
BS2	0.157129	1699	5.570596
Diopside	0.134328	1664	1.5068
LS2	0.170384	1307	22.198
OTP	0.069685	329.35	0.02954
Salol	0.087192	315	0.008884
Anorthite	0.131345	1823	39.81072
$Zr_{57}Ni_{43}$	0.234171	1450	0.01564
$Pd_{40}Ni_{40}P_{20}$	0.154701	1030	0.030197
$Zr_{74}Rh_{26}$	0.187851	1350	0.03643
$Pd_{77.5}Cu_6Si_{16.5}$	0.124879	1058	0.0446
Albite	0.103344	1393	24154952.8
$Cu_{64}Zr_{36}$	0.142960	1230	0.021
$Ni_{34}Zr_{66}$	0.209359	1283	0.0269
$Zr_{50}Cu_{48}Al_2$	0.167270	1220	0.0233
$Ni_{62}Nb_{38}$	0.109488	1483	0.042
Vit106a	0.133724	1125	0.131
$Cu_{55}Zr_{45}$	0.144521	1193	0.0266
H_2O	0.133069	273.15	0.001794
Glucose	0.079455	419	0.53
Glycerol	0.108834	290.9	1.9953
$Ti_{40}Zr_{10}Cu_{30}Pd_{20}$	0.185389	1279.226	0.01652
$Zr_{70}Pd_{30}$	0.21073	1350.789	0.02288
$Zr_{80}Pt_{20}$	0.169362	1363.789	0.04805
NS2	0.134626	1147	992.274716
$Cu_{60}Zr_{20}Ti_{20}$	0.103380	1125.409	0.04516
$Cu_{69}Zr_{31}$	0.157480	1313	0.01155
$Cu_{46}Zr_{54}$	0.156955	1198	0.02044535
$Ni_{24}Zr_{76}$	0.244979	1233	0.02625234
$Cu_{50}Zr_{42.5}Ti_{7.5}$	0.148249	1152	0.0268
D Fructose	0.050124	418	7.31553376
TNB1	0.07567	472	0.03999447
Selenium	0.130819	494	2.9512
CN60.40	0.149085	1170	186.2087
CN60.20	0.161171	1450	12.5887052
$Pd_{82}Si_{18}$	0.137623	1071	0.03615283
$Cu_{50}Zr_{45}Al_5$	0.118631	1173	0.03797
$Ti_{40}Zr_{10}Cu_{36}Pd_{14}$	0.137753	1185	0.0256
$Cu_{50}Zr_{50}$	0.166699	1226	0.02162
Isopropylbenzene	0.073845	177	0.086
ButylBenzene	0.085066	185	0.0992
$Cu_{58}Zr_{42}$	0.131969	1199	0.02526
Vit 1	0.111185	937	36.59823
Trehalose	0.071056	473	2.71828
Sec-Butylbenzene	0.080088	190.3	0.071
SiO_2	0.090948	1873	1.196×10^8

References

1. V. L. Berezinski, Destruction of Long-Range Order in One-Dimensional and Two-Dimensional Systems Having a Continuous Symmetry Group I. Classical Systems, *Soviet Physics, JETP* **32**, 493 (1970).
2. V. L. Berezinski, Destruction of Long-Range Order in One-Dimensional and Two-Dimensional Systems Possessing a Continuous Symmetry Group II. Quantum Systems, *Soviet Physics, JETP* **34**, 610 (1971).
3. J. M. Kosterlitz and D. J. Thouless, Ordering, metastability and phase transitions in two-dimensional systems, *Journal of Physics C: Solid State Physics* **6 (7)**, 1181 (1973).
4. J. M. Kosterlitz and D. J. Thouless, Long range order and metastability in two dimensional solids and superfluids. (Application of dislocation theory), *J. Phys. C* **5**, L124 (1972).
5. J. V. Jose, 40 Years of Berezinskii-Kosterlitz-Thouless Theory (World Scientific, 2013).
6. N. D. Mermin and H. Wagner, Absence of Ferromagnetism or Antiferromagnetism in One- or Two-Dimensional Isotropic Heisenberg Models, *Phys. Rev. Lett.* **17**, 1133 (1966).
7. S. Coleman, There are no Goldstone bosons in two dimensions, *Commun. Math. Phys.* **31**, 259 (1973).
8. O. McBryan and T. Spencer, On the decay of correlations in SO(n)-symmetric ferromagnets, *Commun. Math. Phys.* **53**, 299 (1977).
9. Z. Nussinov, Commensurate and Incommensurate O(n) Spin Systems: Novel Even-Odd Effects, A Generalized Mermin-Wagner-Coleman Theorem, and Ground States, `http://arxiv.org/pdf/cond-mat/0105253.pdf` (2001).
10. Xiao-Gang Wen, *Quantum Field Theory of Many Body Systems- From the Origin of Sound to an Origin of Light and Electrons* (Oxford Univ. Press, Oxford, 2004).
11. Z. Nussinov and G. Ortiz, A symmetry principle for topological quantum order', *Annals of Physics* **324**, 977 (2009); Z. Nussinov and G. Ortiz, Sufficient symmetry conditions for topological quantum order, *Proceedings of the National Academy of Sciences of the United States of America* **106**, 16944 (2009).
12. F. D. M. Haldane, Nonlinear field theory of large-spin Heisenberg antiferromagnets: semiclassically quantized solitons of the one-dimensional easy-axis Neel state, *Phys. Rev. Lett.* **50**, 1153 (1983); F. D. M. Haldane, Model for a quantum Hall effect without Landau levels: Condensed-matter realization of the "parity anomaly", *Physical Review Letters* **61**, 2015 (1988).
13. D. J. Thouless, M. Kohmoto, M. P. Nightingale, and M. den Nijs, Quantized Hall Conductance in a Two-Dimensional Periodic Potential, *Physical Review Letters* **49**, 405 (1982).
14. H. Kleinert, Gauge Fields in Condensed Matter, Volume II: Stress and Defects (World Scientific, Singapore, 1989).
15. H. Kleinert, Multivalued Fields in Condensed Matter, Electromagnetism, and Gravitation (World Scientific, Singapore, 2008).
16. Aron J. Beekman, Jaakko Nissinen, Kai Wu, Ke Liu, Robert-Jan Slager, Zohar Nussinov, Vladimir Cvetkovic, and Jan Zaanen, Dual gauge field theory of quantum liquid crystals in two dimensions, *Physics Reports* **683**, 1 (2017).
17. F. Nabarro, Theory of Dislocations (Dover, New York, 1982).
18. P. Ronhovde, S. Chakrabarty, M. Sahu, K. F. Kelton, N. A. Mauro, K . K. Sahu, and Z. Nussinov, Detecting hidden spatial and spatio-temporal structures in glasses and complex physical systems by multiresolution network clustering, *The European Physics Journal E* **34**, 105 (2011).

19. P. Ronhovde, S. Chakrabarty, M. Sahu, K. K. Sahu, K. F. Kelton, N. Mauro, and Z. Nussinov, Detection of hidden structures on all scales in amorphous materials and complex physical systems: basic notions and applications to networks, lattice systems, and glasses, *Scientific Reports* **2**, 329 (2012).

20. S. S. Schoenholz, E. D. Cubuk, D. M. Sussman, E. Kaxiras, and A. J. Liu, A structural approach to relaxation in glassy liquids, *Nature Physics* **12**, 469 (2016).

21. N. B. Weingartner, R. Soklaski, K. F. Kelton, and Z. Nussinov, "Dramatically growing shear rigidity length scale in the supercooled glass former NiZr2", *Phys. Rev. B* **93**, 214201 (2016).

22. Nicholas B. Weingartner and Zohar Nussinov, "Probing local structure in glass by the application of shear", *J. Stat. Mech.* **094001** (2016).

23. M. Telford, The case for bulk metallic glass, *Materials Today* **7**, 36 (2004).

24. C. A. Angell, Formation of glasses from liquids and biopolymers, *Science* **267**, 1924 (1995).

25. E. Perim, D. Lee, Y. Liu, C. Toher, P. Ging, Y. Li, W. N. Simmons, O. Levy, J. J. Vlassak, J. Schoers, and S. Curtarolo, Spectral descriptors for bulk metallic glasses based on the thermodynamics of competing crystalline phases, *Nature Communications* **7**, 12315 (2016).

26. A. L. Greer, Confusion by design, *Nature* **366**, 303 (1993).

27. L. Berthier and M. D. Ediger, Facets of glass physics, *Physics Today* **69**(1), 40 (2016).

28. R. Zallen, *The Physics of Amorphous Solids*, John Wiley & Sons, Inc., pages 23-32 (1983).

29. A. L. Greer and E. Ma, Bulk metallic glasses– at the cutting edge of metals research, *MRS Bulletin* **32**, 611 (2007).

30. B. C. Hancock and M. Parks, What is the true solubility advantage for amorphous pharmaceuticals?, *Pharmaceutical Research* **17**, 397 (2000).

31. M. Wuttig, and N. Yamada, Phase-change materials for rewriteable data storage, *Nature Materials* **6**, 824 (2007).

32. P. W. Anderson, Through the glass lightly, *Science* **267**, 1615 (1995).

33. L. Berthier and G. Biroli, Theoretical perspective on the glass transition and amorphous materials, *Rev. Mod. Phys.* **83**, 587 (2011).

34. F. H. Stillinger and P. G. Debenedetti, Glass Transition Thermodynamics and Kinetics, *Annu. Rev. Condens. Matter Phys.* **4**, 263 (2013).

35. C. A. Angell, Formation of glasses from liquids and biopolymers, *Science* **267**, 1924 (1995).

36. H. Vogel, The law of the relationship between viscosity of liquids and the temperature, *Z. Phys.* **22**, 645 (1921); G. S. Fulcher, Analysis of recent measurements of the viscosity of glasses, *J. Am. Ceram. Soc.* **8**, 339 (1925); G. Tammann and W. Z. Hesse, Die Abhangigkeit der Viscositat von der Temperatur bie unterkhlten Flussigkeiten, *Anorg. Allgem. Chem.* **156**, 245 (1926).

37. G. Adam and J. H. Gibbs, On the Temperature Dependence of Cooperative Relaxation Properties in Glass-Forming Liquids, *J. Chem. Phys.* **43**, 139 (1965).

38. T. R. Kirkpatrick, D. Thirumalai, and P. G. Wolynes, Scaling concepts for the dynamics of viscous liquids near an ideal glassy state, *Phys. Rev. A* **40**, 1045 (1989).

39. G. Parisi and M. Mezard, A first-principle computation of the thermodynamics of glasses *J. Chem. Phys.* **111**, 1076 (1999).

40. V. Lubchenko and P. G. Wolynes, Theory of structural glasses and supercooled liquids, *Ann. Rev. of Phys. Chemistry* **58**, 235 (2007).

41. M. H. Cohen and G. S. Grest, Liquid-glass transition, a free-volume approach, *Phys. Rev. B* **20**, 1077 (1979).

42. E. Leutheusser, Dynamical model of the liquid-glass transition, *Phys. Rev. A* **29**, 2765 (1984).

43. U. Bengtzelius, W. Gotze, and A. Sjolander, Dynamics of supercooled liquids and the glass transition, *J. Phys. C* **17**, 5915 (1984).

44. W. Gotze, *Complex dynamics of glass-forming liquids: A mode-coupling theory* (Oxford University Press, Oxford, 2008).

45. D. Kivelson, S. A. Kivelson, X. Zhao, Z. Nussinov, and G. Tarjus, A Thermodynamic Theory of Supercooled Liquids, *Physica A*, **219**, 27 (1995).

46. G. Tarjus, S. A. Kivelson, Z. Nussinov, and P. Viot, The frustration based approach to supercooled liquids and the glass transition: a review and critical assessment, *J. Phys: Condens Matter*, **17**, R1143 (2005).

47. Z. Nussinov, Avoided Phase Transitions and Glassy Dynamics In Geometrically Frustrated Systems and Non-Abelian Theories, *Phys. Rev. B* **69**, 014208 (2004).

48. Y. S. Elmatad, D. Chandler, and J. P. Garrahan, Corresponding states of structural glass formers. II., *J. Phys. Chem. B* **113**, 5563 (2009).

49. Y. S. Elmatad, R. L. Jack, D. Chandler, and J. P. Garrahan, Finite-temperature critical point of a glass transition, *PNAS* **107**, 12793 (2010).

50. M. Blodgett, T. Egami, Z. Nussinov, and K. F. Kelton, Scientific Reports, **5**, 13837 (2015).

51. J. C. Mauro, Y. Yue, A. J. Ellison, P. K. Gupta, and D. C. Allan, Viscosity of glass-forming liquids, *PNAS*, **106**, 19780 (2009).

52. Z. Nussinov, "A one parameter fit for glassy dynamics as a quantum corollary of the liquid to solid transition", https://arxiv.org/pdf/1510.03875.pdf (2015), *Philosophical Magazine* **97**, 1509 (2017); Z. Nussinov, "A one parameter fit for glassy dynamics as a quantum corollary of the liquid to solid transition (vol 97, pg 1509, 2017)", *Philosophical Magazine* **97**, 1567 (2017).

53. Nicholas B. Weingartner, Chris Pueblo, Flavio Nogueira, Kenneth F. Kelton, and Zohar Nussinov, A phase space approach to supercooled liquids and a universal collapse of their viscosity, https://arxiv.org/pdf/1611.03018.pdf, *Frontiers in Materials* **3**, 50 (2016).

54. Z. Nussinov, Infinite Range Correlations in Non-Equilibrium Quantum Systems and Their Possible Experimental Realizations, https://arxiv.org/pdf/1710.06710.pdf (2017).

55. Nicholas B. Weingartner, Chris Pueblo, F. S. Nogueira, K. F. Kelton, and Zohar Nussinov, https://arxiv.org/pdf/1512.04565.pdf (2015).

56. E. A. Guggenheim, The Principle of Corresponding States, *J. Chem. Phys.* **13**, 253 (1945).

57. R. L. McGreevy, Understanding liquid structures, *J. Phys.: Condens. Matter* **3**, F9 (1991).

58. T. H. Kim and K. F. Kelton, Structural study of supercooled liquid transition metals, *The Journal of Chemical Physics* **126**, 054513 (2007).

59. J. D. Honeycutt and H. C. Andersen, Molecular dynamics study of melting and freezing of small Lennard-Jones clusters, *J. Phys. Chem.* **91**, 4950 (1987).

60. H. Sillescu, Heterogeneity at the glass transition: a review, *Journal of Non-Crystalline Solids* **243**, 81 (1999).

61. M. D. Ediger, Spatially heterogeneous dynamics in supercooled liquids, *Annual Review of Physical Chemistry* **51**, 99 (2000).

62. R. Richert, Heterogeneous dynamics in liquids: fluctuations in space and time, *Journal of Physics: Condensed Matter* **14**, R 703 (2002).

63. W. Kob, C. Donati, S. J. Plimpton, P. H. Poole, and S. C. Glotzer, Dynamical het-

erogeneities in a supercooled Lennard-Jones liquid, *Physical Review Letters* **79**, 2827 (1997).

64. C. Donati, J. F. Douglas, W. Kob, S. J. Plimpton, P. H. Poole, and S. C. Glotzer, Stringlike cooperative motion in a supercooled liquid, *Physical Review Letters* **80**, 2338 (1998).

65. Hsin-Hua Lai and Kun Yang, Entanglement entropy scaling laws and eigenstate typicality in free fermion systems, *Phys. Rev. B* **91**, 081110(R) (2015).

66. Hiroyuki Fujita, Yuya O. Nakagawa, Sho Sugiura, and Masataka Watanabe, Universality in volume law entanglement of pure quantum states, https://arxiv.org/pdf/1703.02993.pdf (2017).

67. Lev Vidmar, Lucas Hackl, Eugenio Bianchi, and Marcos Rigol, Entanglement Entropy of Eigenstates of Quadratic Fermionic Hamiltonians, *Phys. Rev. Lett.* **119**, 020601 (2017).

68. Adam M. Kaufman, M. Eric Tai, Alexander Lukin, Matthew Rispoli, Robert Schittko, Philipp M. Preiss, and Markus Greiner, Quantum thermalization through entanglement in an isolated many-body system, *Science* **353**, 794 (2016); Anatoli Polkovnikov and Dries Sels, Chaos and thermalization in small quantum systems, *Science* **353**, 752 (2016).

69. D. Basko, I. Aleiner, and B. Altshuler, Metal-insulator transition in a weakly interacting many-electron system with localized single-particle states, *Annals of Physics* **321**, 1126 (2006); D. Basko, I. Aleiner, and B. Altshuler, Possible experimental manifestations of the many-body localization, *Physical Review B* **76**, 052203 (2007).

70. V. Oganesyan and D. A. Huse, Localization of interacting fermions at high temperature, *Physical Review B* **75**, 155111 (2007).

71. R. Vosk and E. Altman, Many-Body Localization in One Dimension as a Dynamical Renormalization Group Fixed Point, *Physical Review Letters* **110**, 067204 (2013); E. Altman and R. Vosk, Universal Dynamics and Renormalization in Many-Body-Localized Systems, *Annual Review of Condensed Matter Physics* **6**, 383 (2015).

72. J. Z. Imbrie, On Many-Body Localization for Quantum Spin Chains, *Journal of Statistical Physics* **163**, 998 (2016).

73. Rahul Nandkishore and David A. Huse, Many-Body Localization and Thermalization in Quantum Statistical Mechanics, *Annual Review of Condensed Matter Physics* **6**, 15 (2015).

74. M. Schreiber, S. S. Hodgman, P. Bordia, H. P. Luschen, M. H. Fischer, R. Vosk, E. Altman, U. Schneider, and I. Bloch, Observation of many-body localization of interacting fermions in a quasirandom optical lattice, *Science* **349**, 842 (2015).

75. Wojciech De Roeck and Francois Huveneers, Stability and instability towards delocalization in MBL systems, *Phys. Rev. B* **95**, 155129 (2017).

76. B. Derrida, "Non-Self-Averaging Effects in Sums of Random Variables, Spin Glasses, Random Maps, and Random Walks" in *On Three Levels*, Edited by M. Fannes *et al.*, Plenum Press, New York (1994).

77. A. Aharony and A.B. Harris, Absence of Self-Averaging and Universal Fluctuations in Random Systems near Critical Points, *Phys. Rev. Lett.* **77**, 3700 (1996).

78. Shai Wiseman and Eytan Domany, Finite-Size Scaling and Lack of Self-Averaging in Critical Disordered Systems, *Phys. Rev. Lett.* **81**, 22 (1998).

79. P. H. Lundow and I. A. Campbell, Non-self-averaging in Ising spin glasses and hyperuniversality, *Physical Review E* **93**, 012118 (2016).

Surface ferromagnetism in a topological crystalline insulator

Sahinur Reja, H. A. Fertig*, and Shixiong Zhang

Department of Physics, Indiana University
Bloomington, Indiana 47401, USA
*E-mail: hfertig@indiana.edu

Luis Brey

Instituto de Ciencia de Materiales de Madrid, (CSIC),
Cantoblanco, 28049 Madrid, Spain

Topological crystalline insulators support gapless surface states even in the presence of a bulk gap, in situations where the surface preserves some crystalline symmetry of the bulk. In most realizations this is a mirror symmetry. We study the effect of magnetic impurities in such a system, for which ordering at the surface lowers the electronic energy by spontaneous breaking the crystalline symmetry. This is shown to induce surface ferromagnetism, with favored directions and numbers of energetically equivalent groundstates that are sensitive to the precise density of electrons at the surface. In particular for a (111) surface of a SnTe model in the topological state, magnetic states will have either a two-fold or six-fold symmetry. We compute spin stiffnesses within the model to demonstrate the stability of ferromagnetic states, and consider their ramifications for thermal disordering. Possible experimental consequences of the surface magnetism are discussed.

Keywords: Topological Crystalline Insulator, Magnetism, Phase Transitions.

1. Introduction

In 2016 the Nobel Prize was awarded to Duncan Haldane, Mike Kosterlitz, and David Thouless for their pioneering work topology in condensed matter systems. Among the accomplishments recognized in this award was the discovery that the Hall conductance of a two-dimensional electron gas in a magnetic field is related to a topological invariant of the system[1]. This non-trivial topology necessitates that quantum Hall edges support conducting states, even when the Fermi energy lies in a gap of the bulk spectrum[2]. For a long time it was believed that this type of behavior was very special to the quantum Hall system, but eventually – indeed, more than two decades later – it was recognized that this type of behavior does not require a magnetic field, or even that the system be two-dimensional[3,4].

Among such systems are topological crystalline insulators (TCI's)[5]. For these systems, gapless surface states are present and protected from becoming gapped when there is an unbroken crystalline symmetry[6]. An important three-dimensional realization of these materials is found in (Sn,Pb)Te and related alloys[7-12]. For these and other topological materials, interesting effects may result when a protecting symmetry is broken. For example, in topological insulators protected by time-reversal symmetry (TRS), magnetic impurities on a surface may spontaneously

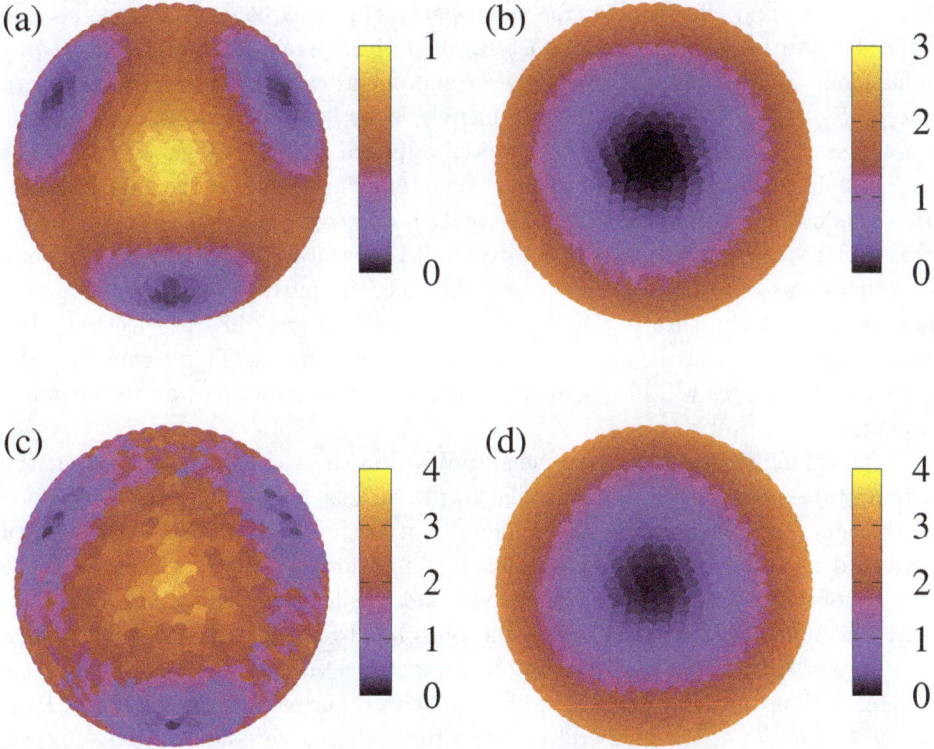

Fig. 1. Total electronic free energy per surface atom as a function of magnetization orientation on the Bloch sphere. Color scale is in units of $10^{-4}t$, with t=nearest neighbor hopping energy. Note that only one hemisphere is shown in each figure, with the $\mathbf{k}_1$ direction represented by the center. In (a) and (c) the Fermi energy is close to $E_{\bar{M}}$; in (b) and (d), it is close to $E_{\bar{\Gamma}}$. Results in (a) and (b) are for fixed particle number; in (c) and (d) they are for fixed chemical potential.

order, or form glassy states[13-17]. This can be understood in terms of gap formation of the surface spectrum[18] which would not be possible if the impurity perturbation left TRS intact. By contrast, because TCI's are protected by a crystalline symmetry, the TRS-breaking associated with polarized magnetic impurities does not by itself energetically favor magnetic ordering[19,20]. However, magnetically ordered states *can* break the underlying crystalline symmetry[21,22], gapping the surface state spectrum and potentially favoring the state energetically. In this work we discuss how symmetries of the system lead to a manifold of degenerate ferromagnetic surface states, whose size is dictated by both the surface symmetries and the system doping, as well as the interesting behaviors expected when these states become disordered. Many of the results discussed here have been presented in Ref. 23.

As a paradigm of this physics we consider in detail the behavior of a (111) surface of $(\text{Sn},\text{Pb})\text{Te}^{12,24-26}$. We generalize a method developed for computing surface states of TRS topological insulators[27-29] to the present situation, from which it is possible to find effective electron spin operators and explore surface magnetism. The (111) surface states are characterized in this system by four surface Dirac points, one at the $\bar{\Gamma}$ point with energy $E_{\bar{\Gamma}}$ and one at each of three $\bar{M}$ points[7] with energy $E_{\bar{M}} \neq E_{\bar{\Gamma}}$. The system further includes substitutional isolectronic magnetic impurities, but with the chemical potential adjusted to the bulk gap one does not expect bulk magnetization do develop. The interesting feature of the TCI system is that conducting surface states in this situation effectively couple the impurities, so that surface magnetism *can* be stable even if it is absent in the bulk. Ferromagnetic ordering turns out to be favored when it breaks the mirror symmetry protecting one or more of the surface Dirac points, by opening gaps in the spectrum that ultimately push down the energies of filled electronic states. This means that the lowest energy states will have magnetization directions dependent on the chemical potential μ.

Figure 1 illustrates this phenomenon. When in the vicinity of the $\bar{\Gamma}$ point, the groundstates are oriented perpendicular to the surface, and the system represents a ferromagnet with Ising symmetry. When μ is near the $\bar{M}$ point energies, there is a six-fold degenerate manifold of groundstate orientations, all of which lie closer to the equator of the Bloch sphere. In principle the μ dependence of these orderings implies that the magnetization direction can be controlled by a gate near the surface. The existence of these states might be detected experimentally via the behavior of their domain wall excitations – which proliferate at the thermal disordering transitions, or can be present at low temperature when the system is zero-field cooled – and we discuss several potential effects from them in what follows.

This papers is organized as follows. In Section II we introduce the bulk Hamiltonian used to model our system, and show how it may be reduced to describe the states in the energy range of interest. Section III explains our method for constructing surface states from the initial bulk Hamiltonian, and shows how this leads to surface (Dirac) Hamiltonians with gapless spectra. In Section IV we introduce our model of magnetic degrees of freedom, and discuss how this impacts the surface Hamiltonians introduced in the prior section, in particular arguing that surface ferromagnetism is likely to result. Numerical calculations supporting our expectations are presented in Section V, demonstrating in particular that the number of such groundstates is determined by the chemical potential of the system. In Section VI we present results of spin-stiffness calculations, demonstrating that there are effective inter-spin interactions that further enhance surface ferromagnetism, but in qualitatively different ways depending on the chemical potential. This turns out to have important consequences for a variety of experiments through its impact on domain walls of the system, which we discuss in Section VII.

2. Bulk Hamiltonian

Our analysis begins with a tight-binding Hamiltonian for materials in the (Sn,Pb)Te class, which is a rocksalt structure. Its explicit form[7] is given by $H_{bulk} = H_m +$

$H_{nn} + H_{nnn} + H_{so}$, with

$$H_m = \sum_j m_j \sum_{\mathbf{R},s} \vec{c}^\dagger_{j,s}(\mathbf{R}) \cdot \vec{c}_{j,s}(\mathbf{R}),$$

$$H_{nn} = t \sum_{(\mathbf{R},\mathbf{R}'),s} \vec{c}^\dagger_{a,s}(\mathbf{R}) \cdot \vec{d}_{\mathbf{R},\mathbf{R}'}\vec{d}_{\mathbf{R},\mathbf{R}'} \cdot \vec{c}_{b,s}(\mathbf{R}') + h.c.,$$

$$H_{nnn} = \sum_j t'_j \sum_{((\mathbf{R},\mathbf{R}')),s} \vec{c}^\dagger_{j,s}(\mathbf{R}) \cdot \vec{d}_{\mathbf{R},\mathbf{R}'}\vec{d}_{\mathbf{R},\mathbf{R}'} \cdot \vec{c}_{j,s}(\mathbf{R}') + h.c.,$$

$$H_{so} = i \sum_j \lambda_j \sum_{\mathbf{R},s,s'} \vec{c}^\dagger_{j,s}(\mathbf{R}) \times \vec{c}_{j,s'}(\mathbf{R}) \cdot (\vec{\sigma})_{s,s'}. \tag{1}$$

In these equations $\mathbf{R}$ labels the sites of a cubic lattice, $j = a, b$ are the species type (Sn/Pb or Te), with on-site energies $m_{a,b}$, and $s = \uparrow, \downarrow$ is the electron spin. The operator-vector $\vec{c}_{j,s}(\mathbf{R})$ annihilates electrons in p_x, p_y and p_z orbitals, and there is a local spin-orbit coupling strength λ_j on each site. As usual, $\vec{\sigma}$ is the vector of Pauli matrices. The unit vectors $\vec{d}_{\mathbf{R},\mathbf{R}'}$ point from $\mathbf{R}$ to $\mathbf{R}'$, and, finally, the sum over $(\mathbf{R}, \mathbf{R}')$ denotes positions which are nearest neighbors, while $((\mathbf{R}, \mathbf{R}'))$ denotes next nearest neighbors. In all, for each unit cell there are twelve basis states.

For appropriately chosen parameters, the model is a direct gap semiconductor with smallest gaps at the L points $[\mathbf{k} = \mathbf{k}_1, \mathbf{k}_2, \mathbf{k}_3, \mathbf{k}_4 \equiv (\frac{\pi}{2}, \frac{\pi}{2}, \frac{\pi}{2}), (-\frac{\pi}{2}, \frac{\pi}{2}, \frac{\pi}{2}),$ $(\frac{\pi}{2}, -\frac{\pi}{2}, \frac{\pi}{2}), (\frac{\pi}{2}, \frac{\pi}{2}, -\frac{\pi}{2})$ in units of the inverse nearest neighbor separation]. Its applicability to SnTe has been confirmed by photoemission experiments[30]. For plane-wave states $|\mathbf{k}, \alpha\rangle$ with wavevector $\mathbf{k}$ precisely at an L-point and α all other quantum numbers needed to specify a state, the nearest neighbor hopping integral vanishes so that the states at these points have well-defined sublattice index. Because of this, adjusting $m_b - m_a \equiv m$ continuously allows two states to cross in energy, leading to band inversion and a transition from trivial to topological bands[7].

Because the nearest neighbor coupling term H_{nn} vanishes at an L point, at these wavevectors each sublattice supports an independent Hamiltonian, which may be diagonalized analytically. The resulting eigenvalues are $E_{0,j} = 4t'_j - \lambda_j + m_j$, $E_{\pm,j} = -2t'_j + \lambda_j/2 \pm R_j + m_j$, with $R_j \equiv \sqrt{144t'^2_j + 24t'_j\lambda_j + 9\lambda^2_j}/2$, with $j = a, b$. Each of these three energy states are doubly degenerate. For large enough m, the lowest eigenvalue for the b sites approaches the highest eigenvalue for the a sites, and one enters the regime in which system can transition between trivial and topological as a function of m. This important behavior is captured by projecting the Hamiltonian onto the four states whose energies are in the range of the band inversion, those with energies $E \sim E_{+,a} \sim E_{-,b}$. The resulting Hamiltonian is a good approximation for the full tight-binding model in this range of energies when $\mathbf{k}$ is near an L point.

The projected Hamiltonian is conveniently written in terms of states which have well-defined quantum numbers upon $2\pi/3$ rotation around a $\Gamma - L$ direction. For example, for the L point $\mathbf{k_1} \equiv (\frac{\pi}{2}, \frac{\pi}{2}, \frac{\pi}{2})$, we define states

$$|m, s, j\rangle \equiv \frac{1}{\sqrt{3}}\left[|p_x, j\rangle + w^m|p_y, j\rangle + w^{2m}|p_z, j\rangle\right] \otimes |s\rangle$$

where $|p_{x\,(y,z)}, j\rangle$ is a $p_{x\,(y,z)}$-orbital on a site of type j, $|s = \pm\rangle$ is a spin state with quantization axis parallel to $\mathbf{k_1}$, and $w = e^{2\pi i/3}$. The relevant eigenstates of the Hamiltonian are then

$$|1, j\rangle = -u_j w|m = 0, s = +, j\rangle + v_j|m = 2, s = -, j\rangle,$$
$$|2, j\rangle = u_j w^*|m = 0, s = -, j\rangle + v_j|m = 1, s = +, j\rangle,$$

$$(2)$$

with

$$u_j = \frac{\sqrt{2}\lambda_j}{\sqrt{(8t'_j + \varepsilon_j)^2 + 2\lambda_j^2}}, \qquad v_j = \frac{8t'_j + \varepsilon_j}{\sqrt{(8t'_j + \varepsilon_j)^2 + 2\lambda_j^2}}, \qquad (3)$$

$\varepsilon_a = -2t'_a + \lambda_a/2 + R_a$, and $\varepsilon_b = -2t'_b + \lambda_b/2 - R_b$. These states have the properties that, under a $2\pi/3$ rotation around the (111) direction, $|1, j\rangle \to e^{-i\pi/3}|1, j\rangle$ and $|2, j\rangle \to e^{i\pi/3}|2, j\rangle$.

To proceed, we form the $\mathbf{k}$-dependent bulk Hamiltonian $H_b(\mathbf{k}) = e^{-i\mathbf{k}\cdot\mathbf{R}}H_{bulk}e^{i\mathbf{k}\cdot\mathbf{R}}$ which acts on states of definite crystal momentum $\mathbf{k}$, and project this into the 4×4 space defined by the states in Eq. (2). For example, writing $\mathbf{k} = \mathbf{k_1} + \mathbf{q}$, for small q one finds, after considerable algebra and up to an overall constant,

$$\bar{H}_1 = Aq_3\tau_x - B[q_1\tilde{\sigma}_2 + q_2\tilde{\sigma}_1]\tau_y + [m + C_{12}^{(-)}(q_1^2 + q_2^2) + C_3^{(-)}q_3^2]\tau_z + C_{12}^{(+)}(q_1^2 + q_2^2) + C_3^{(+)}q_3^2.$$

$$(4)$$

In this expression, $\tilde{\sigma}_1 \equiv \frac{\sqrt{3}}{2}\sigma_x - \frac{1}{2}\sigma_y$ and $\tilde{\sigma}_2 \equiv \frac{\sqrt{3}}{2}\sigma_y + \frac{1}{2}\sigma_x$ are 2×2 matrices acting on the 1,2 indices of the basis states (Eq. (2)), and τ_x, τ_y, τ_z are standard Pauli matrices acting on the sublattice index. The coefficients in Eq. (4) are explicitly given by $A = -2\sqrt{3}(u_a u_b + v_a v_b)$, $B = \sqrt{6}(-u_a v_b + v_a u_b)$, $C_{12}^{(\pm)} = \frac{1}{2}[(C_a - D_a/2) \pm (C_b - D_b/2)]$, and $C_3^{(\pm)} = -\frac{1}{2}(D_a \pm D_b)$, with $C_j = 2t'_j(\frac{1}{3} + u_j^2 - v_j^2)$ and $D_j = 8t'_j/3$. Finally, the wavevector coordinates (q_1, q_2, q_3) are given in terms of the wavevector components of $\mathbf{q}$ by

$$\begin{pmatrix} q_1 \\ q_2 \\ q_3 \end{pmatrix} = \begin{pmatrix} \frac{1}{\sqrt{2}} & -\frac{1}{\sqrt{2}} & 0 \\ \frac{1}{\sqrt{6}} & \frac{1}{\sqrt{6}} & -\frac{2}{\sqrt{3}} \\ \frac{1}{\sqrt{3}} & \frac{1}{\sqrt{3}} & \frac{1}{\sqrt{3}} \end{pmatrix} \begin{pmatrix} q_x \\ q_y \\ q_z \end{pmatrix}.$$

Two comments are in order here. First, the general form of Eq. (4) to linear order in q has been obtained previously[7]; the quadratic terms that we have retained turn out to be important in our approach to generating explicit surface states. Second, the matrix $\tilde{\sigma}_1$ in Eq. (4) carries out a mirror reflection which is a symmetry of

H_{bulk}; perturbations which do not spoil this symmetry will result in an effective Hamiltonian near an L point of a similar form[5].

Analogous approximate forms for H_{bulk} near the other three L points can be obtained by appropriate symmetry operations: the states and effective Hamiltonian in the vicinity of $\mathbf{k_4} \equiv (\frac{\pi}{2}, \frac{\pi}{2}, -\frac{\pi}{2})$ are related to Eqs. 2 and 4 by a mirror reflection across the $x - y$ plane, and these quantities for the other two L points can be constructed by $2\pi/3$ rotations around the (111) direction from those of $\mathbf{k_4}$. For example, for the $\mathbf{k_4}$ point we denote the mirror operation by M_z, and note the fact that $H_{bulk}(k_x, k_y, -k_z) = M_z H_{bulk}(k_x, k_y, k_z) M_z^{-1}$. Writing $\mathbf{k} = \mathbf{k_4} + \mathbf{q'}$, one finds a long-wavelength form for $H_{bulk}(\mathbf{k})$ in the vicinity of $\mathbf{k_4}$, $\bar{H}_4$, which is formally the same form as Eq. (4), but with the caveat that the $\tilde{\sigma}_i$ operators act on the mirror reflected states $|1, j\rangle_4 \equiv M_z |1, j\rangle$ and $|2, j\rangle_4 \equiv M_z |2, j\rangle$, and $\mathbf{q} \to \mathbf{q'}$ with

$$\begin{pmatrix} q'_1 \\ q'_2 \\ q'_3 \end{pmatrix} = \begin{pmatrix} \frac{1}{\sqrt{2}} & -\frac{1}{\sqrt{2}} & 0 \\ \frac{1}{\sqrt{6}} & \frac{1}{\sqrt{6}} & \frac{2}{\sqrt{3}} \\ \frac{1}{\sqrt{3}} & \frac{1}{\sqrt{3}} & -\frac{1}{\sqrt{3}} \end{pmatrix} \begin{pmatrix} q_x \\ q_y \\ q_z \end{pmatrix}.$$

Rewriting $\bar{H}_4$ in terms of $\mathbf{q}$, one finds

$$\begin{aligned} \bar{H}_4 = {}& A \left(\eta q_2 + q_3 \right) \tau_x - B \left[q_1 \tilde{\sigma}'_2 + (q_2 + \eta q_3) \, \tilde{\sigma}'_1 \right] \tau_y \\ & + \left\{ m + C_{12}^{(-)} \left[q_1^2 + (q_2 + \eta q_3)^2 \right] + C_3^{(-)} (q_2 + \eta q_3)^2 \right\} \tau_z \\ & + C_{12}^{(+)} \left[q_1^2 + (q_2 + \eta q_3)^2 \right] + C_3^{(+)} (q_2 + \eta q_3)^2, \end{aligned} \tag{5}$$

with $\eta = 2\sqrt{2}/3$.

3. Surface Hamiltonian

For $m < 0$, on the (111) surface there are gapless surface states[6] located at the four different projections of the L points onto the surface Brillouin zone. These may be constructed individually[27–29], by searching for eigenstates of the bulk Hamiltonian which are evanescent deep inside the system bulk, and vanish on the surface itself. Focusing on $\mathbf{k_1}$, the direction perpendicular to the surface is parameterized by q_3, so we set $q_1 = q_2 = 0$, $q_3 \to i\kappa$ in $\bar{H}_b$ to find surface states exactly at the $\bar{\Gamma}$ point. Such an evanescent state must satisfy

$$\left[i\kappa A \tau_x + \left(m - C_3^{(-)} \kappa^2 \right) \tau_z \right] |\kappa\rangle = \left(E + C_3^{(+)} \kappa^2 \right) |\kappa\rangle. \tag{6}$$

Note that arriving at this equation is only possible because we have retained quadratic terms in $\bar{H}_b$[27–29]. In order to satisfy Eq. (6), κ must obey

$$\left(E + C_3^{(+)} \right)^2 = \left(m - C_3^{(-)} \kappa^2 \right)^2 - A^2 \kappa^2. \tag{7}$$

For a given value of E, one finds two values of κ^2 that satisfy Eq. (7), $\kappa_\pm^2$. With the proper choice of overall sign for $\kappa_\pm$, the eigenstates $|\kappa_+\rangle$ and $|\kappa_-\rangle$ vanish in

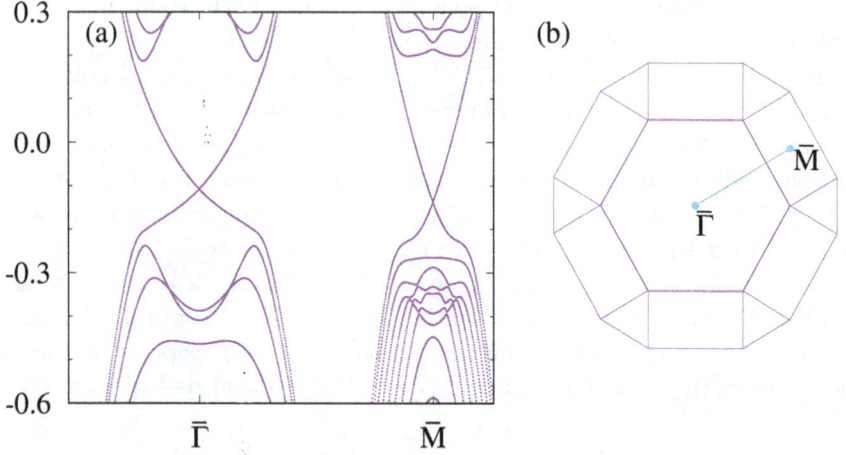

Fig. 2. (a) Band structure for a SnTe slab as a function of wavevector, along a direction connecting the $\bar{\Gamma}$ point to a $\bar{M}$ point, for tight-binding parameters such that the system is topological. Note the two surface Dirac points are at different energies. (b) Line of k-values followed in (a).

real space deep in the bulk of the system, but generically neither vanishes at the surface. However if $|\kappa_+\rangle \propto |\kappa_-\rangle$, we can form a linear combination of these that does[27-29]. One may show that for $\mathrm{sgn}(mC_3^{(-)}) < 0$, there is a unique value of E for which this is possible, given by $E = E_{\bar{\Gamma}} \equiv -\frac{C_3^{(+)}}{C_3^{(-)}}m$. Because the states $|\kappa_\pm\rangle$ have no dependence on the $\tilde{\sigma}_i$ operators we can form two independent pairs of these. It is convenient to write these as eigenstates of the operator $\tilde{\sigma}_3 \equiv -i\tilde{\sigma}_1\tilde{\sigma}_2$. Thus we have two degenerate surface states at the $\bar{\Gamma}$ point, forming a surface Dirac point with energy $E_{\bar{\Gamma}}$ within the bulk gap. The corresponding analysis for the $\bar{M}$ points yields Dirac point energies $E_{\bar{M}} = -m\frac{C_{12}^{(+)}\eta^2+C_3^{(+)}}{C_{12}^{(-)}\eta^2+C_3^{(-)}}$, so that $E_{\bar{\Gamma}} \neq E_{\bar{M}}$, in agreement with tight-binding calculations. This is illustrated in Fig. 2. (See also Ref. 31.)

To simplify subsequent discussion, we drop the particle-hole symmetry-breaking terms (i.e., set $C_{12}^{(+)}$, $C_3^{(+)} = 0$) in $\bar{H}_1$, $\bar{H}_4$, and the long-wavelength Hamiltonians for the other L points, except to note that the surface Dirac point energies at the $\bar{\Gamma}$ and $\bar{M}$ points are different, so that we add $E_{\bar{\Gamma}}$ and $E_{\bar{M}}$ to to $\bar{H}_1$ and $\bar{H}_4$, respectively. Thus we make the replacements

$$\bar{H}_1 \to Aq_3\tau_x - B[q_1\tilde{\sigma}_2 + q_2\tilde{\sigma}_1]\tau_y + [m + C_{12}^{(-)}(q_1^2 + q_2^2) + C_3^{(-)}q_3^2]\tau_z + E_{\bar{\Gamma}},$$

$$\bar{H}_4 \to A\left(\eta q_2 + q_3\right)\tau_x - B\left[q_1\tilde{\sigma}_2' + (q_2 + \eta q_3)\,\tilde{\sigma}_1'\right]\tau_y$$

$$+ \left\{m + C_{12}^{(-)}\left[q_1^2 + (q_2 + \eta q_3)^2\right] + C_3^{(-)}\left(q_2 + \eta q_3\right)^2\right\}\tau_z + E_{\bar{M}}.$$

$$(8)$$

These Hamiltonians are next projected into space of surface states derived above. The explicit form of these that are relevant for the $\bar{\Gamma}$ point (with effective Hamiltonian $\bar{H}_1$) are

$$|u_1\rangle = \frac{1}{\sqrt{2}} \begin{pmatrix} 1 \\ 0 \\ -i\,\mathrm{sgn}(A) \\ 0 \end{pmatrix} \mathcal{N}_z \left(e^{-\kappa_+ z} - e^{-\kappa_- z} \right),$$

$$|u_2\rangle = \frac{1}{\sqrt{2}} \begin{pmatrix} 0 \\ 1 \\ 0 \\ -i\,\mathrm{sgn}(A) \end{pmatrix} \mathcal{N}_z \left(e^{-\kappa_+ z} - e^{-\kappa_- z} \right).$$

In these expressions, the four entries are coefficients of the $(|1, a\rangle, |2, a\rangle, |1, b\rangle, |2, b\rangle)$ states at the $\mathbf{k}_1$ point, z denotes the direction perpendicular to the surface (with $z > 0$ being points inside the system), $\mathcal{N}_z$ is a normalization constant, and $\kappa_\pm^2 = m + \left[A^2 \pm \sqrt{4mC_3^{(-)}A^2 + A^4} \right]/2C_3^{(-)}$. Note that the wavefunctions have equal weight on the two sublattices, and moreover are eigenstates of the $\tilde{\sigma}_3$ operator. The corresponding results for the $\mathbf{k}_4$ point may also be written as eigenstates of the relevant $\tilde{\sigma}_3$ operator (i.e., acting on states near the $\mathbf{k}_4$ point), and also have equal weight on the two sublattices, although the relative phase is more complicated and has the form

$$e^{i\theta_4} \equiv \frac{A - i\eta B}{\sqrt{A^2 + \eta^2 B^2}}.$$

With the explicit surface wavefunctions in hand it is straightforward to project $\bar{H}_1$ and $\bar{H}_4$ onto the space of surface states, yielding surface Hamiltonians for the $\bar{\Gamma}$ and $\bar{M}$ points. These take the form

$$H_{\bar{\Gamma}} \equiv B \left[q_1 \tilde{\sigma}_2 + q_2 \tilde{\sigma}_1 \right] + E_{\bar{\Gamma}} \tag{9}$$

and

$$H_{\bar{M}} \equiv \frac{AB}{\sqrt{A^2 + \eta^2 B^2}} \left[(\eta^2 - 1)\, q_2 \tilde{\sigma}_1 + q_1 \tilde{\sigma}_2 \right] + E_{\bar{M}}. \tag{10}$$

Note that the states which the $\tilde{\sigma}_i$ operators act upon in $H_{\bar{M}}$ are different than those of $H_{\bar{\Gamma}}$. Analogous results may be obtained for the other two $\bar{M}$ points by $2\pi/3$ rotations of Eq. (10).

4. Surface Magnetism

It has long been known that metals in the (Sn/Pb)Te class[32–40], may be doped with magnetic ions which in some circumstances order ferromagnetically at low temperature. In these systems the magnetic ions enter substitutionally for Sn/Pb atoms, and the coupling of the magnetic moments with the conduction electrons can be understood rather well using an $s - d$ model[41], $H_{sd} = J \sum_i \vec{S}(\mathbf{r}_i) \cdot \vec{s}_i$, where $\vec{s}_i$ represents an impurity spin at location $\mathbf{r}_i$ and $\vec{S}(\mathbf{r}_i)$ is the conduction electron spin density at that location. Here we consider a situation in which the chemical potential is in a gap of the bulk spectrum, so that free carriers are not present and bulk magnetic ordering is not expected. In the TCI state, however, surface electrons couple the magnetic moments of the substitutional impurities near the surface, and may lead to ferromagnetic ordering. Analogous physics occurs, for example, in graphene[42]. We will model this by assuming magnetic impurities are present in the system, on one sublattice, near the surface.

The explicit surface wavefunctions derived above (Eq. (9)) provides one way to explore this: the electron spin operators in H_{sd} may be projected onto the spaces of surface states for the $\bar{\Gamma}$ and $\bar{M}$ points. The spin operators on a single (say, the a) sublattice may be written for either the $\bar{\Gamma}$ or an $\bar{M}$ point as

$$\vec{S}^{(a)} = \frac{1}{4}\left(u_a^2\tilde{\sigma}_2, u_a^2\tilde{\sigma}_1, (u_a^2 - v_a^2)\tilde{\sigma}_3\right), \tag{11}$$

where the quantities u_a, v_a are given in Eq. (3). Note that the $\tilde{\sigma}_i$ operators act on a different pair of surface states for each of the four surface Dirac points, specifically eigenstates of $\tilde{\sigma}_3$ defined with respect to the relevant Γ-L direction. To further simplify the model, we assume the impurity spins ferromagnetically order and treat the Hamiltonian in mean-field theory; the linear stability of the state against formation of a spin-density wave can then be checked, as we show below. Projecting H_{sd} onto the subspace of surface states for a Dirac point using Eq. (11) leads to an effective Hamiltonian of the form

$$H_i \approx E_i + \alpha_i(q_2 - b_2)\tilde{\sigma}_1 + \beta_i(q_1 - b_1)\tilde{\sigma}_2 + \Delta_i\tilde{\sigma}_3, \tag{12}$$

where i denotes either $\bar{\Gamma}$ or one of the $\bar{M}$ points, and the relationships between (q_1, q_2) and (q_x, q_y, q_z) depend on the specific Dirac point i. As expected on general symmetry grounds, $\alpha_{\bar{\Gamma}} = \beta_{\bar{\Gamma}}$, but $\alpha_{\bar{M}} \neq \beta_{\bar{M}}$. The offsets b_1 and b_2 are proportional to components of the impurity magnetization perpendicular to $\mathbf{k}_i$, while Δ is proportional to the component along it. The resulting spectra, $\varepsilon_i = E_i \pm \sqrt{\alpha_i^2(q_1 - b_1)^2 + \beta_i^2(q_2 - b_2)^2 + \Delta_i^2}$, provides the important observation that when the moments align along a $\Gamma - L$ direction and $\mu \sim E_i$, a gap opens in the corresponding surface spectrum that lowers its contribution to the total electron energy[18].

It is interesting to consider how the surface symmetries determine the number of magnetic groundstates one expects due to this mechanism. In particular the (111) surface supports a 3-fold symmetry rotational symmetry (C_3), of which the surface states with $\mathbf{k}$ at the $\bar{\Gamma}$ point represents a two-fold representation, while those at the $\bar{M}$ points form a six-fold representation. When the energetics is dominated by the former (as we expect if $\mu \sim E_{\bar{\Gamma}}$), then there will be two magnetic groundstates, associated with maximal breaking of the two-fold degeneracy, with one of the two degenerate states forming the 2-fold representation being maximally pushed down in energy. On the other hand, one expects a corresponding 6-fold degeneracy for $\mu \sim E_{\bar{\Gamma}}$. We next show these expectations to be borne out using numerical studies of the system.

5. Numerical Studies

To test these ideas we have numerically computed the electronic energy of a TCI slab with (111) surfaces, using the tight-binding model H_{bulk} as our Hamiltonian, and adding an effective magnetic field $\vec{b}$ near the surface on only the a sublattice, in such a way that their coupling to the two states associated with the surface Dirac cones is the same. For the present purpose it is convenient to view the fcc lattice as stacked, two-dimensional triangular lattices, i.e., a close packed structure, with ABC stacking. Triangular layers of the a and b sublattices are arranged alternately along the stacking direction.

The electronic structure calculations involve taking a finite but relatively large number of layers. We used 47 layers which we found to be sufficient to avoid significant admixture of states of the two slab surfaces. The model parameters we used[43] in H_{bulk} for these calculations are $t = 0.9, t_a^{'} = -t_b^{'} = -0.5, \lambda_a = \lambda_b = -0.7, m_a = -m_b = 3.5$. These place the system in the topological regime, as can be seen from the band structure of the slab, shown in Fig. 2, which demonstrate the presence of gapless states in the bulk gap around the $\bar{\Gamma}$ and $\bar{M}$ points in the surface Brillouin zone. Again, it is important to note that the energy of the Dirac point at $\bar{\Gamma}$ is slightly higher than that of $\bar{M}$.

As mentioned above, we have modeled the coupling of the electrons to magnetic impurities by an $s - d$ model. In real systems these will be randomly located (on a single sublattice) at different depths from the surface, but on average the coupling of the spins to both states should be the same for both the $\bar{\Gamma}$ and $\bar{M}$ points. To model this, we choose *two* layers, say L_1, L_2, near each surface, on which to locate the magnetic moments. We then choose effective local Zeeman fields on each layer of magnitudes $(J|\vec{s}|)_{L_1}$ and $(J|\vec{s}|)_{L_2}$ such that they satisfy the equation

$$\frac{(J|\vec{s}|)_{L_1}}{(J|\vec{s}|)_{L_2}} = \frac{|\psi_{\bar{\Gamma}}(L_2)|^2 - |\psi_{\bar{M}}(L_2)|^2}{|\psi_{\bar{M}}(L_1)|^2 - |\psi_{\bar{\Gamma}}(L_1)|^2}, \tag{13}$$

where $\psi_{\bar{\Gamma}}$ and $\psi_{\bar{M}}$ are the (numerically generated) wavefunctions at the $\bar{\Gamma}$ and $\bar{M}$ points, respectively. This guarantees that the net coupling of the impurity spins to electron spins in the $\bar{\Gamma}$ and $\bar{M}$ states will have the same overall strength. In practice, in our calculations the magnetic impurities reside on layer 3 and 5 near one surface and on layer 43 and 45 near the other.

Initially we consider a slab with primitive unit cell presenting only a single site of one sublattice on the surface, and introduce a surface magnetization as described above. While this represents a relatively large density of impurities, it captures the correct qualitative physics, and allows us to study a wide enough slab that the surfaces are effectively decoupled. As expected from the above discussion, among the four the Dirac cones the one with largest magnetization projection along its corresponding Γ - L direction develops the largest gap. Figure 1 illustrates the resulting total electronic energy for two different scales of Fermi energy as a function of the magnetic orientation. Panels (a) and (b) show results for particle numbers such that the Fermi energy is near $E_{\bar{M}}$ and $E_{\bar{\Gamma}}$ respectively in the absence

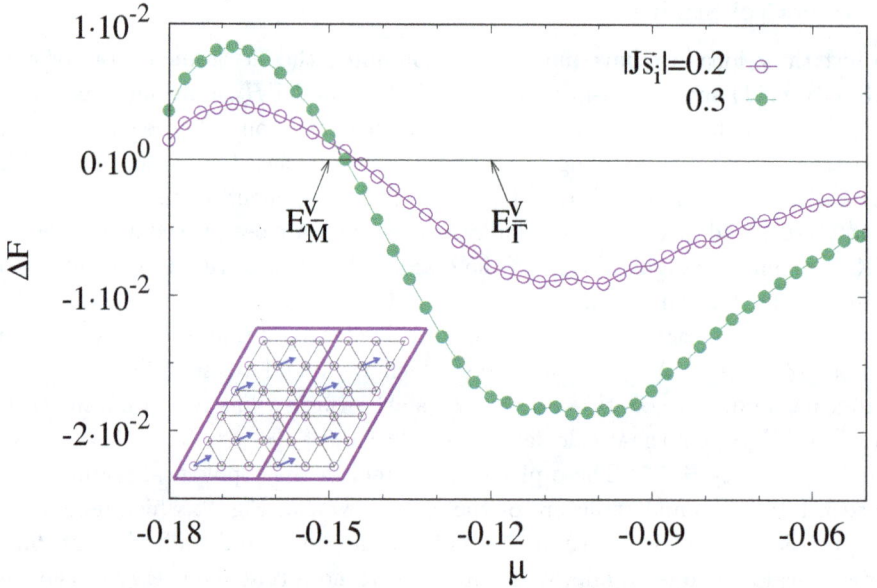

Fig. 3. Difference in free energy per surface atom in units of nearest neighbor hopping t, when magnetic moments are oriented in the (111) direction and in the (11$\bar{1}$) direction, as a function of chemical potential, for two different strengths of $J|\bar{s}|$ in H_{sd}. $E_{\bar{\Gamma}}^V$ and $E_{\bar{M}}^V$ indicate valence band tops for $J|\bar{s}| = 0.3$ with $\bar{s}$ in each direction respectively. Here there are 2 magnetic ions for 9 atoms on the surface in the unit cell (inset).

of magnetic impurities. With them present, the energy is minimized in the former case for $\vec{s}$ along a $\Gamma - L$ direction associated with an $\bar{M}$ point, while in the latter minimization occurs for $\vec{s}$ along $\mathbf{k}_1$. Thus in the $\mu \sim E_{\bar{\Gamma}}$ case one finds two (equal) energy minima on the Bloch sphere, while for $\mu \sim E_{\bar{M}}$ there are six. (Only half of these can be seen in the figures.) Analogous results are found when the chemical potential rather than particle number is fixed, as illustrated in panels (c) and (d). As expected from the above discussion, minimal energy directions are picked out by the orientations that maximize gap openings near the Fermi energy.

To further substantiate this, we also studied a more dilute magnetic moment model, in which the impurities are present only for 2/9 of the atoms of one sublattice near the surface, as illustrated in the inset of Fig. 3. The main panel shows the difference in Gibbs free energy of the system ($\langle H_{bulk} \rangle - \mu N$ with N the number of electrons) when the magnetic moments are oriented in the (111) direction and in the (11$\bar{1}$) direction, as a function of chemical potential. The results again demonstrate that energetically favored directions are determined by μ.

6. Spin Stiffness

In describing this system as a ferromagnet, one might expect that interactions favor alignment of the impurity spins even independent of the directions that are

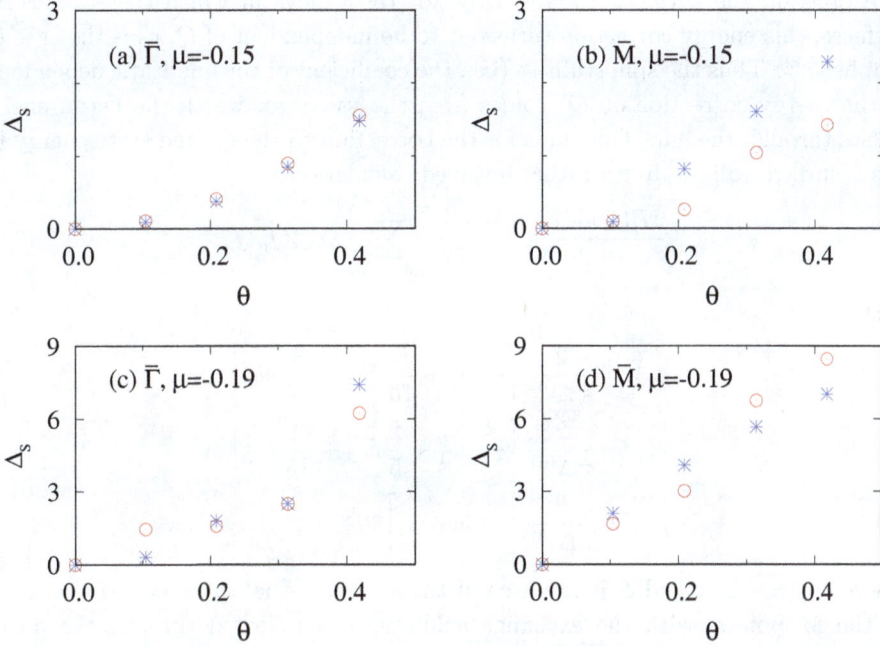

Fig. 4. $\Delta_s(\theta, \phi)$ (see text) for a few values of θ and ϕ. This energy scale is always positive, indicating that ferromagnetic ordered states are stable with respect to small anti-ferromagnetic distortions.

explicitly favored by the symmetry-breaking of mirror symmetries that pick out certain magnetization directions. We can establish that there is in fact such an effective spin-stiffness associated with magnetization gradients. As a first check, we can verify this numerically, using the geometry shown in the inset of Fig. 3. We consider a realization of the two spins in the unit cell in which they are tilted in equal but opposite directions along the Bloch sphere away from the local minimum energy direction for the uniformly magnetized state. Writing $E_{++(--)}(\theta, \phi)$ as the total electronic energy for $\vec{s}_1 = \vec{s}_0 + (-)\delta \vec{s}(\theta, \phi)$ and $\vec{s}_2 = \vec{s}_0 + (-)\delta \vec{s}(\theta, \phi)$, and E_{+-} analogously defined, we numerically compute $\Delta_s(\theta, \phi) \equiv E_{+-} - (E_{++} + E_{--})/2$, which is a measure of the energy cost for the two spins to be unequally aligned. As shown in Fig. 4, this always appears to be positive. This indicates an effective interaction energy between nearby spins that tends to keep them aligned.

A second approach to computing effective inter-impurity interactions is to consider what happens to the electronic energy when the exchange field is allowed to oscillate with some wavevector $\vec{Q}$ along an average direction, $b_{1,2}(\mathbf{r}) = b_{1,2}^{(0)} + \delta b_{1,2} \cos(\mathbf{Q} \cdot \mathbf{r})$, $\Delta(\mathbf{r}) \equiv \Delta^{(0)} + \delta \Delta \cos(\mathbf{Q} \cdot \mathbf{r})$ and compute the correction to the energy at second order in $\delta b_{1,2}$, $\delta \Delta$, and to lowest non-trivial order in Q. To compute this we adopt as our basic Hamiltonian Eq. (12), assuming for simplicity $\alpha_i = \beta_i \equiv \alpha$, and use the directions associated with q_1 and q_2 to define x and y

directions on the surface. Interestingly, for the valleys in which there are Fermi surfaces, this energy correction turns out to be independent of Q, as is the case for graphene[42]. Thus the spin stiffness (i.e., the coefficient of the quadratic dependence of the energy correction on Q) comes from the valley for which the Fermi energy passes through the gap. One finds for the correction to the ground state energy for the $\bar{\Gamma}$ surface valley, after a rather involved calculation,

$$\frac{\delta E(Q) - \delta E(0)}{S} = \frac{1}{2} \sum_{\mu,\nu=x,y,z} \rho_{\mu,\nu} Q_\mu Q_\nu, \tag{14}$$

with

$$\rho_{xx} = \frac{2}{\pi\Delta^{(0)}} \left[\frac{2}{5} \delta b_1^2 + \frac{8}{15} \delta b_2^2 + \frac{4}{15} \delta b_3^2 \right],$$

$$\rho_{yy} = \frac{2}{\pi\Delta^{(0)}} \left[\frac{8}{15} \delta b_1^2 + \frac{2}{5} \delta b_2^2 + \frac{4}{15} \delta b_3^2 \right],$$

$$\rho_{xy} = \frac{2}{\pi\Delta^{(0)}} \left[\frac{8}{15} \delta b_1 \delta b_2 \right],$$

where $\delta b_3 = \delta\Delta/\alpha$ and S is the area of the surface. Analogous expressions apply to the $\bar{M}$ points, with the exchange field vector $\vec{b}$ reflected through the mirror plane that transforms the L_1 point to the L point associated with a given $\bar{M}$ point. Eq. (14) demonstrates that if the exchange field from the surface magnetization has a spatial oscillation, the resulting energy necessarily increases with increasing oscillation wavevector. This is consistent with inter-spin interactions which tend to ferromagnetically align them. It is notable that the stiffnesses associated with this diverge as $1/\Delta^{(0)}$. Likely effects of this are discussed in the next section.

7. Experimental Consequences

We next consider some physical consequences of the surface magnetism discussed above, focusing on temperature ranges where the impurity magnetic moments may be treated classically. As remarked upon above, the sensitivity of the low-energy directions of the magnetization to the chemical potential should allow it to be controlled via a gate potential, which in principle would be observable in direct magnetization measurements. Another basic observation is that the gap openings induce a Berry's curvature in the surface bands, which generically induces an anomalous Hall effect. For this system we do not expect it to be quantized[21,22], since the chemical potential typically cannot pass through a gap for all the surface Dirac species at the same time.

It is interesting to consider possible consequences of $\rho_{\mu\nu} \sim 1/\Delta^{(0)}$ as discussed above. In particular we expect that the multiple minima presented in Fig. 1 imply that there should be domain wall (DW) excitations in the system, with energy per unit length scaling as $\sqrt{\Delta^{(0)}}\rho_0$, with ρ_0 an appropriate average of $\rho_{\mu\nu}$'s. This remains *finite* even as $\Delta^{(0)}$ vanishes, as should happen at high enough temperature.

The divergence of the spin stiffness, $\rho_0 \sim 1/\Delta^{(0)}$, reflects the fact that as the gap vanishes, the quantity $\delta E(\mathbf{Q})$ is no longer analytic in $\mathbf{Q}$, and in particular rises *linearly* with Q in the long wavelength limit[42], suggesting a non-local interaction among spin gradients. Presuming μ ends up at the Dirac point as $\Delta^{(0)}$ vanishes, the simplest model of the system is a clock model with long-range interactions, which in the Ising case would approach the transition with mean-field exponents[44]. For other values of μ it is possible that DW's with finite energy per unit length can be stabilized in this circumstance, but if so would not have the simplest structure[45]. In this situation, for the $\mu \sim E_{\bar{\Gamma}}$ case [cf. Fig. 1(a)], the system presumably will undergo a second order phase transition in the Ising universality class. In the $\mu \sim E_{\bar{M}}$ case [Fig. 1(b)], with six different minima one expects that a simpler theory with the same long-wavelength physics will be a six-state clock model. In this case, the phase transition is known to be in the Kosterlitz-Thouless universality class[46].

We also note that DW excitations in this system may accumulate charge, both due to mid-gap states[47,48], as well as from the surface valleys which have Fermi surfaces allowing low-energy scattering. At the critical temperature T_c where the transition occurs, one expects DW's to proliferate, opening a channel for conduction which is absent below T_c. This could lead to singular behavior (e.g., a cusp) in the conductivity of the surface at the transition[49,50], and should also have a signature when the surface is probed via tunneling. A further possibility is to probe the system by looking for differences in surface conduction when the system is field-cooled through its critical temperature from when it is zero field-cooled. The latter leads to nucleation of groundstate domains with random orientation, and DW's between them, which cannot relax on the time scale of an experiment. Thus one expects stronger surface conduction from a zero field-cooled sample[20,51,52].

Finally, the presence of charged DW's on the surface might be detected directly via coupling to electromagnetic waves[53]. One interesting possibility is to probe diffuse scattering in reflectance from the surface, which should be sensitive to DW proliferation because of the charge associated with them. In a simple model we can associate charge with gradients of the spin parallel to the ordering axis, which in the $\bar{\Gamma}$ case would be the surface normal $\hat{z}$. Assuming the scattering amplitude is proportional to the charge density leads to an amplitude $A_{\vec{q}} \sim \hat{z} \cdot \vec{\mathcal{E}} \times \mathbf{q}\, s_z(\mathbf{q})$, where $\vec{\mathcal{E}}$ is the electric field of the light, for plane waves scattered by a wavevector $\vec{q}$ away from the in-plane component of the incident direction. The scattering intensity then reveals information about $\chi = \langle s_z(-\vec{q}) s_z(\vec{q}) \rangle$ which in principle will allow one to identify the universality class of the transition.

In summary, the surface of a magnetically-doped TCI hosts magnetic ordering in the topological state even when the bulk is magnetically disordered. The unique electronic structure of a TCI surface leads to a rich set of possible ordered states, with defect structures – domain walls – that reflect how the symmetry of a surface is realized at the chemical potential of the conducting surface electrons. This allows thermal disordering transitions of a number of different universality classes to be realized in this single system, simply by adjusting the electron doping near the surface. In this way the magnetically-doped TCI system allows for an exploration of a diverse set of effective magnetic systems, all in a single setting.

Acknowledgements

The authors thank Fernando de Juan for helful comments, and U. Nitzsche for technical assistance. This work was supported by the NSF through Grant Nos. DMR-1506263 and DMR-1506460, by the US-Israel Binational Science Foundation, and by MEyC-Spain under grant FIS2015-64654-P. Computations were carried out on the ITF/IFW and IU Karst clusters.

References

1. D. J. Thouless, M. Kohmoto, M. P. Nightingale and M. den Nijs, Quantized hall conductance in a two-dimensional periodic potential, *Phys. Rev. Lett.* **49**, 405 (Aug 1982).
2. B. I. Halperin, Quantized hall conductance, current-carrying edge states, and the existence of extended states in a two-dimensional disordered potential, *Phys. Rev. B* **25**, 2185 (Feb 1982).
3. M. Z. Hasan and C. L. Kane, Colloquium, *Rev. Mod. Phys.* **82**, 3045 (Nov 2010).
4. X.-L. Qi and S.-C. Zhang, Topological insulators and superconductors, *Rev. Mod. Phys.* **83**, 1057 (Oct 2011).
5. L. Fu, *Phys. Rev. Lett.* **106**, p. 106802 (2011).
6. J. Liu, W. Duan and L. Fu, *Phys. Rev. B* **88**, p. 241303(R) (2013).
7. T. Hsieh, H. Lin, J. Liu, W. Duan and L. Fu, *Nat. Commun.* **3**, p. 982 (2012).
8. Y. Tanaka and et al., *Nat. Phys.* **8**, p. 800 (2012).
9. S. Xu and et al., *Nat. Phys.* **3**, p. 1192 (2012).
10. P. Dziawa and et al., *Nat. Comm.* **11**, p. 1023 (2012).
11. Y. Okada and et al., *Science* **341**, p. 6153 (2013).
12. C. Yan and et al., *Phys. Rev. Lett.* **112**, p. 186801 (2014).
13. A. Liu, C. Liu, C. Xu, X. Qu and S. Zhang, Magnetic impurities on the surface of a topological insulator, *Phys. Rev. Lett.* **102**, p. 156603 (2009).
14. R. R. Biswas and A. V. Balatsky, Impurity-induced states on the surface of three-dimensional topological insulators, *Phys. Rev. B* **81**, p. 233405 (Jun 2010).
15. I. Garate and M. Franz, Magnetoelectric response of the time-reversal invariant helical metal, *Phys. Rev. B* **81**, p. 172408 (May 2010).
16. D. Abanin and D. Pesin, Ordering of magnetic impurities and tunable electronic properties of topological insulator, *Phys. Rev. Lett.* **106**, p. 136802 (2011).
17. C.-X. Liu, B. Roy and J. D. Sau, Ferromagnetism and glassiness on the surface of topological insulators, *Phys. Rev. B* **94**, p. 235421 (Dec 2016).
18. D. Efimkin and V. Galitski, *Phys. Rev. B* **89**, p. 115431 (2014).
19. J. Shen and J. J. Cha, *Nanoscale* **6**, p. 14133 (2014).
20. B. A. Assaf, F. Katmis, P. Wei, C.-Z. Chang, B. Satpati, J. S. Moodera and D. Heiman, Inducing magnetism onto the surface of a topological crystalline insulator, *Phys. Rev. B* **91**, p. 195310 (May 2015).
21. F. Zhang, X. Li, J. Feng, C. Kane and E. Mele, arXiv:1309.7682.
22. C. Fang, M. J. Gilbert and B. A. Bernevig, Large-chern-number quantum anoma-

lous hall effect in thin-film topological crystalline insulators, *Phys. Rev. Lett.* **112**, p. 046801 (Jan 2014).

23. S. Reja, H. Fertig, L. Brey and S. Zhang, *Phys. Rev. B* **96**, p. 201111 (R) (2017).
24. Z. Li, S. Shao, N. Li, K. McCall, J. Wang and S. X. Zhang, Single crystalline nanostructures of topological crystalline insulator snte with distinct facets and morphologies, *Nano Letters* **13**, 5443 (2013).
25. A. A. Taskin, F. Yang, S. Sasaki, K. Segawa and Y. Ando, Topological surface transport in epitaxial snte thin films grown on bi₂te₃, *Phys. Rev. B* **89**, p. 121302 (Mar 2014).
26. J. Shen, Y. Jung, A. S. Disa, F. J. Walker, C. H. Ahn and J. J. Cha, Synthesis of snte nanoplates with 100 and 111 surfaces, *Nano Letters* **14**, 4183 (2014).
27. C.-X. Liu, X.-L. Qi, H. Zhang, X. Dai, Z. Fang and S.-C. Zhang, Model hamiltonian for topological insulators, *Phys. Rev. B* **82**, p. 045122 (Jul 2010).
28. P. G. Silvestrov, P. W. Brouwer and E. G. Mishchenko, Spin and charge structure of the surface states in topological insulators, *Phys. Rev. B* **86**, p. 075302 (Aug 2012).
29. L. Brey and H. A. Fertig, *Phys. Rev. B* **89**, p. 084305 (2014).
30. P. Littlewood and et al., *Phys. Rev. Lett.* **105**, p. 086404 (2010).
31. C. Polley and et al., *Phys. Rev. B* **89**, p. 075317 (2014).
32. M. Inoue, K. Ishii and T. Tatsukawa, *J. Low Temp. Phys.* **23**, 785 (1975).
33. M. Inoue, K. Ishii and H. Yagi, *J. Phys. Soc. Japan* **43**, 903 (1977).
34. M. Inoue, T. Tanabe, H. Yagi and T. Tatsukawa, *J. Phys. Soc. Japan* **47**, 1879 (1979).
35. T. Story, R. R. Gałązka, R. B. Frankel and P. A. Wolff, Carrier-concentration induced ferromagnetism in pbsnmnte, *Phys. Rev. Lett.* **56**, 777 (Feb 1986).
36. G. Karczewski, J. K. Furdyna, D. L. Partin, C. N. Thrush and J. P. Heremans, Far-infrared investigation of band-structure parameters and exchange interaction in $pb_{1-x}eu_x te$ films, *Phys. Rev. B* **46**, 13331 (Nov 1992).
37. F. Geist, H. Pascher, N. Frank and G. Bauer, Interband magnetotransmission and coherent raman spectroscopy of spin transitions in diluted magnetic $pb_{1-x}mn_x Se$, *Phys. Rev. B* **53**, 3820 (Feb 1996).
38. F. Geist, W. Herbst, C. Mejía-García, H. Pascher, R. Rupprecht, Y. Ueta, G. Springholz, G. Bauer and M. Tacke, Magneto-optical investigations of eu-based diluted magnetic lead chalcogenides, *Phys. Rev. B* **56**, 13042 (Nov 1997).
39. A. Prinz, G. Brunthaler, Y. Ueta, G. Springholz, G. Bauer, G. Grabecki and T. Dietl, Electron localization in $n - pb_{1-x}eu_x Te$, *Phys. Rev. B* **59**, 12983 (May 1999).
40. A. Łusakowski, A. Jędrzejczak, M. Górska, V. Osinniy, M. Arciszewska, W. Dobrowolski, V. Domukhovski, B. Witkowska, T. Story and R. R. Gałązka, Magnetic contribution to the specific heat of $pb_{1-x}mn_x Te$, *Phys. Rev. B* **65**, p. 165206 (Apr 2002).
41. T. Dietl, C. Śliwa, G. Bauer and H. Pascher, Mechanisms of exchange interactions between carriers and mn or eu spins in lead chalcogenides, *Phys. Rev. B* **49**, 2230 (Jan 1994).
42. L. Brey, H. A. Fertig and S. Das Sarma, Diluted graphene antiferromagnet, *Phys. Rev. Lett.* **99**, p. 116802 (Sep 2007).
43. I. Fulga, N. Avraham, H. Beidenkopf and A. Stern, *Phys. Rev. B* **94**, p. 125405 (2016).
44. M. F. Paulos, S. Rychkov, B. C. van Rees and B. Zan, *Nuc. Phys. B* **902**, 246 (2016).
45. R. Rajaraman, *Solitons and Instantons* (North-Holland, New York, 1989).
46. J. V. José, L. P. Kadanoff, S. Kirkpatrick and D. R. Nelson, Renormalization, vortices, and symmetry-breaking perturbations in the two-dimensional planar model, *Phys. Rev. B* **16**, 1217 (Aug 1977).
47. R. Jackiw and C. Rebbi, *Phys. Rev. D* **13**, p. 3398 (1976).
48. A. Schaakel, *Boulevard of Broken Symmetries* (World Scientific, 2008).

49. T. Jungwirth and A. H. MacDonald, Resistance spikes and domain wall loops in ising quantum hall ferromagnets, *Phys. Rev. Lett.* **87**, p. 216801 (Oct 2001).
50. K. Dhochak, E. Shimshoni and E. Berg, Spontaneous layer polarization and conducting domain walls in the quantum hall regime of bilayer graphene, *Phys. Rev. B* **91**, p. 165107 (Apr 2015).
51. K. Ueda, J. Fujioka, B.-J. Yang, J. Shiogai, A. Tsukazaki, S. Nakamura, S. Awaji, N. Nagaosa and Y. Tokura, Magnetic field-induced insulator-semimetal transition in a pyrochlore $nd_2ir_2o_7$, *Phys. Rev. Lett.* **115**, p. 056402 (Jul 2015).
52. Z. Tian and et al., Field-induced quantum metal-insulator transition in the pyrochlore iridate $nd_2ir_2o_7$, *Nat. Phys.* **12**, 134 (2016).
53. E. Y. Ma, Y.-T. Cui, K. Ueda, S. Tang, K. Chen, N. Tamura, P. M. Wu, J. Fujioka, Y. Tokura and Z.-X. Shen, Mobile metallic domain walls in an all-in-all-out magnetic insulator, *Science* **350**, 538 (2015).

Majorana quasiparticles in ultracold one-dimensional gases

F. Iemini

Abdus Salam International Center for Theoretical Physics, Strada Costiera 11, Trieste, Italy

L. Mazza

*Departement de Physique, Ecole Normale Superieure/PSL Research University,
CNRS, 24 rue Lhomond, F-75005 Paris, France*

L. Fallani

*Department of Physics and Astronomy, University of Florence, I-50019 Sesto Fiorentino, Italy
LENS European Laboratory for Nonlinear Spectroscopy, I-50019 Sesto Fiorentino, Italy*

P. Zoller

*Institute for Theoretical Physics, University of Innsbruck, A-6020 Innsbruck, Austria
Institute for Quantum Optics and Quantum Information of the Austrian Academy of Sciences,
A-6020 Innsbruck, Austria*

R. Fazio

*Abdus Salam International Center for Theoretical Physics, Strada Costiera 11, Trieste, Italy
NEST, Scuola Normale Superiore and Istituto Nanoscienze-CNR, I-56126 Pisa, Italy*

M. Dalmonte

Abdus Salam International Center for Theoretical Physics, Strada Costiera 11, Trieste, Italy

In this Proceedings we report some recent results on Majorana quasi-particles published in Phys. Rev. Lett. 118, 200404 (2017). We show how angular momentum conservation can stabilise a symmetry-protected quasi-topological phase of matter supporting Majorana quasi-particles as edge modes in one-dimensional cold atom gases. We investigate a number-conserving four-species Hubbard model in the presence of spin-orbit coupling. The latter reduces the global spin symmetry to an angular momentum parity symmetry, which provides an extremely robust protection mechanism that does not rely on any coupling to additional reservoirs. The emergence of Majorana edge modes is elucidated using field theory techniques, and corroborated by density-matrix-renormalization-group simulations. Our results pave the way toward the observation of Majorana edge modes with alkaline-earth-like fermions in optical lattices, where all basic ingredients for our recipe - spin-orbit coupling and strong inter-orbital interactions - have been experimentally realized over the last two years.

1. Introduction

The past two decades have witnessed an impressive progress in understanding how to harness quantum systems supporting topological order, one of the ultimate goals being the observation of quasi-particles with non-Abelian statistics – non-Abelian anyons[1–5]. A pivotal role in this search has been the formulation of a model for one-dimensional (1D) p-wave superconductors[6], the so-called Kitaev model, that supports a symmetry-protected topological phase with Majorana quasi-particles (MQPs) as edge modes. The key element for the stability of such edge modes is the presence of a Z_2 parity symmetry. At the mean-field level, the model can be realized via proximity-induced superconductivity in solid-state settings[7–12], or via coupling to molecular Bose-Einsten condensates in cold atoms[13].

Kitaev's model, however, is an effective mean-field model and its Hamiltonian does not commute with the particle number operator. Considerable activity has been devoted to understanding models supporting Majorana edge modes in a number-conserving setting[14–23], as in various experimental platforms (e.g. solid state[16,17] or ultracold atoms[18–23]) this property is naturally present. It was realised that a simple way to promote particle number conservation to a symmetry of the model, while keeping the edge state physics intact, was to consider at least two coupled wires rather than a single one. In this case the parity symmetry emerges via, e.g., engineered pair-tunneling between pairs of wires[19–21]. However, it is important to understand whether, in these number-conserving setups, there exist *fundamental* microscopic symmetries that can serve as a pristine mechanism for the realization of MQPs[23].

Another challenge in the study of such canonical systems arises from its intrinsic many-body character. Since attractive interactions are pivotal to generate superconducting order in the canonical ensemble, one usually faces a complex interacting many-body problem. In this way, approximations such as bosonization[16–18], or numerical approaches[19] are usually invoked. The existence of a few exactly solvable interacting models in number conserving systems with non-trivial topological properties[20,21] are therefore enlightening for such questions, providing the community with an easy-to-handle model where an intuition of the peculiarities of MQP's could be developed through back-of-the-envelope calculations. Recall that one of the reasons of the importance of the one-dimensional Kitaev model is to provide a particularly intuitive understanding of the basic physics of zero-energy Majorana modes in topological systems via simple algebraic manipulations. The knowledge of the exact ground-state wave-function and the possibility to compute various correlation functions is important to grasp the essence of Majorana physics.

In this Proceedings we report some recent results on Majorana quasi-particles published in Phys. Rev. Lett. 118, 200404 (2017), in which a physically motivated canonical Hamiltonian was shown to support MQP's. The model, a number-conserving four-species Hubbard model in the presence of spin-orbit coupling, has angular momentum conservation which can stabilise a symmetry-protected quasi-topological phase of matter supporting MPQ's as edge modes.

The article is organized as follows. In Section 2 we review some basic concepts of the Kitaev model. In Section 3 we present our model Hamiltonian, a number-conserving four-species Hubbard model in the presence of spin-orbit coupling. In Section 4, we present our analysis of the model within the bosonization framework, allowing us to identify a sector in the low-energy field theory which displays the same physics of Kitaev's chain. In Section 5 we employ DMRG simulations in order to demonstrate the existence of a symmetry-protected quasi-topological phase supporting MQPs as edge modes. In Section 6 we provide further (direct) evidence for the MQP nature of the edge states by showing how the quasi-topological phase is adiabatically connected to an exactly solvable model. In Section 7 we discuss about the physical realization of the model using alkaline-earth-like atoms in optical lattices. We present our conclusions in Section 8.

2. Kitaev model

We begin by recapitulating some properties of the Kitaev chain, whose Hamiltonian reads[6]

$$\hat{H}_K = \sum_j \left[-J\hat{a}_j^\dagger \hat{a}_{j+1} - \Delta \hat{a}_j \hat{a}_{j+1} + \text{H.c.} - \mu(\hat{n}_j - 1/2) \right].$$

Here, $J > 0$ denotes the hopping amplitude, μ and Δ the chemical potential and the superconducting gap, respectively; $\hat{a}_j^{(\dagger)}$ are fermionic annihilation (creation) operators on site j, and $\hat{n}_j \equiv \hat{a}_j^\dagger \hat{a}_j$. This model has (i) two density-driven phase transitions from finite densities to the empty and full states at $|\mu| = 2J$ (for $\Delta \neq 0$), and (ii) a transition driven by the competition of kinetic and interaction energy (responsible for pairing) at $\Delta/J = 0$ (for $|\mu| < 2J$). For $|\mu| < 2J$ and $\Delta \neq 0$, the ground state is unique for periodic boundary conditions, but twofold degenerate for open geometry, hosting localized zero-energy Majorana modes. This topological phase is symmetry protected by total fermionic parity $\hat{P} = (-1)^{\hat{N}}$, where $\hat{N} \equiv \sum_j \hat{n}_j$.

In order to exemplify the distinct phases of the model, let us consider two particular cases: (i) at $\Delta = J = 0$, representing a trivial phase; (ii) at $\mu = 0$, $\Delta = J$, usually called as "sweet point", representing a topological phase with zero energy edge Majorana modes. Firstly, we define the Majorana operators $\{\gamma_j\}_{j=1}^{2L}$ satisfying the Clifford algebra,

$$\hat{\gamma}_{2j-1} = i(\hat{a}_j - \hat{a}_j^\dagger), \qquad \hat{\gamma}_{2j} = \hat{a}_j + \hat{a}_j^\dagger, \tag{1}$$

with $\hat{\gamma}_j^\dagger = \hat{\gamma}_j$, $\{\hat{\gamma}_j, \hat{\gamma}_l\} = 2\delta_{j,l}$, and L the total number of sites in the chain. The Hamiltonian in the cases (i) and (ii) are described in terms of Majorana operators as follows,

$$\hat{H}_{(i),\text{trivial}} = -\frac{i}{2}\mu \sum_{j=1}^{L} \hat{\gamma}_{2j-1}\hat{\gamma}_{2j}, \qquad \hat{H}_{(ii),\text{topological}} = iJ \sum_{j=1}^{L-1} \hat{\gamma}_{2j}\hat{\gamma}_{2j+1} \tag{2}$$

In the trivial Hamiltonian, Majorana operators from the *same physical site* are paired together forming a ground state with completely filled or empty occupation, depending on the sign of μ. On the other hand, in the topological Hamiltonian, Majorana operators from *distinct physical sites* are paired together, forming in this case a ground state with p-wave superconductivity. Moreover, in this case, due to the open boundary conditions, the system will have two unpaired Majoranas, $\hat{\gamma}_1$ and $\hat{\gamma}_{2L}$, localised at the edges and with zero energy, thus inducing a two-fold degeneracy in the ground state.

3. Majorana quasi-particles protected by $\mathbb{Z}_2$ angular momentum conservation

A main obstacle in order to implement the Kitaev model is the engineering of pairing correlations. In this Section we show how can we deal with such concept in a number conserving setting, directly applicable in cold atom settings. We show how angular momentum conservation enables the realization of a symmetry-protected quasi-topological phase supporting MQPs in one-dimensional number-conserving systems[25]. In particular, we show how a combination of spin-exchange interactions and *crossed* spin-orbit couplings in orbital Hubbard models (see Fig. 1a-b) naturally gives rise to a $\mathbb{Z}_2$ spin symmetry. This symmetry serves as the enabling tool to realize MQPs, and, as we discuss below, its robustness is guaranteed by the fact that all terms breaking it are not present in the microscopic dynamics, as they would violate angular momentum conservation. Remarkably, these models find direct and natural realization using Alkaline-earth-like atoms (AEAs) in optical lattices[26-34]: in these settings, both spin-exchange interactions[35,36] and spin-orbit couplings[37-39] have already been demonstrated, providing an ideal setting to realize MQPs using state-of-the-art experimental platforms within the paradigm described in the present work.

Model Hamiltonian. Our starting point is a one-dimensional Hubbard model describing four fermionic species, with annihilation operators $c_{j,\alpha,p}$, with $j \in [1, L]$ a site index, L the length of the system, $\alpha \in [\uparrow, \downarrow]$ describing a pseudo-spin encoded in a pair of Zeeman states $m_F, m_F + 1$ (depicted in Fig. 1a-b as arrows), and $p \in [-1, 1]$ describing orbital degrees of freedom, encoded in the electronic state ground ($^1S_0$, blue) and meta-stable ($^3P_0$, orange) states. The system Hamiltonian reads:

$$H = \sum_j (H_{t,j} + H_{U,j} + H_{W,j} + H_{so,j}); \tag{3}$$

(in the following we also use the notation $H_x = \sum_j H_{x,j}$). The first two terms represents tunneling along the wire, $H_{t,j} = -\sum_{\alpha,p} t(c^\dagger_{j,\alpha,p} c_{j+1,\alpha,p} + \text{h.c.})$, and diagonal interactions, $H_{U,j} = \sum_p U_p n_{j,\uparrow,p} n_{j,\downarrow,p} + U \sum_{\alpha,\beta} n_{j,\alpha,-1} n_{j,\beta,1}$. The third term, visualized by grey arrows in Fig. 1, describes spin-exchange interactions[40]:

$$H_{W,j} = W(c^\dagger_{j,\uparrow,-1} c^\dagger_{j,\downarrow,1} c_{j,\downarrow,-1} c_{j,\uparrow,1} + \text{h.c.}). \tag{4}$$

The last term describes a generalized spin-orbit coupling:

$$H_{so,j} = \sum_p (\alpha_R + b)\hat{c}^\dagger_{j,\uparrow,p} c_{j+1,\downarrow,-p} + (b - \alpha_R)\hat{c}^\dagger_{j+1,\uparrow,p} c_{j,\downarrow,-p} + \text{h.c.}, \tag{5}$$

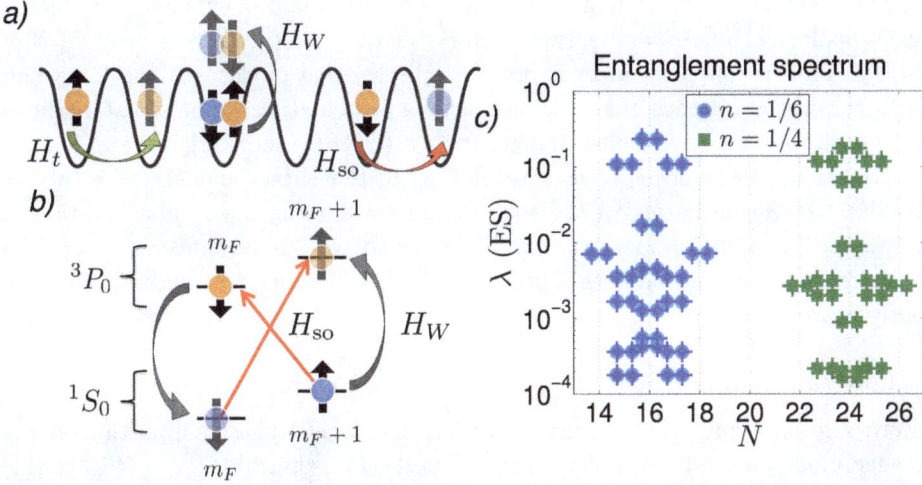

Fig. 1. Schematics of the orbital Hubbard model in the presence of spin-orbit coupling as realized with alkaline-earth-like atoms. *a-b)* The model we consider in Eq. (3) describes tunneling (H_t), spin-orbit-coupling (H_{so}), and spin-exchange processes (H_W). In cold atom settings, the spin degree of freedom is represented by different Zeeman states with nuclear spin m_F, m_F+1, while the orbital degree of freedom is encoded in different electronic states, $^1S_0$ and $^3P_0$. In these systems, H_W and H_{so} are described by the grey and red arrows, respectively. *c)* In the quasi-topological phase of the model, the entanglement spectrum displays a characteristic two-fold degeneracy: eigenvalues of the reduced density matrix with the same number of particles come in pairs with opposite parities (see text).

where α_R denotes the Rashba velocity, and the b term may be seen as momentum-dependent Zeeman field[41-43].

In microscopic implementations, the last two terms in H are embodied by strong inter-orbital spin-exchange interactions (grey arrows)[35,36], and by the possibility of engineering *crossed* spin-orbit couplings (red arrows) via clock lasers[37-39]. The combination of these two ingredients breaks explicitly the global spin-symmetry from $SU(2) \times SU(2)$ down to $\mathbb{Z}_2$ – namely, the number of states in each pair of states coupled by spin-orbit coupling is conserved *modulo* 2, due to the presence of the spin-exchange interactions. Indeed, while for $\alpha_R = b = 0$ the Hamiltonian has a $SU(2) \times SU(2)$ spin symmetry[26,27], for generic values of $\alpha_R, b \neq 0$, the spin symmetry is reduced to $\mathbb{Z}_2$, whose correspondent conserved charge is the mutual parity between the two subsets $[(\uparrow, 1), (\downarrow, -1)]$ and $[(\uparrow, -1), (\downarrow, 1)]$ connected by the spin exchange interaction H_W, i.e., $P_m = \mod 2[(\sum_j (n_{j,\uparrow,1} + n_{j,\downarrow,-1}) - (n_{j,\uparrow,-1} + n_{j,\downarrow,1}))/2]$. The robustness of this emergent parity symmetry stems from from angular momentum conservation: this symmetry may only be broken in the presence of terms such as, e.g., $c_{j\uparrow,-1}^\dagger c_{j,\uparrow,1}$, which generate a quantum of electronic angular momentum while preserving nuclear spin. The mechanism of establishing a $\mathbb{Z}_2$ symmetry is reminiscent of pair hopping of coupled wires[19], although here it emerges naturally, and thus it is experimentally accessible in a physical setting.

This symmetry is the building tool for the realization of a symmetry-protected quasi-topological phase whose spin sector has the same universal properties of Kitaev's model – in particular, it hosts MQPs as edge modes. In the following, we discuss the emergence of such phase using a combination of analytical methods and density-matrix-renormalization-group[44,45] (DMRG) simulations (see Fig. 1c for typical entanglement spectrum results, as in the Kitaev model). We further elucidate the anyon nature of the edge modes by showing how, upon addition of additional four-body interactions, Eq. (3) can be adiabatically connected to a model with exactly soluble ground state properties[20,21], where braiding statistics was recently proved[21].

4. Low-energy field theory

In order to underpin the existence of a quasi-topological phase supporting MQPs as edge modes, we rely on a field theory based on bosonization[47,48]. Within this framework, the essential point is to identify a sector in the low-energy field theory which displays the same physics of Kitaev's chain. Here, we outline the main steps of our treatment. In principle, it is possible to employ as a starting point either H_t, or $H_t + H_{SO}$: in both cases, the topological phase is found. In the following, we follow the former approach, which has the advantage of providing a simpler description of the effects of the diagonal interaction on the stability of the phase.

We first replace the fermionic operators with right and left-movers:

$$c_{j,\alpha,p} = \psi_{\alpha,p;R}(x = ja) + \psi_{\alpha,p;L}(x = ja) \tag{6}$$

where a is the lattice spacing, and then introduce the conventional bosonization representation:

$$\psi_{\alpha,p;r}(x) = \frac{\eta_{\alpha,p;r}}{\sqrt{2\pi a}} e^{irk_{F,\alpha,p}x} e^{-i(r\varphi_{\alpha,p} - \vartheta_{\alpha,p})} \tag{7}$$

with $r = (-1, 1)$ for L/R, and $(\varphi_{\alpha,p}, \vartheta_{\alpha,p})$ being conjugated bosonic operators describing density and phase fluctuations, respectively, and $\eta_{\alpha,p;r}$ are Klein factors (neglected in the following, they play a similar role as in Ref. 19). For our purposes, it would be useful to consider the generic form of the operators including all harmonics, i.e.:

$$\psi_{\alpha,p;r}(x) = \frac{\eta_{\alpha,p;r}}{\sqrt{2\pi a}} e^{i\vartheta_{\alpha,p}} \sum_q e^{irqk_{F,\alpha,p}x} e^{-iqr\varphi_{\alpha,p}} \tag{8}$$

We take as our starting point H_t. For convenience, we introduce two independent sector $f = -1, 1$, with two pair of symmetric and antisymmetric fields, $\varphi_{f,S/A} = (\varphi_{\uparrow,f} \pm \varphi_{\downarrow,f})/\sqrt{2}$. In each of these two sectors, the low-energy physics is described by the conventional theory of spin-orbit coupled gases, given by

$$H_f = \sum_{Z=S,A} \frac{v_Z}{2} \int dx \left[\frac{(\partial_x \varphi_{f,Z})^2}{K_{f,Z}} + K_{f,Z}(\partial_x \vartheta_{f,Z})^2 \right]$$

$$+ g_{SO} \int dx \cos[\sqrt{2\pi}(\vartheta_{f,A} + \kappa\varphi_{f,S})] \tag{9}$$

where the Luttinger parameters are all set to 1, and κ denotes the first harmonic commensurate with spin orbit coupling, which is a function of k_F.

After defining new fields,

$$\varphi_{f,I} = (\kappa\varphi_{f,S} + \vartheta_{f,A})/\sqrt{2\kappa}, \quad \vartheta_{f,I} = (\vartheta_{f,S} + \kappa\varphi_{f,A})/\sqrt{2\kappa} \tag{10}$$

and

$$\varphi_{f,II} = (\kappa\varphi_{f,S} - \vartheta_{f,A})/\sqrt{2\kappa}, \quad \vartheta_{f,II} = (\vartheta_{f,S} - \kappa\varphi_{f,A})/\sqrt{2} \tag{11}$$

one can infer from Eq. (9) that the $\varphi_{f,I}$ are gapped by the cosine terms, while the $\varphi_{f,II}$ remain gapless. In this regime, we can define collective charge and spin fields:

$$\vartheta_\rho = \frac{\vartheta_{1,II} + \vartheta_{-1,II}}{\sqrt{2}}, \quad \vartheta_\sigma = \frac{\vartheta_{1,II} - \vartheta_{-1,II}}{\sqrt{2}} \tag{12}$$

whose physics is, in the absence of interactions, described by two decoupled Tomonaga-Luttinger liquids (TLLs). In this basis, the spin-exchange term reads:

$$\begin{aligned}
H_{W,j} &\simeq W \int dx \left[e^{i\sqrt{2\pi}\vartheta_{1,S}} \sum_q \sin(\sqrt{2\pi}q\varphi_{1,A}) \right] \\
&\quad \times \left[e^{-i\sqrt{2\pi}\vartheta_{-1,A}} \sum_q \sin(\sqrt{2\pi}q\varphi_{-1,S}) \right] + \text{h.c.} \\
&\equiv W \int dx\, e^{i\sqrt{2\pi}(\vartheta_{1,S} - \kappa\varphi_{1,A})} e^{-i\sqrt{2\pi}(\vartheta_{-1,A} - \kappa\varphi_{-1,S})} + \text{h.c.} + \dots \\
&= W \int dx\, e^{i\sqrt{4\pi}\vartheta_{1,II}} e^{-i\sqrt{4\pi}\vartheta_{-1,I}} + \text{h.c.} + \dots \\
&\simeq W \int dx \cos\left[\sqrt{8\pi}\vartheta_\sigma\right]
\end{aligned} \tag{13}$$

where at each stage, the dots indicate the same contributions which are either oscillating, or contain fields which are gapped away from the dynamics, and thus can be neglected. The free Hamiltonian reads:

$$\begin{aligned}
H_{\text{free}} &= \sum_f \frac{v_{II}}{2} \int dx \left[\frac{(\partial_x \varphi_{f,II})^2}{K_{f,II}} + K_{f,II}(\partial_x \vartheta_{f,II})^2 \right] \\
&= \frac{v_\rho}{2} \int dx \left[\frac{(\partial_x \varphi_\rho)^2}{K_\rho} + K_\rho(\partial_x \vartheta_\rho)^2 \right] + \frac{v_\sigma}{2} \int dx \left[\frac{(\partial_x \varphi_\sigma)^2}{K_\sigma} + K_\sigma(\partial_x \vartheta_\sigma)^2 \right]
\end{aligned}$$

with:

$$v_{II}K_{II} = \frac{v_S + v_A/\kappa^2}{2}, \quad v_{II}/K_{II} = \frac{v_S + v_A\kappa^2}{2}, \tag{14}$$

which implies:

$$v_\rho K_\rho = \frac{v_S + v_A/\kappa^2}{2}, \quad v_\rho/K_\rho = \frac{v_S + v_A\kappa^2}{2}, \tag{15}$$

$$v_\sigma K_\sigma = \frac{v_S + v_A/\kappa^2}{2}, \quad v_\sigma/K_\sigma = \frac{v_S + v_A\kappa^2}{2} \tag{16}$$

For the spin sector, the spin Luttinger parameter is:

$$K_\sigma^{\text{free}} = \sqrt{\frac{1 + v_A/v_S}{1 + v_A\kappa^4/v_S}} \tag{17}$$

which implies that, since $\kappa > 1$, the Luttinger parameter gets small for equal velocities. We note here that the scaling dimension of the spin-exchange operator is:

$$d_W = 2/K_\sigma \tag{18}$$

We remark here that both H_W and the diagonal interactions can be used to drastically enhance K_σ. In particular, one has:

$$K_\sigma = K_\sigma^{\text{free}} - a_1 W - a_2 U \tag{19}$$

with a_1, a_2 being non-universal, positive prefactors which depend on κ, α_R, b.

In this way, the charge sector remains gapless (away from commensurability points which can introduce additional umklapp terms), while the spin sector is described by the same low-energy field theory of the Kitaev chain. In the bosonized language, the relevance of the $\mathbb{Z}_2$ symmetry is clear: in the absence of it, terms of the form $\cos(\sqrt{4\pi}\vartheta_\sigma)$ would appear at low energies, and immediately spoil the link to the Kitaev chain (for an exception to this mechanism, see Ref. 46, where specific choices of gauge fields are shown to prevent those terms from appearing). Furthermore, at the Luther-Emery point $K_\sigma = 2$, it is possible to re-fermionize the model so that the correspondence is even clearer - see, e.g, Refs. 19, 24.

In summary, for sufficiently large W/t, the model in Eq. (3) supports a quasi-topological phase, with gapless charge excitations, and decoupled gapped spin-excitations describing a ground state of a Kitaev model, thus supporting MQPs. The role of additional, diagonal interactions can affect the spin sector[23]: since $K_\sigma - 1 \simeq -(W + U)$, attractive interactions further stabilize the quasi-topological phase, while repulsive interactions require larger values of W to open a gap in the spin sector. Equipped with the guideline provided by the low-energy field theory, we present in the following a non-perturbative analysis of the model based on numerical simulations.

5. DMRG results

In order to demonstrate the existence of a symmetry-protected quasi-topological phase supporting MQPs as edge modes, we employ DMRG simulations based on a rather general decimation prescription for an efficient truncation of the Hilbert space. Typically, we use up to $m = 140$ states, which ensure converge on all observables of interest over all parameter regimes[23]. Following the theoretical discussion above, our analysis is based on four observables: *(i)* degeneracies in the entanglement spectrum; *(ii)* finite-size scaling of energy gaps; *(iii)* bulk decay of correlation

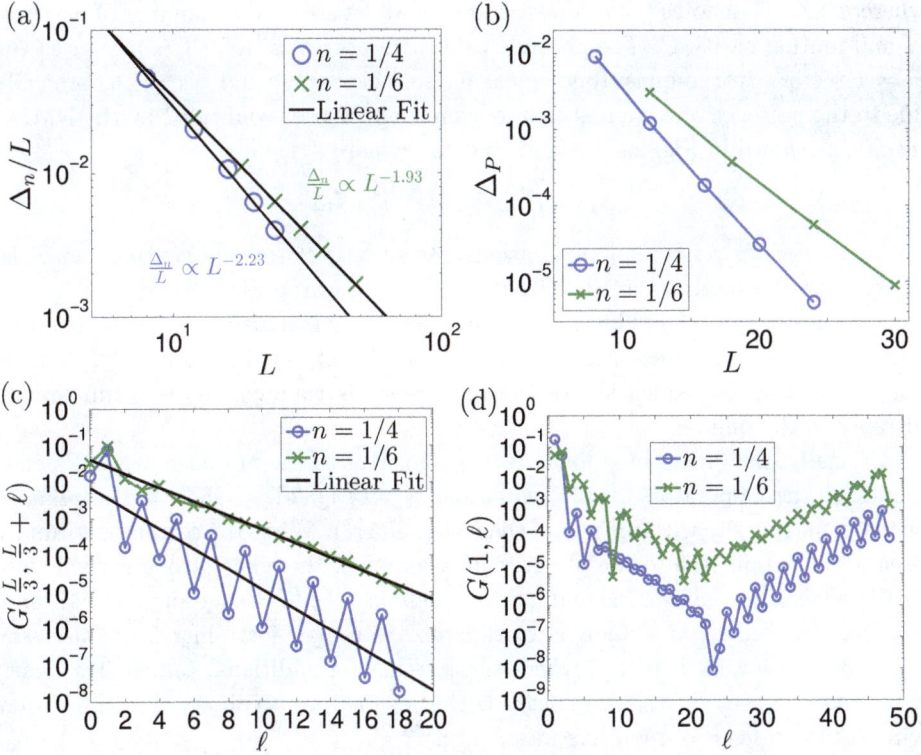

Fig. 2. DMRG analysis of the topological properties of the ground state for model H, at fixed parity, with parameters $W = -8$, $\alpha_R = b = 4$, $U = 0$, at distinct fillings $n = 1/4$ and $n = 1/6$. **(a)** Algebraic scaling of the gap computed at fixed parity, compatible with $\sim L^{-1}$. **(b)** Exponential scaling of the gap between the distinct parity sectors. **(c-d)** Single particle correlations $G(j, \ell) = \langle c^{\dagger}_{j,\uparrow,1} c_{\ell,\uparrow,1} \rangle$ at the bulk (c) and at the edges (d), for a system with $L = 48$ sites.

functions; and *(iv)* edge-to-edge correlations. For convenience, we set $t = 1$ as energy unit.

Given the reduced density matrix ρ_ℓ with respect of a bipartition of the system cutting the ℓ-th link of the lattice, the entanglement spectrum is the collection of its eigenvalues $\{\lambda_\alpha\}$, and is known to provide striking signatures of topological order via degeneracies[51,52]. In Fig. 1c, we show typical results for the entanglement spectrum in the quasi-topological phase at the representative point $W = -8$, $\alpha_R = b = 4$, $U = 0$ (these features are stable in a broad parameter range[49]). Indeed, the low-lying spectrum displays robust degeneracies for both $n = N/4L = 1/4$ and $n = 1/6$ (with N and L total numbers of particles and sites, respectively), as expected for a topological phase supporting MQPs edge modes.

In Fig. 2a, we show the decay of the fixed parity gap with open boundary conditions (OBCs), defined as:

$$\Delta_n = E_L^1[N, P] - E_L^0[N, P] \tag{20}$$

where $E_L^n[N, P]$ denotes the n-lowest-energy state at size L with number of particles N and mutual parity P. The ground state, with energy $E_L^0[N, P]$, is always in the $P = 1$ sector. In the quasi-topological phase, this gap should decay algebraically due to the presence of a gapless charge excitation. This is confirmed by the DMRG results, as shown in Fig. 2a. Instead, the parity gap:

$$\Delta_P = E_L^0[N, -1] - E_L^0[N, 1] \tag{21}$$

is sensitive exclusively to spin excitations. As such, it closes exponentially with the system size L, exactly as in the Kitaev chain, as shown in Fig. 2b.

The presence of a finite bulk gap in the spin sector is signalled by an exponential decay of the Green functions, e.g. $G(j, \ell) = \langle c_{j,\uparrow,1}^\dagger c_{\ell,\uparrow,1} \rangle$, in the bulk[48]. This is portrait in Fig. 2c, which shows that coherence is rapidly lost as a function of distance in the bulk.

Crucially, the Green functions are also sensitive to the presence of MQPs edge modes, as these operators locally switch parity. In Fig. 2d, we show the correlation of one boundary site with the rest of the chain, $G(1, \ell)$. While the correlation rapidly decays in the bulk due to the presence of a spin gap, there is a strong revival close to the edge of the system, signalling the presence of MQP edge modes. We note that the edge-edge correlation is considerably stronger for filling fractions away from commensurate densities, where the presence of additional (albeit irrelevant) operators is expected to slightly degrade the edge modes, as observed in the Kitaev wire in the presence of repulsive interactions[50,53].

6. Adiabatic continuity of the ground state to an exactly-solvable point

Remarkably, it is possible to provide direct evidence for the MQP nature of the edge states by showing how the quasi-topological phase discussed above is adiabatically connected to toy model of spinless fermions with exactly solvable ground state properties[20,21], where Ising anyon braiding was recently demonstrated[21].

The strategy to show adiabatic continuity, which will be discussed in detail at the end of this Section, consists of three steps. First, for each pair of states coupled by spin-orbit coupling, we restrict the dynamics to the lower band, following a procedure introduced in Refs. 41, 54. Then, coupling between the lowest bands is introduced via the spin-exchange interaction, and additional four-Fermi couplings. This enlarged Hamiltonian is characterized by two parameters (β, λ): the point $(0, 0.9)$ represent the model studied in the previous section, while the points $(1, 0.9)$ and $(1, 1)$ represents two points in the phase diagram of the exactly solvable model[20]. We note that all symmetries of the problem are kept for arbitrary (β, λ).

Within this enlarged parameter space, we have carried out DMRG simulations to show that the gap in the spin sector does not close. The latter was extracted from the decay of the Green function in the bulk, $G(j, \ell) \simeq e^{-\alpha_{sp}|j-\ell|}$, and is depicted in Fig. 3a. Along the full path in parameter space, the gap stays open, implying

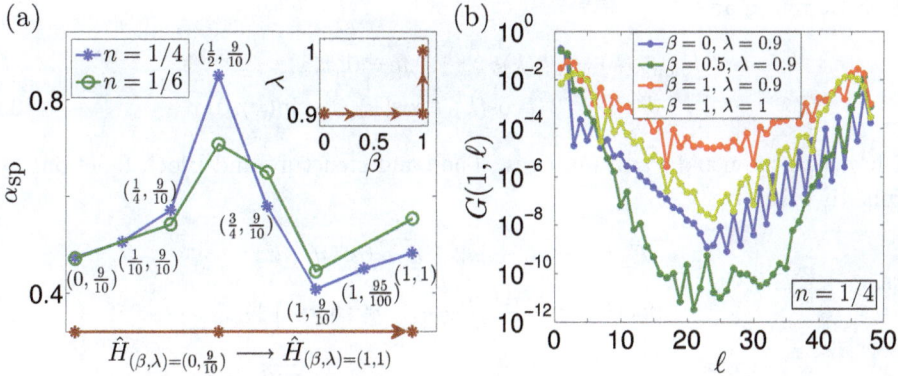

Fig. 3. DMRG analysis for the adiabatic continuity of model H to an exactly solvable model. **(a)** Parity gap "$\alpha_{\rm sp}$" along the adiabatic continuity path $(\beta = 0, \lambda = 9/10) \rightarrow (\beta = 1, \lambda = 1)$. Inset illustrates the Hamiltonian parameters varied along the path. **(b)** Single particle edge correlations along the adiabatic path - similar behavior follows for $n = 1/6$. In all plots we consider a system with $L = 48$ sites and use $m = 140$ number of kept states in the DMRG simulations.

that out quasi-topological state is the same phase as in Ref. 20. Another striking signature of adiabatic continuity is the fact that all diagnostics applied before signal topological order all along the path. This is illustrated in Fig. 3b, where we plot the edge-edge Green function at several points along the path itself.

Adiabatic continuity construction in detail. The adiabatic continuity can be studied in the lower band projected Hamiltonian in the limit of strong spin-orbit coupling. Specifically, we employ below the generalized spin-orbit coupling introduced in Refs. 41, 43, which allows a straightforward, yet exact projection into the lower band of each pair of coupled states. In this regime, the non-interacting part of the Hamiltonian,

$$
\begin{aligned}
H_{\rm non} &= \sum_j H_{t,j} + H_{so,j}, \\
&= \sum_p \left[\sum_{j,\alpha} t(c^\dagger_{j,\alpha,p} c_{j+1,\alpha,p} + {\rm h.c.}) \right. \\
&\quad \left. + \sum_j \left\{ (\alpha_R + b)\hat{c}^\dagger_{j,\uparrow,p} c_{j+1,\downarrow,-p} + (b - \alpha_R)\hat{c}^\dagger_{j+1,\uparrow,p} c_{j,\downarrow,-p} + {\rm h.c.} \right\} \right] \\
&= \sum_p h_{\rm non,p}
\end{aligned}
\tag{22}
$$

describes two well separated Bloch bands by a Fourier transformation. Let us focus for simplicity in only one of the two decoupled parts of the non-interacting Hamiltonian, *i.e.*, a single term "$h_{\rm non,p}$". On Fourier transform we easily obtain its

Bloch Hamiltonian,

$$h_{\text{non,p}}(k) = d^\mu(k)\sigma_\mu, \qquad \mu = 0, x, y, z$$
$$d^\mu(k) = 2\left(t\cos(k), b\cos(k), \alpha_R\sin(k), 0\right) \qquad (23)$$

with σ_μ are the usual Pauli matrices. The band structure and Bloch functions are explicitly given by,

$$E_\pm(k) = d^0 \pm |\vec{d}|, \quad \vec{d} = (d^x, d^y, d^z)$$
$$|u_\pm(k)\rangle = \frac{P_\pm(k)|\uparrow\rangle}{|P_\pm(k)|\uparrow\rangle|}, \quad \sigma^z|\uparrow\rangle = |\uparrow\rangle$$
$$P_\pm(k) = \frac{1}{2}\left(1 \pm d(k)\cdot\sigma\right), \quad d = \frac{\vec{d}}{|\vec{d}|} \qquad (24)$$

Considering that the ordinary hopping term does not influence the Bloch states due to its spin independence, the effective Hamiltonian in the lower band, at $\alpha_R = b$, is described by a flat band,

$$h_{\text{non,p}} = 2b\sum_j \left[\gamma^\dagger_{j,p,+}\gamma_{j,p,+} - \gamma^\dagger_{j,p,-}\gamma_{j,p,-} + \text{H.c.}\right]$$
$$\sim -2b\sum_j \left[\gamma^\dagger_{j,p,-}\gamma_{j,p,-} + \text{H.c.}\right] \qquad (25)$$

with a gap $4b$ from the upper band, and Bloch states $\gamma_{j,p,\pm} = (c_{j,\downarrow,-p} \pm c_{j+1,\uparrow,p})/\sqrt{2}$; inversely,

$$c_{j,\downarrow,-p} = \frac{1}{\sqrt{2}}(\gamma_{j,p,+} + \gamma_{j,p,-}), \qquad c_{j,\uparrow,p} = \frac{1}{\sqrt{2}}(\gamma_{j-1,p,+} - \gamma_{j-1,p,-}) \qquad (26)$$

We notice now that interacting terms, such as spin-exchange interactions, are effectively described in the lower band picture as pairing interactions (for simplicity of notation we use hereafter the operators $\eta_j \equiv \gamma_{j,1,-}$ and $\chi_j \equiv \gamma_{j,-1,-}$ to explicitly represent the lower band subspace),

$$H_{W,j} = W_{\text{ex}}(c^\dagger_{j,\uparrow,-1}c^\dagger_{j,\downarrow,1}c_{j,\downarrow,-1}c_{j,\uparrow,1} + \text{h.c.})$$
$$= (W_{\text{ex}}/4)(\gamma^\dagger_{j-1,-1,+} - \gamma^\dagger_{j-1,-1,-})(\gamma^\dagger_{j,-1,+} + \gamma^\dagger_{j,-1,-})$$
$$\times(\gamma_{j,1,+} + \gamma_{j,1,-})(\gamma_{j-1,1,+} - \gamma_{j-1,1,-})$$
$$\sim (W_{\text{ex}}/4)(\chi^\dagger_j\chi^\dagger_{j-1}\eta_{j-1}\eta_j + \text{h.c.}) \qquad (27)$$

a key element in order to generate MQPs according to previously studied models [19–21].

Within such a lower band picture one may also introduce some additional Hamiltonian terms $H_{\text{locU},j} = U_{\text{loc}}\sum_p(n_{j,\uparrow,p} + n_{j,\downarrow,-p})^2$, and $H_{\text{lcso},j} = h_{\text{lcso}}(\hat{c}^\dagger_{j,\uparrow,1}c_{j,\downarrow,-1} + \text{H.c.})(\hat{c}^\dagger_{j,\uparrow,-1}c_{j,\downarrow,1} + \text{H.c.})$ describing local diagonal interactions and a local coherent spin-orbit coupling, respectively, in such a way to reproduce the exactly solvable model H_λ proposed in[20]. Briefly recalling, the model in Ref. 20 is given by,

$$\frac{H_\lambda}{4} = - \sum_{j=1,\alpha=\chi,\eta}^{L-1} \left[(\alpha_j^\dagger \alpha_{j+1} + \text{H.c.}) - (n_j^\alpha + n_{j+1}^\alpha) + \lambda n_j^\alpha n_{j+1}^\alpha \right] +$$

$$- \frac{\lambda}{2} \sum_{j=1}^{L-1} \left[(n_j^\chi + n_{j+1}^\chi)(n_j^\eta + n_{j+1}^\eta) - (\chi_j^\dagger \chi_{j+1} + \chi_{j+1}^\dagger \chi_j)(\eta_j^\dagger \eta_{j+1} + \eta_{j+1}^\dagger \eta_j) \right.$$

$$\left. + 2(\eta_j^\dagger \eta_{j+1}^\dagger \chi_{j+1} \chi_j + \text{H.c.}) \right] \tag{28}$$

where $n_j^\chi = \chi_j^\dagger \chi_j$, $n_j^\eta = \eta_j^\dagger \eta_j$. The model has an exactly solvable line at $\lambda = 1$, on varying the density of particles, described by a topological nontrivial ground state. Away from the exactly solvable line the system is topological for $\lambda < 1$, while describing a phase separation (PS) state for $\lambda > 1$.

We will now explicitly show that the additional Hamiltonian terms $H_{\text{locU},j}$, $H_{\text{lcso},j}$ with the diagonal and spin-exchange interactions, H_U and $H_{W_{\text{ex}}}$, indeed reproduce the exact solvable model in their lower band picture.

The diagonal interactions $H_{U,j} = \sum_p U_p n_{j,\uparrow,p} n_{j,\downarrow,p} + U \sum_{\alpha,\beta} n_{j,\alpha,-1} n_{j,\beta,1}$, are described in the lower band at $U_p = U$ as follows,

$$H_{U,j} = (U/2) \sum_{\alpha,p} \left\{ (n_{j,\alpha,p})^2 - (n_{j,\alpha,p}) \right\}$$

$$\sim \frac{U}{4} \left[(n_j^\chi n_{j-1}^\chi + n_j^\eta n_{j-1}^\eta) + (n_j^\chi + n_{j-1}^\chi)(n_j^\eta + n_{j-1}^\eta) \right] \tag{29}$$

The two additional Hamiltonian terms are described in the lower band as,

$$H_{\text{locU},j} = \frac{U_{\text{loc}}}{4}(n_j^\chi + n_{j-1}^\chi + n_j^\eta + n_{j-1}^\eta) + \frac{U_{\text{loc}}}{2}(n_j^\chi n_{j-1}^\chi + n_j^\eta n_{j-1}^\eta) \tag{30}$$

$$H_{\text{cso},j} \sim \frac{h_{\text{cso}}}{4} (\chi_j^\dagger \chi_{j-1} + \text{H.c.})(\eta_j^\dagger \eta_{j-1} + \text{H.c.}) \tag{31}$$

Thus, one can see by Eqs. (27), (29), (30) that the total Hamiltonian, $H = \sum_j (H_{\text{so},j} + H_{t,j} + H_{U,j} + H_{W,j} + H_{\text{lcso},j} + H_{\text{locU},j})$, in the lower band picture ($\alpha_R = b \gg$ other terms), describes the exactly solvable model (Eq. (28)) with: $W_{\text{ex}}/t = -4\lambda$, $U_p = U = 2U_{\text{loc}} = -2\lambda t$, and $h_{\text{clso}}/t = 2\lambda$.

The DRMG analysis for the adiabatic continuity was performed in two steps: (i) first connect our initial *physical* Hamiltonian with $\alpha_R = b = 8t$, $W_{\text{ex}}/t = -8$, $U_p = U = U_{\text{loc}} = h_{\text{clso}} = 0$ to the model $\hat{H}_\lambda$ at $\lambda = 0.9$; (ii) connect $\hat{H}_\lambda$ from $\lambda = 0.9$ to the exactly solvable line $\lambda = 1$. We choose to perform the adiabatic continuity in such two steps in order to avoid possible phase transitions along the path to the exactly solvable line $\lambda = 1$. The total Hamiltonian is thus parametrized by $\hat{H}_{(\beta,\lambda)}$

as follows:

$$W_{\text{ex}}/t = (-8 + 4.4\beta)\,(\lambda/0.9)$$
$$U_p/t = U/t = 2U_{\text{loc}}/t = -2\beta\lambda \tag{32}$$
$$h_{\text{clso}}/t = 2\beta\lambda$$

where $(\beta = 0,\ \lambda = 0.9) \to (1, 0.9)$ describes the first step, and $(1, 0.9) \to (1, 1)$ the second step.

7. Realization using alkaline-earth-like atoms in optical lattices

The model discussed above finds a natural implementation using fermionic isotopes of AEA in optical lattices[28], such as ^{171}Yb, ^{173}Yb, and ^{87}Sr. As illustrated in Fig. 1, the orbital degree of freedom is encoded in the electronic state: the ^{1}S$_0$ ground state manifold representing $p = -1$, and the long- lived excited state manifold ^{3}P$_0$ representing $p = 1$. The spin degree of freedom is instead encoded in the nuclear spin state, which for AEA is basically decoupled from the electronic degree of freedom for both ground and low-lying excited states. In ^{171}Yb, the nuclear spin is $I = 1/2$, so all degrees of freedom are immediately available as required here. For ^{173}Yb and ^{87}Sr, which do have $I = 5/2$ and 9/2, respectively, unwanted Zeeman states can be excluded from the dynamics either employing state-dependent light-shifts[33], or by exploiting the fact that the clock frequency is m_F-dependent due to different linear Zeeman shifts in the ^{1}S$_0$ and ^{3}P$_0$ manifold.

The two key elements of our proposal, large spin-exchange interactions and spin-orbit couplings on the so-called clock transition, build upon state-of-the-art experimental progresses in AEA physics. As demonstrated in recent experiments with ^{173}Yb[35,36], the spin-exchange interaction in these settings can be extremely large, of the order of 5/10 kHz in optical lattices, guaranteeing that the driven interaction strength in our system is considerably larger than typical temperatures. Moreover, the ratio W/t can be tuned via modifying either the optical lattice depth or the trapping in the transverse direction. Spin-orbit coupling between ground and excited states has been recently demonstrated both at JILA[39] and at LENS[38] realizing single-particle band structures akin to the one employed here, albeit with slightly different microscopic Hamiltonians (following Ref. 41, the precise form we use here requires a tilting of the lattice or a superlattice structure). In concrete, considering typical tunneling rates of order $t \simeq 100h$ Hz, spin-orbit couplings of order $400h$ Hz and spin-exchange interactions of order $800h$ Hz would give direct access to the quasi-topological phase. In these experimental settings, the quasi-topological phase can be characterized using both correlation function and spectral properties, as discussed above. The nature of the edge modes can be demonstrated using a variety of techniques[55]. In particular, time-of-flight imaging and edge spectroscopy can be used to demonstrate the existence of zero-energy modes and their inherent correlations. Moreover, the fact that our model is adiabatically connected to an exactly-solvable point provides a qualitative guidance on the shape of the MQP wave-function - generically hard to analytically access in interacting systems -, opening up a concrete perspective to realize braiding operations in such settings.

8. Conclusions

In this Proceedings we reported some recent results on Majorana quasi-particles published in Phys. Rev. Lett. 118, 200404 (2017). We have shown how Majorana quasi-particles can emerge as edge modes of orbital Hubbard models in the presence of spin-orbit interactions. The key element for the realization of the quasi-topological phase supporting them is angular momentum conservation, an epitome building block of atomic physics experiments. The stability of the mechanism we propose paves the way toward the investigation of interacting topological states and Majorana edge modes in both atomic clocks and optical lattice experiments, where the main ingredients of our proposal are naturally realized and have been experimentally demonstrated over the last two years.

References

1. C. Nayak, S. H. Simon, A. Stern, M. Freedman, and S. Das Sarma, Rev. Mod. Phys. **80**, 1083 (2008).
2. F. Wilczek, Nat. Phys. **5**, 614 (2009).
3. J. Alicea, Rep. Prog. Phys. **75**, 076501 (2012).
4. C. W. J. Beenakker, Annu. Rev. Con. Mat. Phys. **4**, 113 (2013).
5. N. Goldman, J. C. Budich, and P. Zoller, Nat. Phys. **12**, 639 (2016).
6. A. Kitaev, Phys. Usp. **44**, 131 (2001).
7. V. Mourik, K. Zou, S. M. Frolov, S. R. Plissard, E. P. A. M. Bakkers, and L. P. Kouwenhoven, Science **336**, 1003 (2012).
8. M. T. Deng, C. L. Yu, G. Y. Huang, M. Larsson, P. Caroff, and H. Q. Xu, Nano Lett. **12**, 6414 (2012).
9. A. Das, Y. Ronen, Y. Most, Y. Oreg, M. Heiblum, and H. Shtrikman, Nat. Phys. **8**, 887 (2012).
10. L. P. Rokhinson, X. Liu, and J. K. Furdyna, Nat. Phys. **8**,795 (2012).
11. S. Nadj-Perge *et al.*, Science **346**, 602 (2014).
12. S. M. Albrecht *et al.*, Nature **531**, 206 (2016).
13. L. Jiang, T. Kitagawa, J. Alicea, A. R. Akhmerov, D. Pekker, G. Refael, J. I. Cirac, E. Demler, M. D. Lukin, and P. Zoller, Phys. Rev. Lett. **106**, 220402 (2011).
14. G. Ortiz, J. Dukelsky, E. Cobanera, C. Esebbag, and C. Beenakker, Phys. Rev. Lett. **113**, 267002 (2014).
15. C. Chen, W. Yan, C. S. Ting, Y. Chen, and F. J. Burnell, arXiv:1701.01794 (2017).
16. L. Fidkowski, R. M. Lutchyn, C. Nayak, and M. P. A. Fisher, Phys. Rev. B **84**, 195436 (2011).
17. J. D. Sau, B. I. Halperin, K. Flensberg, and S. Das Sarma, Phys. Rev. B, **84**, 144509 (2011).
18. M. Cheng and H.-H. Tu, Phys. Rev. B., **84**, 094503 (2011).
19. C. V. Kraus, M. Dalmonte, M. A. Baranov, A. M. Läuchli, and P. Zoller, Phys. Rev. Lett., **111**, 173004 (2013).

20. F. Iemini, L. Mazza, D. Rossini, R. Fazio, and S. Diehl, Phys. Rev. Lett. **115**, 156402 (2015).
21. N. Lang and H. P. Büchler, Phys. Rev. B **92**, 041118 (2015R).
22. Fernando Iemini, Davide Rossini, Rosario Fazio, Sebastian Diehl and Leonardo Mazza, Phys. Rev. B **93**, 115113 (2016).
23. F. Iemini, L. Mazza, L. Fallani, P. Zoller, R. Fazio, and M. Dalmonte, Phys. Rev. Lett. **118**, 200404 (2017).
24. M. Cheng and H.-H. Tu, Phys. Rev. B **84**, 094503 (2011).
25. Quasi-topological phases include a topological, gapped sector, in addition to decoupled gapless modes, see Refs. 19, 58.
26. M. A. Cazalilla, A. F. Ho, and M. Ueda, New J. Phys. **11**, 103033 (2009).
27. A. V. Gorshkov, M. Hermele, V. Gurarie, C. Xu, P. S. Julienne, J. Ye, P. Zoller, E. Demler, M. D. Lukin, and A. M. Rey, Nat. Phys. **6**, 289 (2010).
28. M. A. Cazalilla and A. M. Rey, Rep. Prog. Phys. **77**, 124401 (2014).
29. S. Stellmer, M. K. Tey, B. Huang, R. Grimm, and F. Schreck, Phys. Rev. Lett. **103**, 200401 (2009).
30. B. J. DeSalvo, M. Yan, P. G. Mickelson, Y. N. Martinez de Escobar, and T. C. Killian, Phys. Rev. Lett. **105**, 030402 (2010).
31. S. Sugawa, K. Inaba, S. Taie, R. Yamazaki, M. Yamashita, and Y. Takahashi, Nat. Phys. **7**, 642 (2011).
32. M. D. Swallows, M. Bishof, Y. Lin, S. Blatt, M. J. Martin, A. M. Rey, and J. Ye, Science **331**, 1043 (2011).
33. M. Mancini, G. Pagano, G. Cappellini, L. Livi, M. Rider, J. Catani, C. Sias, P. Zoller, M. Inguscio, M. Dalmonte, and L. Fallani, Science **349**, 1510 (2015).
34. C. Hofrichter, L. Riegger, F. Scazza, M. Höfer, D. Rio Fernandes, I. Bloch, and S. Fölling, Phys. Rev. X **6**, 021030 (2016).
35. G. Cappellini *et al.*, Phys. Rev. Lett. **113**, 120402 (2014).
36. F. Scazza, C. Hofrichter, M. Höfer, P. C. De Groot, I. Bloch, and S. Fölling, Nat. Phys. **10**, 779 (2014).
37. M. L. Wall *et al.*, Phys. Rev. Lett. **116**, 035301 (2016).
38. L. F. Livi *et al.*, Phys. Rev. Lett. **117**, 220401 (2016).
39. S. Kolkowitz *et al.*, Nature **542**, 66 (2017).
40. These interactions are quasi-resonant up to intermediate values of the external magnetic field.
41. J. C. Budich, C. Laflamme, F. Tschirsich, S. Montangero, and P. Zoller, Phys. Rev. B **92**, 245121 (2015).
42. The b-term is not fundamental to stabilize MQPs, but considerably simplifies a part of the theoretical analysis below.
43. J. C. Budich and E. Ardonne, Phys. Rev. B **88**, 035139 (2013).
44. S. R. White, Phys. Rev. Lett. **69**, 2863 (1992).
45. U. Schollwöck, Rev. Mod. Phys.**77**, 259 (2005).
46. Chun Chen, Wei Yan, C. S. Ting, Yan Chen and F. J. Burnell, arXiv:1701.01794.
47. A.O. Gogolin, A.A. Nersesyan, A.M. Tsvelik, *Bosonization and strongly correlated systems*, (Cambridge University press, Cambridge, 1998).
48. T. Giamarchi, *Quantum Physics in one dimension*, (Oxford University press, Oxford, 2003).
49. F. Iemini *et al.*, in progress.
50. E. M. Stoudenmire, J. Alicea, O. A. Starykh and M. P. A. Fisher, Phys. Rev. B **84**, 014503 (2011).

51. F. Pollmann, E. Berg, A. M. Turner, and M. Oshikawa, Phys. Rev. B **81**, 064439 (2010).

52. A. M. Turner, F. Pollmann, and E. Berg, Phys. Rev. B **83**, 075102 (2011).

53. M. Tezuka and N. Kawakami, Phys. Rev. B **85**, 140508 (R) (2012).

54. This passage is reminiscent of treating the orbital degrees of freedom as a synthetic dimension[33,56,57].

55. C. V. Kraus, S. Diehl, P. Zoller and M. A. Baranov, New J. Phys. **14**, 113036 (2012).

56. A. Celi *et al.*, Phys. Rev. Lett. **112**, 043001 (2014).

57. B. Stuhl *et al.*, Science **349**, 1514 (2015).

58. P. Bonderson and C. Nayak, Phys. Rev. B **87**, 195451(2013).

Bose metal as a disruption of the Berezinskii-Kosterlitz-Thouless transition in 2D superconductors

Philip W. Phillips

Loomis Laboratory of Physics and Institute for Condensed Matter Theory, University of Illinois, Urbana-Champaign, Il. 61801-3080
dimer@illinois.edu

Destruction of superconductivity in thin films was thought to be a simple instance of Berezinskii-Kosterlitz-Thouless physics in which only two phases exist: a superconductor with algebraic long range order in which the vortices condense and an insulator where the vortex-antivortex pairs proliferate. However, since 1989 this view has been challenged as now a preponderance of experiments indicate that an intervening bosonic metallic state obtains upon the destruction of superconductivity. Two key features of the intervening metallic state are that the resistivity turns on continuously from the zero resistance state as a power law, namely $\rho_{\rm BM} \propto (g - g_c)^\alpha$ and the Hall conductance appears to vanish. We review here a glassy model which is capable of capturing both of these features. The finite resistance arises from three features. First, the disordered insulator-superconductor transition in the absence of fermionic degrees of freedom (Cooper pairs only), is controlled by a diffusive fixed point [1] rather than the critical point of the clean system. Hence, the relevant physics that generates the Bose metal should arise from a term in the action in which different replicas are mixed. We show explicitly how such physics arises in the phase glass. Second, in 2D (not in 3D) the phase stiffness of the glass phase vanishes explicitly as has been shown in extensive numerical simulations [2-4]. Third, bosons moving in such a glassy environment fail to localize as a result of the false minima in the landscape. We calculate the conductivity explicitly using Kubo response and show that it turns on as a power law and has a vanishing Hall response as a result of underlying particle-hole symmetry. We show that when particle-hole symmetry is broken, the Hall conductance turns on with the same power law as does the longitudinal conductance. This prediction can be verified experimentally by applying a ground plane to the 2D samples.

Keywords: 2D superconductivity, Bose metal, Vortices.

1. Phenomenology

Probably no other problem better exemplifies the key topic of this conference, the Berezinskii-Kosterlitz-Thouless [5,6] (BKT) transition, than the insulator-superconductor transition [7-11] in thin films. Key observations which helped place this transition within this framework are (1) a zero-resistive state below a critical value of the tuning parameter (either film thickness or magnetic field) interpreted as a condensation of vortex-antivortex excitations, (2) non-linear I-V vortex-antivortex excitations out of the zero-resistance state that give rise to the universal $V \approx I^3$ current-voltage [12] characteristics, and (3) exponential drop [12,13] of the resistance below H_{c2}, indicative of thermal activation of vortex-antivortex motion with a binding energy of the BKT [5,6] form, $U(H) = U_0 \ln H_0/H$ where $H_0 \approx H_{c2}$ and U_0 the binding energy.

However, a key feature which does not fit this scenario is the eventual leveling of the resistance for $T \ll U$. Since this phase obtains below H_{c2}, the excitations are fundamentally bosonic, hence the Bose metal. Within the standard XY[11] modeling of the BKT transition, this state of affairs is not possible. In this scenario, a metallic state only obtains at the critical point separating the ordered and disordered states and the quantum of resistance should be $h/4e^2$. Indeed, the initial experiments[7-9] seemed to be in agreement with the predictions[11] of the phase-only XY model that only at the critical point do bosons exhibit the quantum of resistance of $h/4e^2$. However, further experiments[10,14-16] indicated that there is nothing special about the value of the resistance at the critical point, thereby calling into question the relevance or accuracy of the prediction of the phase-only model. In fact since 1989[12,13,16-27], a state with apparent finite $T \to 0$ resistivity appeared immediately upon the destruction of superconductivity. Although the initial observations were derided as an artifact of failed refrigeration[10], the leveling of the resistance persisted in the magnetic-field tuned transition in MoGe[17,21,25], Ta[18,28], InO$_x$[26,29], and NbSe$_2$[12,13]. As mentioned above, the key contribution of the magnetic-field tuned data was to clarify that the intervening state occurred well below H_{c2} in the regime where $T \ll U$, thereby requiring something beyond the classical physics of the BKT transition.

More recent observations of the Bose metal in cleaner samples with either gate[20] or magnetic-field tuning[12] disclose three facts of the transition. First, in the field-effect transistors[20] composed of ion-gated ZrNCl crystals, the superconducting state that ensues for gate voltages exceeding 4V is destroyed[20] for perpendicular magnetic fields as low as $0.05T$. This behaviour was attributed[20] to weak pinning of vortices and hence the authors reach the conclusion that throughout most of the vortex state, be it a liquid or a glass, a metallic state obtains. This conclusion is particularly telling and of fundamental importance to the eventual construction of the theory of the metallic state. Second, in NbSe$_2$ an essentially crystalline material, the resistance turns on[12] continuously as $\rho \approx (g - g_c)^\alpha$, where g_c is the critical value of the tuning parameter for the onset of the metallic state. Similar results have also been observed in MoGe[22]. Third, in InO$_x$ and TaN$_x$, the Hall conductance is observed[29] to vanish throughout the Bose metallic state, thereby indicating that particle-hole symmetry is an intrinsic feature of this state. A similar vanishing of the Hall resistance below a critical value of the applied field was seen earlier by Paalanen, Hebard and Ruel on amorphous indium oxide films[9]. In strong support of this last claim are the recent experiments demonstrating that the cyclotron resonance vanishes in the Bose metallic state[30].

While there have been numerous proposals for a Bose metal[31-36], a state with a finite resistance at $T = 0$ (demonstrated to exist in only one purely bosonic model[36]), the new experiments highly constrain possible theoretical descriptions. Of particular importance is the observation that in the clean samples[20], the vortex state that ensues once the superconducting state is destroyed is metallic! This appears

116

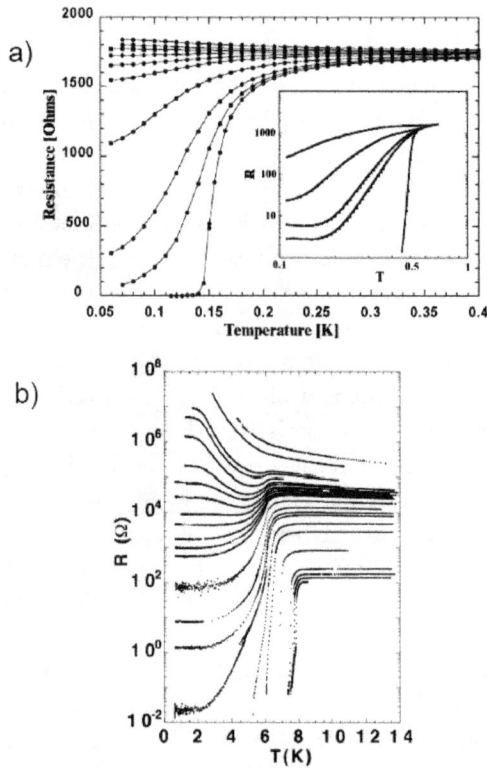

Fig. 1. Resistivity versus temperature for two different systems. (a) Electrical resistance of MoGe thin film plotted vs temperature at B = 0, 0.5, 1.0, 2.0, 3.0, 4.0, 4.4, 4.5, 5.5, 6 kG. The sample becomes a superconductor at 0.15 K in zero field but for fields larger than about 4.4 kG the sample becomes insulating. At fields lower than this but other than zero, the resistance saturates. The saturation behavior is better shown in the inset for another sample with a higher transition temperature. The inset shows data for B = 0, 1.5, 2, 4, and 7 kOe. At higher field, this sample is an insulator. Main figure reprinted from A. Yazdani and A. Kapitulnik, Phys. Rev. Lett. 74, 3037 (1995), while the inset is from, Phys. Rev. Lett. 76. 1529 (1996). (b) Reprinted from C. Christiansen, L. M. Hernandez, and A. M. Goldman, Phys. Rev. Lett., 88, 37004 (2002). Evolution of the temperature dependence of the resistance for a series of Ga films. Film thicknesses range from 12.75 Å to 16.67 Å and increases from top to bottom. The leveling of the resistance once superconductivity is destroyed (zero resistance curves) is not consistent with conventional wisdom. Note that the plateau value of the resistivity increases as the distance from the superconducting phase increases.

to be in potential conflict with the vortex glass state being a superconductor[37,38]. In fact, the broad observation of a metallic state in 2D samples, be they disordered or quite clean, points to a re-examination of the ultimate fate of vortex states in 2D. It is precisely this that we do here as the models my group proposed several years ago[36,39,40] are all based on glassy vortex states in which the conductivity was shown to be finite from an explicit calculation of the conductivity from the Kubo formula. In the collision-dominated (or hydrodynamic) regime, the resistivity has a finite value and turns on as $\rho \approx (g - g_c)^\alpha$, as highlighted in the experiments on NbSe$_2$[12]. While questions[41] regarding the phase stiffness of the phase glass have been raised, numerical simulations all indicate[2-4] that the energy to create a defect in a 2D phase or gauge glass scales as L^θ, where $\theta = -0.39$. Hence, the stiffness is non-existent. In 3D[2-4], $\theta > 0$ and a stiffness obtains. As such glass states are candidates to explain the vortex glass[37,38], that $\theta < 0$ is consistent with the experimental finding[20] in ion-gated ZrNCl, an extreme 2D system, that the resultant vortex state is indeed metallic and not a true superconductor.

2. Preliminaries

In the 80's and '90's, the leading candidate to explain failed superconductivity was the resistively shunted Josephson junction array model[42-44]. All such models are based on a propagator of the form,

$$G_0 = (k^2 + \eta|\omega_n| + m^2)^{-1}, \tag{1}$$

in which the Ohmic dissipative term, $|\omega|$ is an attempt to model the normal electrons. Despite the Ohmic term, all such models yield[44,45] either insulating or superconducting states and hence are not applicable to the metallic state. In computing the conductivity of these models, it is important to note[46] that the conductivity in the vicinity of the transition region is a universal function of the form, $\sigma_Q f(\omega/T)$ where f is a monotonically decreasing function of the frequency, ω, and the temperature, T. The experiments correspond to the limiting procedure $\lim_{T \to 0} \lim_{\omega \to 0}$, that is to $f(0)$ not the inverse limit where $f(\infty)$ enters. The physics of $f(0)$ is pure hydrodynamics in which it is collisions of the quasiparticle excitations of the order parameter that regularize the conductivity. Explicit computation of the conductivity in the resistively shunted Josephson junction array model yielded[45] that although dissipation can drive an intermediate region in temperature where the resistivity levels, ultimately at $T = 0$ a superconductor obtains. Hence, dissipation alone cannot drive the Bose metal.

Before we present the glassy model that has the extra ingredient, it is instructive to look at a simpler model which illustrates the power of the hydrodynamic[47] approach. Our system consists of an array of Josephson junctions, we coarse-grain over the phase associated with each junction and hence use as our starting point the Landau-Ginzburg action,

$$F[\psi] = \sum_{\vec{k},\omega_n} (k^2 + \omega_n^2 + m^2)|\psi(\vec{k},\omega_n)|^2$$

$$+ \frac{U}{2N\beta} \sum_{\omega_1,\ldots,\omega_4;\vec{k}_1,\ldots,\vec{k}_4} \delta_{\omega_1+\cdots\omega_4,0}\delta_{\vec{k}_1+\cdots\vec{k}_4,0}$$

$$\psi_\nu(\omega_1,\vec{k}_1)\psi_\nu(\omega_2,\vec{k}_2)\psi_\mu(\omega_3,\vec{k}_3)\psi_\mu(\omega_4,\vec{k}_4), \tag{2}$$

where $\psi(\vec{r}, \tau)$ is the complex Bose order parameter whose expectation value is proportional to $\langle\exp(i\phi)\rangle$, where ϕ is the phase of a particular junction. The summation in the action over discrete Matsubara frequencies, $\omega_i = 2\pi n_i T$, and integration over continuous wavevectors, $\vec{k}$, is assumed. The parameter m^2 is the inverse square of the correlation length. In writing the action in this fashion, we have already included the one-loop renormalization arising from the quartic term. In the quantum-disordered regime, $m \gg T$ and hence it is the quantum fluctuations that dominate the divergence of the correlation length.

This model has two phases, a superconductor for $m^2 < 0$ and an insulator for $m^2 > 0$. The conductivity in the insulator is expected to vanish. But in fact it does not precisely because the resistivity in the quantum disordered regime is mediated by finite temperature collisions between the bosons. Such events are exponentially activated. However, their lifetime is also exponentially long as can be seen from an explicit calculation[47]. To recount the calculation, we note that the central quantity appearing in the collision integral is the polarization function

$$\Pi(\vec{q}, i\Omega_m) = T \sum_n \int \frac{d^2p}{(2\pi)^2} G_0(\vec{p} + \vec{q}, \omega_n + \Omega_m) G_0(\vec{p}, \omega_n) \tag{3}$$

where the bare field propagator $G_0(\vec{p}, \omega_n) = (p^2 + \omega_n^2 + m^2)^{-1}$. As it is the imaginary part of Π that is required in the collision integral, we must perform an analytical continuation $\Omega_n \to -i\Omega_n - \delta$ with δ a positive infinitesimal. It is the polarization function that appears explicitly in the scattering time,

$$\frac{1}{\tau_{\vec{k}}} = \frac{1}{2(2\pi)^2} \left[\int \frac{d^2q}{\epsilon_{\vec{q}+\vec{k}}\epsilon_{\vec{k}}} \left(\mathrm{Im}\frac{1}{\Pi(\vec{q}, \epsilon_{\vec{q}+\vec{k}} - \epsilon_{\vec{k}})} \right) n(\epsilon_{\vec{q}+\vec{k}} - \epsilon_{\vec{k}}) \right.$$

$$\left. + \int \frac{d^2q}{\epsilon_{\vec{q}}\epsilon_{\vec{k}}} \left(\mathrm{Im}\frac{1}{\Pi(\vec{q} + \vec{k}, \epsilon_{\vec{q}} + \epsilon_{\vec{k}})} \right) n(\epsilon_{\vec{q}}) \right], \tag{4}$$

which is related to the conductivity through

$$\sigma = 2\frac{(e^*)^2}{\hbar} \int \frac{d^2k}{(2\pi)^2} \frac{k_x^2}{\epsilon_{\vec{k}}^2} \tau_{\vec{k}} \left(-\frac{\partial n(\epsilon_{\vec{k}})}{\partial \epsilon_{\vec{k}}} \right). \tag{5}$$

The essence of our central result is that to leading order in T/m, the inverse relaxation time $1/\tau_{\vec{k}}$ is momentum-independent and given by

$$\frac{1}{\tau} = \pi T e^{-m/T}. \tag{6}$$

Substitution of this expression into Eq. (5) leads to the mutual cancellation of the exponential factors yielding the remarkable result

$$\sigma(T \to 0) = \frac{2}{\pi} \frac{4e^2}{h}. \tag{7}$$

It is curious to note[48,49] that a similar cancellation of exponential factors (from the mean free path and the density of states) arises in the context of the quasiparticle conductivity in a dirty d-wave superconductivity yielding the identical numerical prefactor $2/\pi$. In essence, the insulator is a metal because the mean-free path of the bosonic excitations is exponentially small with the same factor that enters the population of bosonic excitations. Since it is the product of the scattering time and the population that leads to the conductivity, the result must be finite. While this result is interesting, this metal state is quite fragile as it is destroyed by any perturbation. Hence, the answer to the experiments lies elsewhere.

3. Bose Metal

As shown previously[50], any amount of dirt in a 2D superconductor induces $\pm J$ disorder, J the Josephson coupling. Consequently, a disordered superconductor is closer to a disordered XY model rather than a dirty superfluid. Justifiably, the starting point for analyzing the experiments is the disordered XY model,

$$
H = -E_C \sum_i \left(\frac{\partial}{\partial \theta_i}\right)^2 - \sum_{\langle i,j \rangle} J_{ij} \cos(\theta_i - \theta_j),
\tag{8}
$$

with random Josephson couplings J_{ij} but fixed on-site energies, E_C. The phase of each island is θ_i. If the Josephson couplings are chosen from a distribution with zero mean, only two phases are possible: (1) a glass arising from the distribution of positive and negative $J'_{ij}s$ and (2) a disordered paramagnetic state. A superconducting phase obtains if the distribution

$$
P(J_{ij}) = \frac{1}{\sqrt{2\pi J^2}} \exp\left[-\frac{(J_{ij} - J_0)^2}{2J^2}\right]
\tag{9}
$$

of $J'_{ij}s$ has non-zero mean, J_0, and J the variance. To distinguish between the phases, it is expedient to introduce[51] the set of variables $\mathbf{S}_i = (\cos\theta_i, \sin\theta_i)$ which allows us to recast the interaction term in the random Josephson Hamiltonian as a spin problem with random magnetic interactions, $\sum_{\langle i,j \rangle} J_{ij} \mathbf{S}_i \cdot \mathbf{S}_j$. Let $\langle ... \rangle$ and $[...]$ represent averages over the thermal degrees of freedom and over the disorder, respectively. Integrating over the random interactions will introduce two auxilliary fields

$$
Q_{\mu\nu}^{ab}(\vec{k}, \vec{k}', \tau, \tau') = \langle S_\mu^a(\vec{k}, \tau) S_\nu^b(\vec{k}', \tau') \rangle
\tag{10}
$$

and $\Psi_\mu^a(\vec{k}, \tau) = \langle S_\mu^a(\vec{k}, \tau) \rangle$, respectively. The superscripts represent the replica indices. A non-zero value of $\Psi_\mu^a(\vec{k}, \tau)$ implies phase ordering of the charge $2e$ degrees of freedom. For quantum spin glasses, it is the diagonal elements of the Q-matrix $D(\tau - \tau') = \lim_{n \to 0} \frac{1}{Mn} \langle Q_{\mu\mu}^{aa}(\vec{k}, \vec{k}', \tau, \tau') \rangle$ in the limit that $|\tau - \tau'| \to \infty$ that serve as the effective Edwards-Anderson spin-glass order parameter[52-54] within Landau theory. The free energy per replica

$$\mathcal{F}[\Psi, Q] = \mathcal{F}_{\text{SG}}(Q) + \sum_{a,\mu,k,\omega_n} (k^2 + \omega_n^2 + m^2) |\Psi_\mu^a(\vec{k}, \omega_n)|^2$$

$$- \frac{1}{\kappa t} \int d^d x \int d\tau_1 d\tau_2 \sum_{a,b,\mu,\nu} \Psi_\mu^a(x, \tau_1) \Psi_\nu^b(x, \tau_2) Q_{\mu\nu}^{ab}(x, \tau_1, \tau_2)$$

$$+ U \int d\tau \sum_{a,\mu} \left[\Psi_\mu^a(x, \tau) \Psi_\mu^a(x, \tau) \right]^2, \tag{11}$$

consists of a spin-glass part which is a third-order functional of the $Q-$ matrices discussed previously[36,52], the Ψ_μ^a terms that describe the charge 2e condensate and the term which couples the charge and glassy degrees of freedom. The parameters, κ, t and U are the standard coupling constants in a Landau theory and m^2 is the inverse correlation length. The essential aspect of the quantum rotor spin glass is that the saddle point solution for the corresponding action is minimized by a solution of the form

$$Q_{\mu\nu}^{ab}(\vec{k}, \omega_1, \omega_2) = \beta(2\pi)^d \delta^d(k) \delta_{\mu\nu} \left[D(\omega_1) \delta_{\omega_1 + \omega_2, 0} \delta_{ab} + \beta \delta_{\omega_1, 0} \delta_{\omega_2, 0} q^{ab} \right], \tag{12}$$

where the diagonal elements are given by

$$D(\omega) = -\sqrt{\omega^2 + \Delta^2}/\kappa, \tag{13}$$

with κ a coupling constant in the Landau free energy for the spin glass. The diagonal elements of the Q-matrices describe the excitation spectrum. Throughout the glassy phase, $\Delta = 0$ and hence the spectrum is ungapped and given by $D(\omega) = -|\omega|/\kappa$. The linear dependence on $|\omega|$ arises because the correlation function $Q_{\mu\mu}^{aa}(\tau)$ decays[52-54] as τ^{-2}. This dependence results in a fundamental change in the dynamical critical exponent from $z = 1$ to $z = 2$.

To compute the conductivity in the glassy phase, we note that near the spin-glass/superconductor boundary, m^2 should be regarded as the smallest parameter. Hence, it is the fluctuations of Ψ_μ^a rather than those of Q^{ab} that dominate. In this regard, we recall the work of Chamon and Nayak[1] who noted that that the disorder at the insulator-superconductor phase transition drives the critical behaviour away from that of the clean system towards a diffusive fixed point in which the critical resistance vanishes. They then advocate that the resistivity should turn on continuously from zero as a power law. It is precisely this behaviour that we prove here.

To compute the conductivity, we focus on the part of the free energy,

$$\mathcal{F}_{\text{gauss}} = \sum_{a,\vec{k},\omega_n} (k^2 + \omega_n^2 + \eta|\omega_n| + m^2) |\psi^a(\vec{k}, \omega_n)|^2$$

$$- \beta q \sum_{a,b,\vec{k},\omega_n} \delta_{\omega_n, 0} \psi^a(\vec{k}, \omega_n) [\psi^b(\vec{k}, \omega_n)]^*, \tag{14}$$

governed by the fluctuations of the superconducting order parameter at the Gaussian level. In the above action, we introduced the effective dissipation $\eta = 1/(\kappa^2 t)$

and rescaled $q \to q\kappa t$. As is evident in this action, disorder appears explicitly as a mixing between the replicas. It is from this term that fundamentally new physics arises and the origin of the transition to the diffusive fixed point. The new physics is captured by the associated Gaussian propagator

$$G_{ab}^{(0)}(\vec{k}, \omega_n) = G_0(\vec{k}, \omega_n)\delta_{ab} + \beta G_0^2(\vec{k}, \omega_n)q\delta_{\omega_n,0} \qquad (15)$$

which is obtained by inverting the free energy in ultrametric space[55] in the $n \to 0$ limit[55] with $G_0(\vec{k}, \omega_n) = (k^2 + \omega_n^2 + \eta|\omega| + m^2)^{-1}$. The first term in Eq. (15) is the standard Gaussian propagator in the presence of Ohmic dissipation. The Ohmic dissipative term in the free-energy arises from the diagonal elements of the $Q-$ matrices. However, it is the $q-$dependent term in the Gaussian free energy, the last term in Eq. (14), that is new and changes fundamentally the form of the propagator. Because of the $\delta_{\omega_n,0}$ factor in the second term in the free energy, the propagator now contains a frequency-independent part, $\beta G_0^2(\vec{k}, \omega_n = 0)q$.

To compute the conductivity, we use the generalization[56] of the Kubo formula for a replicated action and obtain for the conductivity

$$\sigma(i\omega_n) = \frac{2(e^*)^2}{n\hbar\omega_n}T \sum_{a,b,\omega_m} \int \frac{d^2k}{(2\pi)^2} \left[G_{ab}^{(0)}(\vec{k}, \omega_m)\delta_{ab} \right.$$
$$\left. - 2k_x^2 G_{ab}^{(0)}(\vec{k}, \omega_m)G_{ab}^{(0)}(\vec{k}, \omega_m + \omega_n) \right]. \qquad (16)$$

The conductivity contains two types of terms. All terms not proportional to q have been evaluated previously[45,47] and vanish as $T \to 0$. The terms proportional to q^2 vanish in the limit $n \to 0$. The only terms remaining are proportional to q and yield after an appropriate integration by parts

$$\sigma(i\omega_n) = \frac{8qe^{*2}}{\hbar\omega_n} \int \frac{d^2k}{(2\pi)^2} k_x^2 G_0^2(\vec{k}, 0) \left[G_0(\vec{k}, 0) - G_0(\vec{k}, \omega_n) \right].$$

The momentum integrations are straightforward and yield

$$\sigma(\omega = 0, T \to 0) = \frac{8e^2}{\hbar} \frac{q\eta}{2m^4} \qquad (17)$$

a temperature-independent value for the conductivity as $T \to 0$. The dependence on q and η imply that dissipation alone is insufficient to generate a metallic state. What seems to be the case is that a bosonic excitation moving in a dissipative environment in which many false minima exist does not localize because it takes an exponentially long amount of time to find the ground state. This is the physical mechanism that defeats localization in a glassy phase. From the dependence on m^4, we see clearly that the resistivity turns on as a power law

$$\rho \approx (g - g_c)^{2z\nu} \qquad (18)$$

as is seen experimentally[12,22] and consistent with the Chamon/Nayak[1] work that at the superconductor-insulator transition, the resistivity should turn on continuously from zero not $h/4e^2$. As shown elsewhere[36], the finite resistivity obtained here is robust to the quartic interactions in Eq. (11). Hence, the metallic state is not an artifact of the Gaussian approximation. Quantifying how this exponent changes as a function of disorder is of utmost importance as it would determine if all observations of the Bose metal lie in the same universality class.

Aside from the turn-on of the resistivity, the phase glass can also explain the apparent vanishing Hall response in the metallic state. We have recently computed the Hall conductance[57] and found it to vanish as a result of the inherent particle-hole symmetry in this model. Using a model in which the glassy degrees of freedom were generated from a random magnetic field in the form, $\cos(\theta_i - \theta_j - A_{ij})$ (where $A_{ij} = e^*/\hbar \int_i^j \mathbf{A} d\mathbf{l}$, $(e^* = 2e)$) rather than the constant term in Eq. (8) Away from the particle-hole symmetric point, the Hall conductance turns on as

$$\sigma_H(i\omega_\nu) = \frac{\lambda q (e^* m_H^2)^2}{\hbar m^4} \left(\frac{2}{x} - \frac{\Psi(1, \frac{x+2}{2x})}{x^3} \right) \tag{19}$$

where $x = \frac{m_H^2}{m^2}$, $m_H^2 = \frac{e^*}{c\hbar} B$, and $\Psi(1, x)$ is the first digamma function. The compensating effect on the diagonal conductivity is

$$\sigma_{xx}(i\omega_\nu) = \frac{\eta q (e^* m_H)^2}{\hbar m^4} \left(\frac{2}{x} - \frac{\Psi(1, \frac{x+2}{2x})}{x^3} \right). \tag{20}$$

Hence, both have identical trends in a magnetic field. This prediction should serve as a guide to further experiments.

4. Final Remarks

In conclusion, all current experiments on the destruction of superconductivity in thin films in 2D can be understood within a glassy model as the intermediary before the onset of the insulating state. An analogy which might be helpful here is with the Bose-Hubbard model. In the absence of disorder, this model admits a direct transition from a superfluid to a Mott insulator. However, in the presence of disorder, a Bose glass[58] phase intervenes disrupting the direct transition to the Mott insulator. The work presented here implies a similar trend is manifest in the charged case as well. The falsifiable prediction for the turn-on of the Hall conductance can be confirmed by ground-plane experiments and should offer a new window into the true nature of the ground state of the Bose metal.

Acknowledgements

This paper is largely a review of the previous works on this topic I co-authored with D. Dalidovich and J. May-Mann. I thank Steve Kivelson for pointing out the references on the numerical simulations on the stiffness in the 2D spin glass. This works was funded by NSF-DMR-1461952.

References

1. C. Chamon and C. Nayak, Anomalous quantum diffusion at the superfluid-insulator transition, *Phys. Rev. B* **66**, p. 094506 (Sep 2002).
2. H. G. Katzgraber and A. P. Young, Numerical studies of the two- and three-dimensional gauge glass at low temperature, *Phys. Rev. B* **66**, p. 224507 (December 2002).
3. J. M. Kosterlitz and N. Akino, Numerical Study of Spin and Chiral Order in a Two-Dimensional XY Spin Glass, *Physical Review Letters* **82**, 4094 (May 1999).
4. J. M. Kosterlitz and N. Akino, Numerical Study of Order in a Gauge Glass Model, *Physical Review Letters* **81**, 4672 (November 1998).
5. V. L. Berezinskiĭ, Destruction of Long-range Order in One-dimensional and Two-dimensional Systems Possessing a Continuous Symmetry Group. II. Quantum Systems, *Soviet Journal of Experimental and Theoretical Physics* **34**, p. 610 (1972).
6. J. M. Kosterlitz and D. J. Thouless, Ordering, metastability and phase transitions in two-dimensional systems, *Journal of Physics C: Solid State Physics* **6**, p. 1181 (1973).
7. A. F. Hebard and A. T. Fiory, Critical-exponent measurements of a two-dimensional superconductor, *Phys. Rev. Lett.* **50**, 1603 (May 1983).
8. A. T. Fiory, A. F. Hebard and W. I. Glaberson, Superconducting phase transitions in indium/indium-oxide thin-film composites, *Phys. Rev. B* **28**, 5075 (Nov 1983).
9. M. A. Paalanen, A. F. Hebard and R. R. Ruel, Low-temperature insulating phases of uniformly disordered two-dimensional superconductors, *Phys. Rev. Lett.* **69**, 1604 (Sep 1992).
10. H. M. Jaeger, D. B. Haviland, B. G. Orr and A. M. Goldman, Onset of superconductivity in ultrathin granular metal films, *Phys. Rev. B* **40**, 182 (Jul 1989).
11. M. P. A. Fisher, Quantum phase transitions in disordered two-dimensional superconductors, *Phys. Rev. Lett.* **65**, 923 (Aug 1990).
12. A. W. Tsen, B. Hunt, Y. D. Kim, Z. J. Yuan, S. Jia, R. J. Cava, J. Hone, P. Kim, C. R. Dean and A. N. Pasupathy, Nature of the quantum metal in a two-dimensional crystalline superconductor, *Nat Phys* **advance online publication** (12 2015).
13. N. E. Staley, J. Wu, P. Eklund, Y. Liu, L. Li and Z. Xu, Electric field effect on superconductivity in atomically thin flakes of nbse$_2$, *Phys. Rev. B* **80**, p. 184505 (Nov 2009).
14. A. Yazdani and A. Kapitulnik, Superconducting-insulating transition in two-dimensional a-moge thin films, *Phys. Rev. Lett.* **74**, 3037 (Apr 1995).
15. J. A. Chervenak and J. M. Valles, Absence of a zero-temperature vortex solid phase in strongly disordered superconducting bi films, *Phys. Rev. B* **61**, R9245 (Apr 2000).
16. H. Q. Nguyen, S. M. Hollen, J. M. Valles, J. Shainline and J. M. Xu, Disorder influences the quantum critical transport at a superconductor-to-insulator transition, *Phys. Rev. B* **92**, p. 140501 (October 2015).
17. D. Ephron, A. Yazdani, A. Kapitulnik and M. R. Beasley, Observation of quantum dissipation in the vortex state of a highly disordered superconducting thin film, *Phys. Rev. Lett.* **76**, 1529 (Feb 1996).
18. Y. Seo, Y. Qin, C. L. Vicente, K. S. Choi and J. Yoon, Origin of nonlinear transport across the magnetically induced superconductor-metal-insulator transition in two dimensions, *Phys. Rev. Lett.* **97**, p. 057005 (Aug 2006).
19. S. Park, J. Shin and E. Kim, Scaling analysis of field-tuned superconductor–insulator transition in two-dimensional tantalum thin films, **7**, 42969 EP (02 2017).
20. Y. Saito, Y. Kasahara, J. Ye, Y. Iwasa and T. Nojima, Metallic ground state in an ion-gated two-dimensional superconductor, *Science* **350**, 409 (2015).

21. N. Mason and A. Kapitulnik, Dissipation effects on the superconductor-insulator transition in 2d superconductors, *Phys. Rev. Lett.* **82**, 5341 (Jun 1999).
22. N. Mason, Superconductor-metal transition in moge, PhD thesis, Standord Univ.2003.
23. H. Q. Nguyen, S. M. Hollen, J. Shainline, J. M. Xu and J. M. Valles, Driving a Superconductor to Insulator Transition with Random Gauge Fields, *Scientific Reports* **6**, p. 38166 (November 2016).
24. K. A. Parendo, K. H. S. B. Tan, A. Bhattacharya, M. Eblen-Zayas, N. E. Staley and A. M. Goldman, Electrostatic tuning of the superconductor-insulator transition in two dimensions, *Phys. Rev. Lett.* **94**, p. 197004 (May 2005).
25. S. Misra, L. Urban, M. Kim, G. Sambandamurthy and A. Yazdani, Measurements of the magnetic-field-tuned conductivity of disordered two-dimensional $mo_{43}ge_{57}$ and ino_x superconducting films: Evidence for a universal minimum superfluid response, *Phys. Rev. Lett.* **110**, p. 037002 (Jan 2013).
26. W. Liu, L. Pan, J. Wen, M. Kim, G. Sambandamurthy and N. P. Armitage, Microwave spectroscopy evidence of superconducting pairing in the magnetic-field-induced metallic state of ino_x films at zero temperature, *Phys. Rev. Lett.* **111**, p. 067003 (Aug 2013).
27. C. A. Marrache-Kikuchi, H. Aubin, A. Pourret, K. Behnia, J. Lesueur, L. Bergé and L. Dumoulin, Thickness-tuned superconductor-insulator transitions under magnetic field in a-nbsi, *Phys. Rev. B* **78**, p. 144520 (Oct 2008).
28. Y. Qin, C. L. Vicente and J. Yoon, Magnetically induced metallic phase in superconducting tantalum films, *Phys. Rev. B* **73**, p. 100505 (Mar 2006).
29. N. P. Breznay and A. Kapitulnik, Particle-hole symmetry reveals failed superconductivity in the metallic phase of two-dimensional superconducting films, *Science Advances* **3**, p. e1700612 (September 2017).
30. Y. Wang, I. Tamir, D. Shahar and N. P. Armitage, A Bose metal has no cyclotron resonance, *ArXiv e-prints* (August 2017).
31. D. Das and S. Doniach, Bose metal: Gauge-field fluctuations and scaling for field-tuned quantum phase transitions, *Phys. Rev. B* **64**, p. 134511 (October 2001).
32. M. Mulligan and S. Raghu, Composite fermions and the field-tuned superconductor-insulator transition, *Phys. Rev. B* **93**, p. 205116 (May 2016).
33. E. Shimshoni, A. Auerbach and A. Kapitulnik, Transport through quantum melts, *Phys. Rev. Lett.* **80**, 3352 (Apr 1998).
34. B. Spivak, P. Oreto and S. A. Kivelson, Theory of quantum metal to superconductor transitions in highly conducting systems, *Phys. Rev. B* **77**, p. 214523 (Jun 2008).
35. A. Paramekanti, L. Balents and M. P. Fisher, Ring exchange, the exciton Bose liquid, and bosonization in two dimensions, *Phys. Rev. B* **66**, p. 054526 (August 2002).
36. D. Dalidovich and P. Phillips, Phase Glass is a Bose Metal: A New Conducting State in Two Dimensions, *Physical Review Letters* **89**, p. 027001 (June 2002).
37. M. P. A. Fisher, Vortex-glass superconductivity: A possible new phase in bulk high-t_c oxides, *Phys. Rev. Lett.* **62**, 1415 (Mar 1989).
38. D. S. Fisher, M. P. A. Fisher and D. A. Huse, Thermal fluctuations, quenched disorder, phase transitions, and transport in type-ii superconductors, *Phys. Rev. B* **43**, 130 (Jan 1991).
39. P. Phillips and D. Dalidovich, The Elusive Bose Metal, *Science* **302**, 243 (October 2003).
40. J. Wu and P. Phillips, Vortex glass is a metal: Unified theory of the magnetic-field and disorder-tuned Bose metals, *Phys. Rev. B* **73**, p. 214507 (June 2006).
41. P. Phillips and D. Dalidovich, Absence of phase stiffness in the quantum rotor phase glass, *Phys. Rev. B* **68**, p. 104427 (Sep 2003).
42. S. Chakravarty, S. Kivelson, G. T. Zimanyi and B. I. Halperin, Effect of quasipar-

ticle tunneling on quantum-phase fluctuations and the onset of superconductivity in granular films, *Phys. Rev. B* **35**, 7256 (May 1987).

43. S. Chakravarty, G.-L. Ingold, S. Kivelson and G. Zimanyi, Quantum statistical mechanics of an array of resistively shunted josephson junctions, *Phys. Rev. B* **37**, 3283 (Mar 1988).

44. K.-H. Wagenblast, A. van Otterlo, G. Schön and G. T. Zimányi, New universality class at the superconductor-insulator transition, *Phys. Rev. Lett.* **78**, 1779 (Mar 1997).

45. D. Dalidovich and P. Phillips, Fluctuation Conductivity in Insulator-Superconductor Transitions with Dissipation, *Physical Review Letters* **84**, 737 (January 2000).

46. K. Damle and S. Sachdev, Nonzero-temperature transport near quantum critical points, *Phys. Rev. B* **56**, 8714 (Oct 1997).

47. D. Dalidovich and P. Phillips, Interaction-induced bose metal in two dimensions, *Phys. Rev. B* **64**, p. 052507 (Jul 2001).

48. E. Fradkin, Critical behavior of disordered degenerate semiconductors. i. models, symmetries, and formalism, *Phys. Rev. B* **33**, 3257 (Mar 1986).

49. P. A. Lee, Localized states in a d-wave superconductor, *Phys. Rev. Lett.* **71**, 1887 (Sep 1993).

50. B. Spivak, P. Oreto and S. A. Kivelson, Theory of quantum metal to superconductor transitions in highly conducting systems, *Phys. Rev. B* **77**, p. 214523 (Jun 2008).

51. D. Dalidovich and P. Phillips, Landau theory of bicriticality in a random quantum rotor system, *Phys. Rev. B* **59**, 11925 (May 1999).

52. N. Read, S. Sachdev and J. Ye, Landau theory of quantum spin glasses of rotors and ising spins, *Phys. Rev. B* **52**, 384 (Jul 1995).

53. J. Miller and D. A. Huse, Zero-temperature critical behavior of the infinite-range quantum ising spin glass, *Phys. Rev. Lett.* **70**, 3147 (May 1993).

54. A. J. Bray and M. A. Moore, Critical behavior of the three-dimensional Ising spin glass, *Phys. Rev. B* **31**, 631 (January 1985).

55. J. Zinn-Justin, *Quantum Field Theory and Critical Phenomena* (Clarendon Press, Oxford.

56. I. F. Herbut, Dual theory of the superfluid-bose-glass transition in the disordered bose-hubbard model in one and two dimensions, *Phys. Rev. B* **57**, 13729 (Jun 1998).

57. J. May-Mann and P. W. Phillips, Vanishing hall conductance in the phase-glass bose metal at zero temperature, *Phys. Rev. B* **97**, p. 024508 (Jan 2018).

58. M. P. A. Fisher, P. B. Weichman, G. Grinstein and D. S. Fisher, Boson localization and the superfluid-insulator transition, *Phys. Rev. B* **40**, 546 (Jul 1989).

Topological superfluidity with repulsive fermionic atoms

G. Ortiz*

Department of Physics, Indiana University, 727 E Third St.,
Bloomington, Indiana 47405, USA
*E-mail: ortizg@indiana.edu
www.iub.edu/~iubphys/faculty/ortizg.shtml

L. Isaev, A. Kaufman, and A. M. Rey[†]

JILA, NIST, Department of Physics, University of Colorado, 440 UCB
Boulder, Colorado 80309, USA
†E-mail: arey@jilau1.colorado.edu

What is a topological superfluid in a particle-number conserving system? Motivated by this question we report on a novel pathway to topological superfluidity with fermionic atoms in an optical lattice. We consider a situation where atoms in two internal states experience different lattice potentials: one species is localized and the other itinerant, and show how quantum fluctuations of the localized fermions give rise to an attraction and strong effective spin-orbit coupling in the itinerant band. At low temperature, these effects stabilize a topological superfluid of mobile atoms even if their bare interactions are repulsive. This emergent state can be engineered with ^{87}Sr atoms in a dimerized superlattice. To probe its unique properties we describe protocols that use high spectral resolution and controllability of the Sr clock transition, such as momentum-resolved spectroscopy and supercurrent response to a synthetic (laser-induced) magnetic field.

Keywords: Topological Fermion Superfluids; Majorana fermions; Ultracold Atoms; Optical Lattices; Fermion Parity Switches; Magnetoelectric Phenomena.

1. Introduction

Our understanding of many-body systems traditionally relies on the Landau classification of states of matter based on global symmetries spontaneously broken within the ordered phase. This symmetry breaking scenario is accompanied by emergence of an order parameter, i.e. a non-zero expectation value of a local physical observable, that uniquely characterizes the phase. For instance, a hallmark signature of a fermionic superfluid is the spontaneous breaking of the particle-number conservation ($U(1)$) symmetry which occurs as a result of Cooper pairing. The corresponding order parameter plays the role of a Cooper pair wavefunction and defines an energy gap in the excitation spectrum, allowing dissipationless particle currents[1]. However, many phases of matter defy the Landau classification paradigm; these are known as topological ordered phases. An important class of such systems are topological superfluids (TSFs)[2,3], i.e. phases that in addition to $U(1)$, break a residual $\mathbb{Z}_2$ symmetry. The latter symmetry breaking is a global phenomenon that occurs *in the absence of local order parameter* and only for appropriate boundary conditions[4–7].

Although fermion superfluids are theoretically distinguished by spontaneous breaking of the symmetry of particle-number conservation, actual (closed and finite) physical superfluids do not alter the number of their elementary constituents. Despite this simple observation, the most successful description of superfluidity is based on a (Bogoliubov) mean-field approximation that *explicitly* breaks the $U(1)$ symmetry. Whether such mismatch between theory and experiments matters depends on the particular physical property that is being investigated. For instance, a bulk thermodynamic property is inherently less prone to scrutiny. Less clear is its relevance in the manipulation of localized Majorana excitations whose braiding achieve non-Abelian statistics, a hallmark of TSFs. Thus, understanding topological superfluidity at its most fundamental level is a problem that requires exquisite experimental control and novel theoretical tools beyond the standard mean-field approximation. In this respect, ultracold atomic systems (which manifestly preserve their number of atoms) are pristine quantum testbeds for topological superfluidity and for the study of the challenges involved in quantum-controlling emergent Majorana modes in particle-conserving systems.

In this paper we take over where Refs. 4 and 5 left off and propose an experimental realization of a number-conserving TSF with repulsive fermionic cold atoms in an optical lattice. The main goal of our study is to identify signatures of topological superfluidity most readily accessible in cold-atom experiments. Quite counter-intuitively, in this sense the simplest realization of a TSF is served by a system of *repulsively* interacting fermions.

The outline of this paper is as follows. We begin Section 2.1 by considering the problem of characterizing a particle-conserving fermionic superfluid as topologically trivial or non-trivial, in terms of a topological invariant known as *fermion parity switch*. This characterization was shown to have a definite answer only very recently[4]. As explained in this paper, a topological fermion superfluid (or superconductor), in addition to the global $U(1)$ symmetry of particle-number conservation, *spontaneously* breaks the discrete $\mathbb{Z}_2$ symmetry of fermionic parity for particular non-symmetry-breaking boundary conditions. In Section 2.2 we illustrate how to compute fermion parity switches for a paradigmatic exactly-solvable model of a particle-conserving superfluid that, as will be shown explicitly, displays topologically trivial and non-trivial phases[4]. We then proceed to advance a pathway towards topological superfluidity in Section 3. Our proposal[8] considers repulsive alkaline-earth fermionic atoms, like ^{87}Sr, trapped in suitably designed optical lattices. We start in Section 3.1 by introducing main ideas in the simpler context of quasi-one-dimensional optical lattices, although higher-dimensional extensions can also be implemented[8]. We study the quantum phase diagram of the resulting fermionic system, and the stability of its distinct phases in Section 3.2. In Section 3.3 we proceed to describe a series of new experimental probes that allows us to detect the exotic superfluid phases of the model. These probes include the unveiling of a novel magneto-electric phenomenon that emerges in the topological superfluid phase as a result of applying a synthetic magnetic field, which also leads to an asymmetry in the momentum distribution. We also propose to probe the spectrum of Bogoliubov excitations by momentum-resolved spectroscopy. Section 3.4 describes a practical experimental protocol to engineer the quasi-one dimensional optical lattice loaded with fermionic atoms in specific clock states. We close with a discussion about detection of Majorana fermions and outlook in Section 4.

2. Topological superfluidity in particle-number conserving systems

Most investigations of fermionic condensed matter are based on a type of mean-field approximation popularized by Bogoliubov[9]. The approach is mathematically represented by a tractable Lie-algebraic formalism by which the problem of diagonalizing their Hamiltonian in Fock space becomes one of polynomial complexity in the total number of degrees of freedom. Apart from providing an intuitive and natural picture of fermionic systems in terms of Landau quasi-particles, topological invariants are easily evaluated within this framework.

These positive features come at a surprising cost in the context of superfluidity (electrically neutral fermions) or superconductivity (charged fermions): the *explicit* breaking of the global continuous $U(1)$ symmetry of particle-number conservation. Interestingly, a subclass of fermionic (mean-field) superfluids, known as topological superfluids, additionally displays zero-energy quasi-particle modes. These modes are localized on the boundaries of the system or on defects (such as vortices) and have Majorana fermion braiding statistics. At a fundamental level, however, the thermodynamic state of a fermionic superfluid (or superconductor) is characterized by the *spontaneous* breaking of that global $U(1)$ symmetry. Consequences of this mismatch between models and actual systems depend on the particular experimental context and measured physical quantity. Many calculations of thermodynamic and transport properties of fermionic superfluids have firmly established the phenomenological success of particle-non-conserving mean-field theories. But these successes do not imply that every experimentally accessible characteristic of the fermionic superfluid state is well described by breaking particle conservation[10,11]. This issue has recently being acknowledged as pressing[4,12,13] for quantum information processing with Majorana modes[14,15].

2.1. *Fermionic parity switch: A many-body topological invariant*

In light of this discussion, what constitutes a topological superfluid state in closed, hence necessarily number-conserving, physical systems? Reference 4 gave an answer to this question. Before this work, it was not known how to test for topological fermion superfluidity in particle-number conserving systems. The fact that a witness based on detecting switches of fermionic parity works in spite of the conservation of particle number could be firmly established thanks to the realization of an exactly-solvable topological fermion superfluid, the Richardson-Gaudin-Kitaev (RGK) wire (see Sec. 2.2).

In mean-field theories, the emergence of topological order in a superconducting wire, closed in a ring, is associated with switches in the ground-state fermion parity upon increasing the enclosed flux $\Phi = (\phi/2\pi) \times \Phi_0$ ($\Phi_0 = h/2e$ is the superconducting flux quantum)[14,16-21]. Any spin-active superconductor, topologically trivial or not, may experience a crossing of the ground state energies for even and odd number of electrons[22-25]. Regardless of spin, what matters is the number of crossings N_X between $\Phi = 0 = \phi$ and $\Phi = \Phi_0$, $\phi = 2\pi$. The superconductor is topologically non-trivial if N_X is odd, otherwise it is trivial.

In the many-body, number conserving, case we need to identify the relevant parity switches signaling the emergence of a topological fermion superfluid phase. Reference 4 introduced a quantitative criterion for establishing the emergence of topological superfluidity in any, particle-number conserving, attractive or repulsive many-fermion system. The criterion exploits the behavior of the ground state energy of a system of N, $N-1$, and $N+1$ particles, for both periodic and antiperiodic boundary conditions. To identify the fermion parity switches we calculate the ground state energy $\mathcal{E}_0^\phi(N)$ for a given number N of fermions, with periodic ($\phi = 0$) or antiperiodic ($\phi = 2\pi$) boundary conditions, and compare

$$\mathcal{E}_0^{\text{odd}}(\phi) = \tfrac{1}{2}\mathcal{E}_0^\phi(N+1) + \tfrac{1}{2}\mathcal{E}_0^\phi(N-1) \text{ and } \mathcal{E}_0^{\text{even}}(\phi) = \mathcal{E}_0^\phi(N), \tag{1}$$

where we assume N even. The difference $\mathcal{E}_0^{\text{odd}}(\phi) - \mathcal{E}_0^{\text{even}}(\phi)$ (inverse compressibility) determines the fermionic parity switch

$$\mathcal{P}_N(\phi) = \text{sign}\left[\mathcal{E}_0^{\text{odd}}(\phi) - \mathcal{E}_0^{\text{even}}(\phi)\right], \tag{2}$$

that has opposite sign at $\phi = 0$ and $\phi = 2\pi$ in the topologically non-trivial phase. We can now apply these ideas to our exactly-solvable problem, the RGK wire.

2.2. A toy model of a number-conserving topological superfluid

By exploiting the algebraic Bethe ansatz, Ref. 4 succeeded in establishing and characterizing topological fermion superfluidity for a prototypical particle-number conserving, and thus necessarily interacting, superconducting system beyond mean-field theory. The RGK wire of length L, is defined by the Hamiltonian

$$H_{\text{RGK}} = \sum_{k \in \mathcal{S}_k^\phi} \varepsilon_k \, \hat{c}_k^\dagger \hat{c}_k - 8G \sum_{k,k' \in \mathcal{S}_{k+}^\phi} \eta_k \eta_{k'} \hat{c}_k^\dagger \hat{c}_{-k}^\dagger \hat{c}_{-k'} \hat{c}_{k'}. \tag{3}$$

Here, $\hat{c}_k^\dagger$ represent spinless (or fully spin-polarized) fermion creation operators, with momentum k-dependent single-particle spectrum

$$\varepsilon_k = -2t_1 \cos k - 2t_2 \cos 2k, \tag{4}$$

where t_1, t_2 represent the nearest and next-nearest neighbor hopping amplitudes, and $G > 0$ the attractive interaction strength. The interaction strength is modulated by the potential

$$\eta_k = \sin(k/2)\sqrt{t_1 + 4t_2 \cos^2(k/2)} \tag{5}$$

odd in k, $\eta_k = -\eta_{-k}$, as is characteristic of p-wave superconductivity. The pair potential and the single-particle spectrum are connected by the simple relation

$$4\eta_k^2 = \varepsilon_k + 2t_+ \quad (t_+ = t_1 + t_2). \tag{6}$$

In a ring geometry, periodic boundary conditions ($\phi = 0$) correspond to enclosed flux $\Phi = 0$ and antiperiodic boundary conditions ($\phi = 2\pi$) correspond to $\Phi = \Phi_0 = h/2e$. The resulting sets of allowed momenta $\mathcal{S}_k^\phi$ are

$$\mathcal{S}_k^0 = \mathcal{S}_{k+}^0 \oplus \mathcal{S}_{k-}^0 \oplus \{0, -\pi\} \quad \text{and} \quad \mathcal{S}_k^{2\pi} = \mathcal{S}_{k+}^{2\pi} \oplus \mathcal{S}_{k-}^{2\pi}, \tag{7}$$

with

$$\mathcal{S}_{k\pm}^0 = L^{-1}\{\pm 2\pi, \pm 4\pi, \cdots, \pm(\pi L - 2\pi)\} \text{ and}$$
$$\mathcal{S}_{k\pm}^{2\pi} = L^{-1}\{\pm\pi, \pm 3\pi, \cdots, \pm(\pi L - \pi)\}. \tag{8}$$

Eigenstates with exactly $2M + N_\nu$ fermions are given by

$$|\Phi_{M,\nu}\rangle = \prod_{\alpha=1}^{M}\left(\sum_{k \in \mathcal{S}_{k+}^\phi} \frac{\eta_k}{\eta_k^2 - E_\alpha}\hat{c}_k^\dagger\hat{c}_{-k}^\dagger\right)|\nu\rangle, \tag{9}$$

where M is the number of fermion pairs. The state $|\nu\rangle$ with N_ν unpaired fermions satisfies $\hat{c}_{-k}\hat{c}_k|\nu\rangle = 0$ for all k. Moreover, $\frac{1}{2}(\hat{c}_k^\dagger\hat{c}_k + \hat{c}_{-k}^\dagger\hat{c}_{-k} - 1)|\nu\rangle = -s_k|\nu\rangle$, with $s_k = 0$ if the level k is singly-occupied or $s_k = 1/2$ if it is empty. The spectral parameters E_α are determined by the Richardson-Gaudin (Bethe) equations

$$\sum_{k \in \mathcal{S}_{k+}^\phi} \frac{s_k}{\eta_k^2 - E_\alpha} - \sum_{\beta(\neq\alpha)} \frac{1}{E_\beta - E_\alpha} = \frac{Q_\phi}{E_\alpha}, \tag{10}$$

where $Q_\phi = 1/2G - \sum_{k \in \mathcal{S}_{k+}^\phi} s_k + M - 1$. For periodic boundary conditions ($\phi = 0$), the two momenta $k = 0, -\pi$ are inactive, i.e., are not affected by the interactions and must be included separately. Then, all eigenvectors of H_{RGK} are given by

$$|\Psi_N\rangle = |\Phi_{M,\nu}\rangle \otimes |n_0 n_{-\pi}\rangle, \tag{11}$$

where $N = 2M + N_\nu + n_0 + n_{-\pi}$ is the total number of fermions and $n_0, n_{-\pi} \in \{0, 1\}$.

The quantum phase diagram of the RGK chain is determined from the analytical dependence of its ground energy $\mathcal{E}_0^\phi(\rho, g)$ on the density $\rho = N/L$ and scaled coupling strength $g = GL/2$. Depending on the boundary condition ϕ and fermion-number parity, one should consider either $N_\nu = 0$ or 1. The resulting phase diagram is shown in Fig. 1. The RGK chain is gapped for all $g > 0$, except for the Read-Green coupling $g = g_c = 1/(1-2\rho)$ where it becomes critical in the thermodynamic limit, independently of the choice of boundary conditions ϕ. This critical line defines the phase boundary separating weak from strong pairing phases, and thus is a line of non-analyticities. At g_c a cusp develops in the second-order derivative of $e_0 \equiv \lim_{N,L\to\infty} \mathcal{E}_0^\phi/L$, $\rho = $ constant, that leads to singular discontinuous behavior of the third-order derivative. Hence the transition from a weakly-paired to a strongly-paired fermionic superfluid is of third order, just like for the two-dimensional chiral

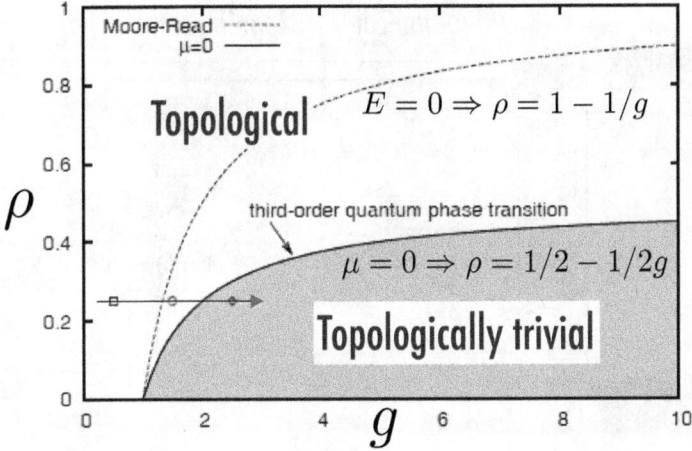

Fig. 1. Quantum phase diagram of the RGK wire in the (ρ, g)-plane. Dashed and full lines represent the Moore-Read ($g^{-1} = 1 - \rho$) and Read-Green ($g^{-1} = 1 - 2\rho$) boundaries, respectively. As shown in the text, the weak-pairing phase is topologically non-trivial while the strong pairing phase is topologically trivial. The horizontal dashed arrow corresponds to the density $\rho = 1/4$ and μ is the chemical potential.

p-wave superconductor[26,27]. Which one of these two superfluid phases, if any, may be properly characterized as topologically non-trivial? We address this question next.

The RGK chain provides analytic access only at $\phi = 0$ and $\phi = 2\pi$, but this is sufficient to determine whether N_X is even or odd. Notice that odd N_X means that the flux Φ should be advanced by $2\Phi_0$, rather than Φ_0, in order to return to the initial ground state. This is the essence of the 4π-periodic Josephson effect[14,28]. The results, shown in Figs. 2, unambiguously demonstrate the topologically non-trivial nature of the superfluid for $g < g_c$ — both in a finite system and in the thermodynamic limit, and without relying on any mean-field approximation.

3. Topological superfluidity with repulsive fermionic atoms

A perfect implementation of a number-conserving system is a cold-atom cloud confined in a magneto-optical trap. Ultracold atomic systems offer a high degree of control and are almost defect-free which makes them pristine testbeds for quantum physics, and convenient quantum simulators. Although, traditionally, superfluidity in cold atoms is achieved via a Feshbach resonance, here we focus on a different class of elements, called alkaline-earth metal atoms (AEAs) that do not support a useful resonant scattering channel, and show that such atoms can support a topological superfluid state without externally-tuned interactions.

Despite all efforts dedicated to the search for topological superfluids in cold-atom systems as well as real materials, they remain elusive with the only confirmed

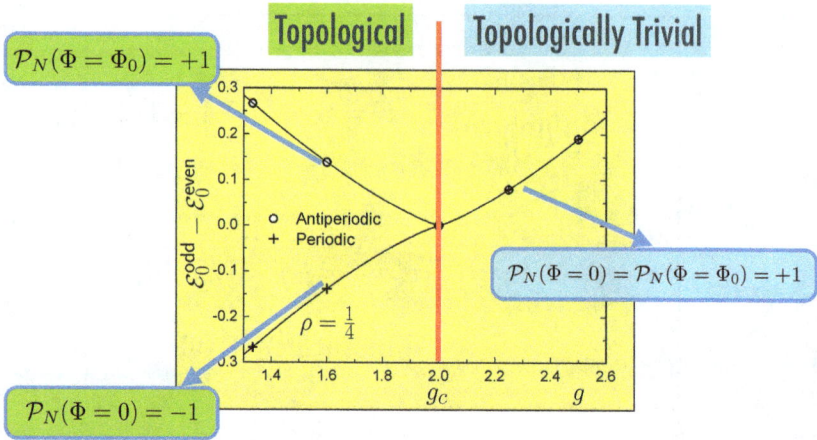

Fig. 2.   Ground state energy differences (in units of $t_1 \equiv 1$, for $t_2 = 0$) for even ($N = 2M$) and odd ($N = 2M \pm 1$) number of fermions, and with periodic ($\phi = 0$) or antiperiodic ($\phi = 2\pi$) boundary conditions. The odd-even difference is shown as a function of the interaction strength g for a finite system (data points, for $N = 512$, $L = 2048$) and in the thermodynamic limit (continuous lines). The topologically non-trivial state is entered for $g < g_c = 2$. Also indicated are the values of the fermion parity switches $\mathcal{P}_N(\Phi)$ across the transition.

realization being liquid ^{3}He[29]. One reason for such scarcity is that topological superfluids require a very particular orbital structure of Cooper pairs, at least p-wave, which originates either from strongly spin-dependent interactions or a large spin-orbit coupling (SOC) that couples particle's motion to its spin. Apparently, the coexistence of strong SOC and attractive interactions (leading to Cooper pairing) is quite unusual in nature[30], fundamentally because SOC and fermion pairing have very different physical origins.

Here we demonstrate how this difficulty can be avoided in a specially-crafted system that uses *the same* ingredients to engineer attractive interactions and effective spin-orbit coupling (SOC). We study a model of repulsive spinless fermions in two bands: one localized and another itinerant, and show that inhomogeneities spanning few lattice sites (e.g. dimerization) in the localized band lead to two profound phenomena. First, they induce *short-range* attractive interactions among the itinerant species, by virtue of local quantum fluctuations. Second, they enlarge the unit cell in accordance with the extent of localized wavefunctions. As a result, itinerant states reside in several Bloch bands, whose index plays the role of a spin degree of freedom. These pseudospins flip whenever a fermion tunnels between unit cells, thus coupling to the orbital motion. The strength of this emergent SOC is determined by the tunneling amplitude and is *of the order of the itinerant bandwidth*. We show that a combination of this *ultra-strong* SOC and attractive interactions gives rise to a robust p-wave topological superfluid in quasi-one dimension (1D) and a chiral $p_x + ip_y$ superfluid in 2D.

Our topological superfluid state can be observed in ultracold *nuclear-spin po-larized* fermionic AEAs[31], e.g. ^{87}Sr[32] or ^{173}Yb[33,34], in an optical superlattice with a few-sites unit cell[35-38]. The localized (itinerant) states can be implemented with atoms in an excited $^3P_0$ (ground state $^1S_0$) clock state (respectively, *e*- and *g*-states), with a *single e-atom per unit cell*. We propose several experimental probes for characterizing the topological superfluids, including momentum-resolved spec-troscopy[39,40] and generation of a particle supercurrent with a laser-induced syn-thetic magnetic field[41-43]. Our approach avoids many known experimental issues: (i) the only relevant interactions occur through the a_{eg}^- channel, and therefore the system is not affected by inelastic *e-e* losses[44,45] or strong scattering in the a_{eg}^+ channel[46,47]; (ii) *p*-wave interactions in our case emerge as a result of quantum fluc-tuations as opposed to a *p*-wave Feshbach resonance, and our setup is free from the three-body losses reported in experiments[48-51]; (iii) the SOC in our system is gener-ated as a result of the lattice structure and hence avoids heating, inherent to earlier proposals to create SOC using near-resonant Raman lasers[52-55]. Our proposal is much simpler than previous works that involve more complicated laser arrays, RF pulses or additional molecular states[56-61]. Finally, our cold-atom system provides a long-sought-after realization of a pairing mechanism in repulsive fermions, which emerges because of nanoscale inhomogeneities[62-65].

3.1. *Topological p-wave superfluidity in a 1D superlattice*

Key aspects of the emergent Cooper pairing and SOC leading to our proposed topological superfluid state can be seen by studying a dimerized 1D optical lattice, shown in Fig. 3(a) and described by the model Hamiltonian:

$$\hat{H} = -J_e \sum_i (\hat{e}_{i1}^\dagger \hat{e}_{i2} + \text{h.c.}) - J_g \sum_i (\hat{g}_{i1}^\dagger \hat{g}_{i2} + \hat{g}_{i+1,1}^\dagger \hat{g}_{i2} + \text{h.c.})$$
$$+ U_{eg}^- \sum_{ia} \hat{n}_{ia}^e \hat{n}_{ia}^g, \tag{12}$$

where $i = x_i = 0, \ldots, N_d - 1$ and $a = 1, 2$ labels dimers and sites within a dimer, respectively. The operator $\hat{e}_{ia}^\dagger$ ($\hat{g}_{ia}^\dagger$) creates a *nuclear-spin polarized e* (*g*) atom at site a within a dimer i ($\hat{n}_{ia}^e = \hat{e}_{ia}^\dagger \hat{e}_{ia}$ and similarly for $\hat{n}_{ia}^g$). The *e*-atoms occupy a dimerized lattice with a large intra-dimer hopping $J_e > 0$ and one atom per dimer (we assume that dimers are decoupled). The *g*-atoms propagate in a simple (non-dimerized) lattice with a nearest-neighbor tunneling J_g. The second term in (12) contains a local *e-g* repulsion of strength $U_{eg}^- > 0$.

We focus on the regime $J_e \gg U_{eg}^-$ and J_g when interactions and *g*-atom kinetic energy in (12) can be considered a perturbation to the *e*-atom kinetic energy. For the *i*-th dimer, the latter has eigenstates $|\lambda\rangle_i = \frac{1}{\sqrt{2}}(\hat{e}_{i1}^\dagger + \lambda \hat{e}_{i2}^\dagger)|\text{vac}\rangle$ ($\lambda = \pm 1$ and $|\text{vac}\rangle$ is the vacuum state without atoms) with energies $-\lambda J_e$ [Fig. 3(b)]. States of the entire *e-g* system can be approximately written as $|\Psi_{eg}\rangle = \prod_i |\lambda\rangle_i \otimes |\Psi_g\rangle$ ($|\Psi_g\rangle$ is a state of only *g*-atoms), thanks to the single-dimer gap $2J_e$.

We next assume that the *e*-subsystem is prepared in the excited state $\prod_i |\lambda = -1\rangle_i$. This configuration is stable because of the large energy penalty $2J_e$ that

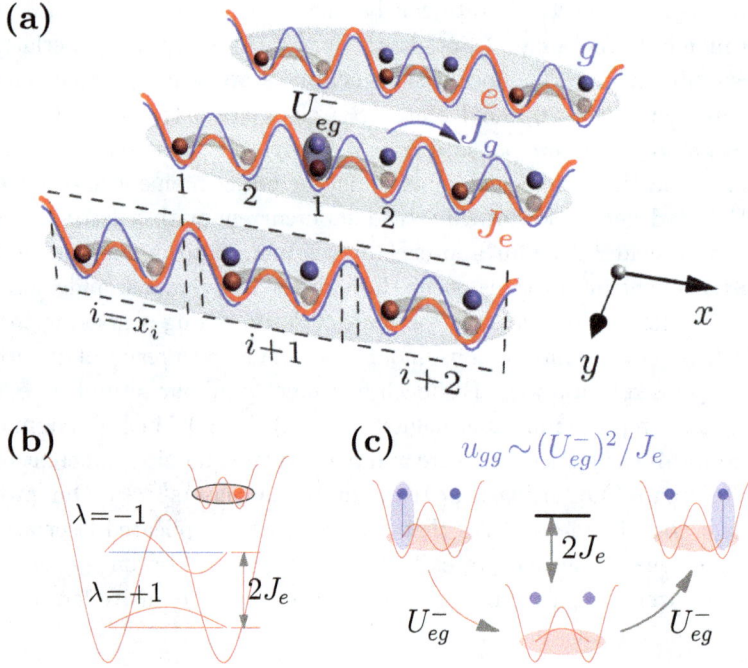

Fig. 3. **(a)** The system described by Eq. (12) can be implemented by tightly confining in an array of 1D tubes an ultra-cold gas of nuclear-spin polarized fermionic alkaline-earth atoms prepared in the clock states g (in blue) and e (red color). Along the tubes, e-atoms experience a superlattice that consists of weakly-coupled double-wells (dimers) with large intra-dimer tunneling J_e. The g-atoms are itinerant and experience a weaker lattice potential along the tube direction with a nearest-neighbor hopping J_g. Within each tube, a unit cell (dashed rectangle) at a position $x_i = i$ includes two lattice sites labeled with $a = 1, 2$ (4 wells overall). There is a e-g repulsive interaction $U_{eg}^- > 0$, assumed small compared to J_e: $U_{eg}^- \ll J_e$. **(b)** Symmetric ($\lambda = +$) and antisymmetric ($\lambda = -$) e-atom kinetic-energy eigenstates within a dimer. **(c)** When e-atom dimers are prepared in the anti-symmetric mode, virtual transitions to the symmetric state, caused by the e-g interaction, induce an effective attraction $u_{gg} = (U_{eg}^-)^2/4J_e$ between two g-fermions within a dimer. These processes are captured by the effective model (13).

suppresses decay of individual dimers to their ground state with $\lambda = +1$ in the absence of decoherence sources (this requirement is well satisfied in cold-atom systems), for instance due to e-g scattering. The weak interactions U_{eg}^- only induce e-atom virtual transitions to dimer states with $\lambda = +1$, which we take into account via 2nd order perturbation theory (the kinetic energy of g-atoms amounts to a 1st order correction because it operates within the degenerate subspace $\{|\Psi_{eg}\rangle\}$). These virtual processes, shown in Fig. 3(c), give rise to an effective Hamiltonian for the g-subsystem

$$\hat{H}_{\text{ef}} = \hat{H}_0^g - u_{gg} \sum_i \hat{n}_{i1}^g \hat{n}_{i2}^g, \tag{13}$$

$$\hat{H}_0^g = -J_g \sum_k [\sigma^x(1 + \cos k) + \sigma^y \sin k]_{ab} \, \hat{g}_{ka}^\dagger \hat{g}_{kb},$$

where $\hat{g}_{ka} = \frac{1}{\sqrt{N_d}} \sum_i e^{-ikx_i} \hat{g}_{ia}$, $\boldsymbol{\sigma}$ are Pauli matrices, momentum $k \in [-\pi, \pi]$ (in units of inverse lattice spacing $1/a_0$) is defined in a dimer Brillouin zone (BZ) with N_d states. $u_{gg} = (U_{eg}^-)^2/4J_e$ is the strength of intra-dimer g-atom attraction mediated by quantum fluctuations of localized e-atoms. If we associate the site index $a = 1$, 2 inside a dimer with a spin-$\frac{1}{2}$ degree of freedom, $\hat{H}_0^g$ contains kinetic energy with a SOC that arises because any tunneling event "flips" pseudospin a. This effective model will give rise to a topological superfluid phase.

The physical origin of the p-wave topological superfluid phase is especially transparent at weak coupling $u_{gg} \ll J_g$ and low filling $n^g \ll 1$, when kinetic energy in (13) dominates and is diagonalized by states $\hat{f}_{k\tau} = \frac{1}{\sqrt{2}}(\hat{g}_{k1} - \tau e^{-i\frac{k}{2}}\hat{g}_{k2})$ with energies $\epsilon_k = 2\tau J_g \cos\frac{k}{2}$ ($\tau = \pm 1$). Because relevant momenta are small, $|k| \ll \pi$, it is allowed to keep only the $\hat{f}_{k,-1} \equiv \hat{f}_k$ mode. As a result, interactions in (13) become manifestly p-wave:

$$\hat{H}_{\text{ef}} \approx \sum_k \epsilon_k \hat{f}_k^\dagger \hat{f}_k - \frac{u_{gg}}{4N_d} \sum_{k'kq} e^{i\frac{q}{2}} \hat{f}_{k+q}^\dagger \hat{f}_{k'-q}^\dagger \hat{f}_{k'} \hat{f}_k. \tag{14}$$

Within the Bogoliubov mean-field theory one introduces a pairing order parameter $\Delta = -\frac{u_{gg}}{4N_d} \sum_q e^{-i\frac{q}{2}} \langle \hat{f}_{-q}\hat{f}_q \rangle$ which parameterizes the single-particle excitation spectrum $E_k = \sqrt{(\epsilon_k - \mu)^2 + |D_k|^2}$ with a p-wave gap $D_k \approx ik\Delta$. μ is the g-atom chemical potential (set by the Fermi energy). Δ is a solution of the self-consistency equation $1 = \frac{u_{gg}}{16N_d} \sum_k \frac{k^2}{E_k}$, $\Delta \approx J_g e^{-2\pi J_g/u_{gg}} \sqrt{1-(\mu/2J_g)^2}$.

3.2. Stability and topological nature of the superfluid state

To assess the stability of the p-wave superfluid state beyond the weak coupling limit, and uncover its topological properties, we compute the phase diagram of $\hat{H}_{\text{ef}}$ within a fully unconstrained Hartree-Fock-Bogoliubov mean-field approach in real space. This variational technique minimizes the grand potential $\langle \hat{H}_{\text{ef}} - \mu \sum_{ia} \hat{g}_{ia}^\dagger \hat{g}_{ia} \rangle$ (μ is the g-atom chemical potential) w.r.t. local order parameters $\Delta_i = -u_{gg}\langle \hat{g}_{i2}\hat{g}_{i1} \rangle$, $\xi_i = \langle \hat{g}_{i2}^\dagger \hat{g}_{i1} \rangle$ and $n_{ia}^g = \langle \hat{g}_{ia}^\dagger \hat{g}_{ia} \rangle$, and includes the competition between superfluid phases with a finite gap Δ_i and various inhomogeneous states, e.g. charge-density waves (CDWs), characterized by a site-dependent n_{ia}^g, and magnetic phases signaled by ξ_i. The minimization is performed at zero temperature $T = 0$ in a system with periodic boundary conditions. Once the ground state is self-consistently determined, we open the chain and diagonalize the Bogoliubov-de Gennes (BdG) mean-field Hamiltonian *with fixed OPs* to determine edge modes: If a superfluid phase displays zero-energy (Majorana) modes, we call it topological[2].

Figure 4(a) shows the phase diagram of model (13) as a function of chemical potential μ and interaction u_{gg}. In agreement with the previous section, the superfluid phase is stable at weak coupling $u_{gg} < J_g$ and low density, and is characterized by the mixing of singlet $\langle \hat{g}_{-k,2}\hat{g}_{k1} \rangle$ and triplet $\langle \hat{g}_{-k,a}\hat{g}_{ka} \rangle$ Cooper pair amplitudes. To better understand this effect, let us consider properties of the model (13) under space inversion I: $x \to -x$. In the dimerized lattice, $I = \sigma^x \otimes I_d$, where I_d: $k \to -k$

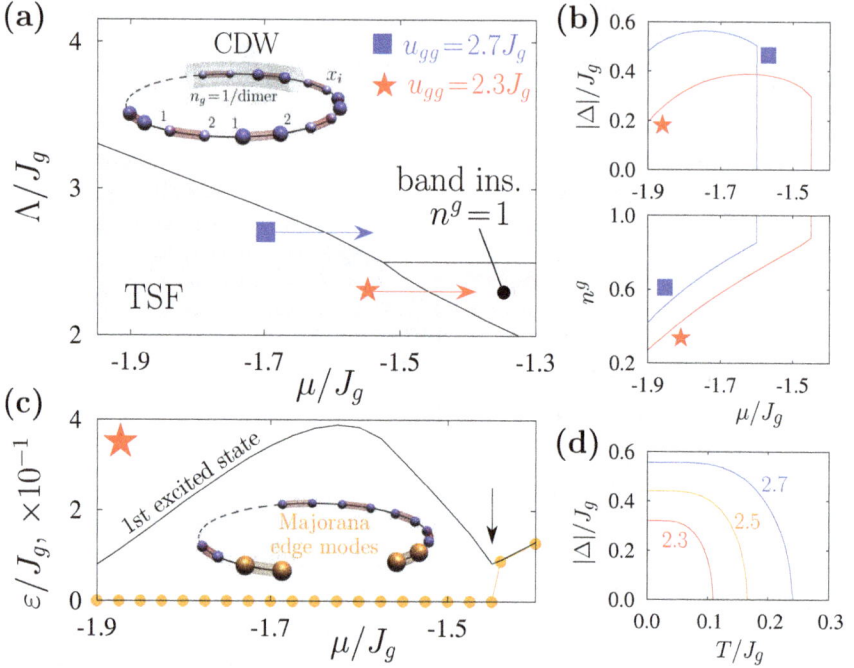

Fig. 4. **(a)** Zero-temperature phase diagram of Eq. (13) computed using an unconstrained Hartree-Fock-Bogoliubov mean-field theory in a system with $N_d = 100$ dimers and periodic boundary conditions. μ is the g-atom chemical potential. Thick [thin] lines indicate 1st order transitions between topological superfluid (TSF) and insulator states [2nd order transition inside the insulating region]. In the charge-density wave (CDW) state the unit cell has two dimers with an average density $n^g = 1$ atom per dimer. For small u_{gg} the CDW undergoes a transition to a band insulator with a single-dimer unit cell. Blue square (red star) corresponds to $u_{gg}/J_g = 2.7$ (2.3). **(b)** Superfluid gap Δ and average density $n^g = \frac{1}{N_d} \sum_i (n_{i1}^g + n_{i2}^g)$ plotted along the arrows shown in (a). **(c)** Two *lowest-magnitude* eigenvalues ε of the BdG Hamiltonian computed in an open chain for $u_{gg} = 2.3 J_g$. The order parameters were taken from a converged solution in (a). Orange circles indicate Majorana edge modes inside the TSF phase. The arrow marks a TSF-band insulator transition. **(d)** Gap Δ as a function of temperature T for $\mu = -1.8 J_g$, and $u_{gg}/J_g = 2.3$, 2.5, 2.7. The Boltzmann constant is $k_B = 1$.

is an inversion acting on the dimer center-of-mass and σ^x appears because I must interchange dimer sites. The SOC $\hat{H}_0^g$ is invariant under I but *manifestly breaks* I_d due to the odd-momentum terms. In a superfluid state, Cooper pair wavefunctions inherit this feature and the system exhibits p-wave pairing between same-flavor g-atoms $\langle \hat{g}_{-k,a} \hat{g}_{ka} \rangle \sim k^n$ (n is odd) despite the s-wave nature of interactions in Eq. (13). This situation is similar to singlet-triplet mixing in non-centrosymmetric superconductors with Rashba SOC[30,41,42].

When u_{gg} or μ is increased, the system undergoes a 1st order transition to a non-superfluid gapped state with an average density $n^g = 1$ [Fig. 4(b)]. This phase is a band insulator for small u_{gg}, and a CDW with two dimers per unit cell for strong interactions. As shown in Fig. 4(c), the superfluid state is *topological* (i.e.

possesses zero-energy edge modes) for all u_{gg} and μ where it is stable. This happens because we used the Hartree-Fock order parameters in our variational scheme: if the minimization were constrained to include only site-independent Δ_i, one would recover a well-known transition[2] from topological superfluid to a non-topological superfluid state. The latter phase is unstable towards CDW formation and the transition never happens.

The phase diagram in Fig. 4(a) remains valid at finite temperature $T > 0$. Indeed, as demonstrated in Fig. 4(d) a typical critical temperature, above which the superfluid phase disappears, is $T_c \sim 0.1 J_{gg}$ (here and below we use the units with Boltzmann constant $k_B = 1$). For $T > T_c$, the system becomes a homogeneous Fermi liquid. Although the presentation above illustrates main ideas behind this emergent phenomenon using mean-field approximations, the topological nature of the superfluid state remains intact beyond mean-field. We confirmed this by performing exact diagonalization in a single tube and verifying that the ground state realizes a fermionic parity switch[4] for all fermion fillings.

3.3. Probing a topological superfluid

The simplest way to validate our theory in cold-atom experiments, is to probe the g-atom attraction via quench dynamics in a *normal* state using the following protocol: *(i)* For times $t < 0$, g-atoms fill a non-interacting Fermi sea. *(ii)* At $t = 0$, J_g is switched off (e.g. by increasing the lattice depth), and g-atoms are brought in contact with e-atoms, thus allowing them to experience the e-g interactions described in Eq. (12). Then, one lets the system evolve for a time t_0. As a result, basis states with doubly occupied dimers accumulate a phase $-Ut_0$, where $U[= -u_{gg}$ in Eq. (13)] is the induced g-atom interaction. *(iii)* At $t = t_0$, the e-atoms are removed, hopping J_g is restored and the system evolves with a non-interacting Hamiltonian [1st term in (13)]. The sign of U can be determined by measuring an average number of doubly occupied dimers:

$$n_2(t) = \langle \psi(t)| \sum_i \hat{n}^g_{i1} \hat{n}^g_{i2} |\psi(t)\rangle \tag{15}$$

in the state $|\psi(t)\rangle$ of the system at time t. For short evolution times, when $|U|t_0 \ll 1$ and $t - t_0 \ll 1/J_g$, $n_2(t) = n_2(0) - \zeta U(t - t_0)$ with $\zeta > 0$. Hence, the average double occupancy decreases (increases) for the repulsive (attractive) interaction U.

The spectrum of Bogoliubov excitations can be probed by momentum-resolved spectroscopy[39,40]. Let us assume that one tube in Fig. 3(a) contains no atoms and is detuned relative to its neighboring tubes. A laser with a wavevector along the x-axis, Rabi frequency Ω, detuned by δ from the atomic e-g transition, transfers g-atoms from the superfluid phase to e-states in the empty tube (the transfer happens along the y-axis and does not change an atom's x-position along the tube). The e-lattice depth in that tube has been reduced to make flat bands at $\pm J_e$ [see Fig. 3(b)] dispersive with a band structure $\epsilon^e_{k\tau} = \tau J_e \sqrt{1 + \eta^2 + 2\eta \cos k}$ where $\tau = \pm 1$ and

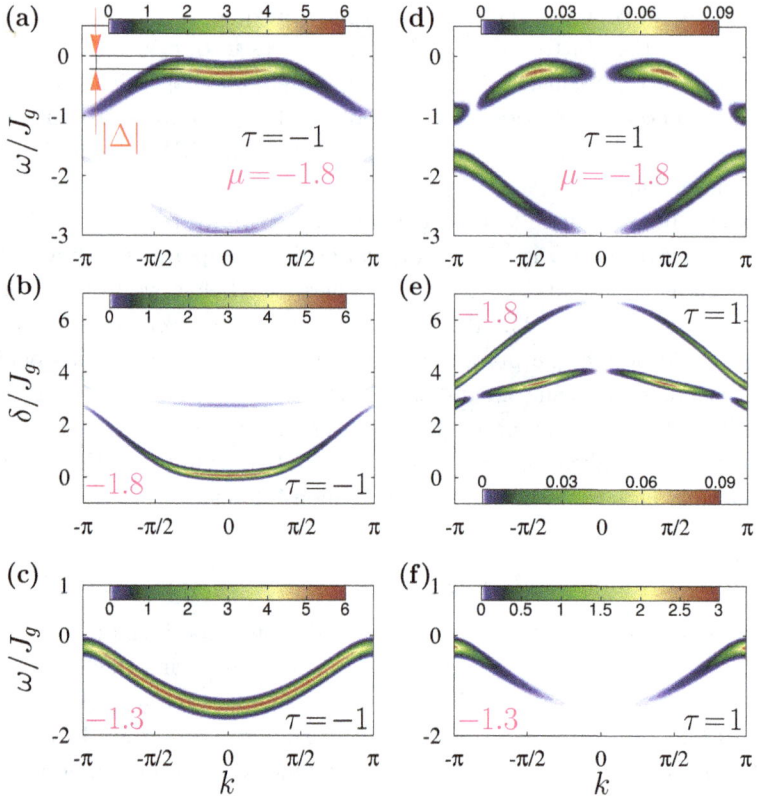

Fig. 5.   The transfer rate $\mathcal{R}_\tau(\delta, k)$ (in arbitrary units) of g-atoms from the topological superfluid with $\mu = -1.8 J_g$, and from the insulator with $\mu = -1.3 J_g$ [see Fig. 4(a)]. These values of μ are indicated by numbers in magenta. The interaction strength is $u_{gg} = 2.3 J_g$. For concreteness, in the empty tube we assume $J_e = J_g$ and $\eta = 1$, so that the e-atom band is $\epsilon^e_{k\tau} = 2\tau J_g \cos \frac{k}{2}$. The left (right) column corresponds to $\tau = \mp 1$. (a) and (d) $\mathcal{R}_\tau$ as a function of the shifted frequency $\omega = \epsilon^e_{k\tau} - \mu - \delta$, which reveals the BdG band structure and allows us to extract the superfluid gap Δ. (b) and (e) The signal in the detuning-momentum plane, as it would be measured in a real experiment. (c) and (f) Same as in panels (a) and (d), but inside the band insulator with a fully filled *single-particle* band $\epsilon_{k,-1}$. A weaker signal for $\tau = 1$ is due to Hartree-Fock corrections.

ηJ_e is the inter-dimer hopping. The g-atom transfer rate to an e-band $\epsilon^e_{k\tau}$, $\mathcal{R}_\tau(\delta, k)$, can be written in terms of the spectral density $\mathcal{A}_{ab}(\omega, k)$ of the single-particle normal Green function[40]:

$$\mathcal{R}_\tau(\delta, k) = \frac{\Omega^2}{2}\left[\mathcal{A}_{11}(\omega, k) + \mathcal{A}_{22}(\omega, k) - 2\tau \frac{\mathrm{Re}\left[(1 + \eta e^{ik})\mathcal{A}_{12}(\omega, k)\right]}{\sqrt{1 + \eta^2 + 2\eta \cos k}}\right], \qquad (16)$$

with $\omega = \epsilon^e_{k\tau} - \mu - \delta$,

$$\mathcal{A}_{ab}(\omega, k) = \mathrm{i}\, f(\omega)[\langle \hat{g}_{ka}\hat{g}^\dagger_{kb}\rangle_{\omega + \mathrm{i}0} - \langle \hat{g}_{ka}\hat{g}^\dagger_{kb}\rangle_{\omega - \mathrm{i}0}], \qquad (17)$$

$f(\omega) = (e^{\omega/T} + 1)^{-1}$ is the Fermi function at temperature T, and $\mathrm{i}0$ is an infinitesimal imaginary number.

The representative signal $\mathcal{R}$ is shown in Fig. 5. Its maximum (for a fixed k) occurs when ω coincides with the highest occupied BdG energy state. Therefore, one can map out the BdG band structure and extract the excitation gap Δ [see panels (a) and (d)]. In Fig. 5(b) and (e) we plot the same signal as a function of the bare laser detuning δ, as it would be observed in an experiment. This spectroscopy technique can also be used to probe the insulating phase in Fig. 3(a). The transfer rate inside the band insulator regime is shown in Fig. 5(c) and (f). In this case, the excitation spectrum is again gapped, but as opposed to the superfluid state, this gap exists because all *single-particle* states below the Fermi level are filled, and not due to fermion pairing. Finally, we note that the signal with $\tau = -1$ [panels (a) – (c)] is significantly stronger than the one with $\tau = 1$ [(d) – (f)], which highlights the validity of the effective model (14).

Due to the SOC inherent in Eq. (13), the topological superfluid state has remarkable features that set it apart from a usual s-wave superfluid and can be used as its "fingerprint". Perhaps its most revealing property is an analog of the spin-galvanic effect, when an applied Zeeman magnetic field induces a bulk supercurrent[42,43]. Since the pseudospin degrees of freedom in $\hat{H}_{\text{ef}}$ correspond to a site index inside the unit cell, this "field" must couple to the motion of g-atoms and can be implemented as a laser-assisted tunneling within dimers[52,53]. This hopping can be set to have an arbitrary phase θ, but we focus on the case $\theta = \frac{\pi}{2}$, i.e. consider a perturbation:

$$\delta\hat{H}_{\text{ef}} = -b_y \sum_{i,\,ab} \sigma^y_{ab}\, \hat{g}^\dagger_{ia}\hat{g}_{ib} \tag{18}$$

and compute the supercurrent response $\langle \hat{K} \rangle = \kappa\, b_y$ [Fig. 6(a)]. We can anticipate this response based on pure symmetry arguments: as explained previously, the SOC, being odd in momentum, breaks the space inversion symmetry in the dimer lattice. On the other hand, the synthetic Zeeman term $\delta\hat{H}_{\text{ef}}$ with $b_y \neq 0$ violates time-reversal symmetry. Breaking of these two symmetries is a necessary condition to stabilize a state with a non-zero current in the system.

The operator $\hat{K}$ is a single-particle mass current, obtained by varying the Hamiltonian $\hat{H}_b = \hat{H}^g_0 + \delta\hat{H}_{\text{ef}}$ [cf Eq. (13)] w. r. t. the flux φ piercing the ring: $\hat{K} = \delta\hat{H}_b/\delta\varphi\big|_{\varphi=0}$. This flux enters via a phase factor $e^{i\varphi}$ on each physical link in the lattice with one-site unit cell, but in the dimerized lattice one must replace $\hat{g}_{i1} \to e^{i2x_i\varphi}\hat{g}_{i1}$ and $\hat{g}_{i2} \to e^{i(2x_i+1)\varphi}\hat{g}_{i2}$, because the position x in a non-dimerized lattice is $x = 2x_i + a - 1 = 0, \ldots, 2N_d - 1$ and $\hat{g}_x = \hat{g}_{ia}$. We obtain

$$\hat{K} = \frac{1}{N_d} \sum_k \left[\sigma^y(1 - \cos k) + \sigma^x\left(\sin k - b_y/J_g\right)\right]_{ab} \hat{g}^\dagger_{ka}\hat{g}_{kb}. \tag{19}$$

In Fig. 6(b) we show the magneto-electric coefficient κ as a function of the chemical potential for several interaction strengths. Remarkably, κ exists *only inside* the superfluid phase and vanishes across the topological superfluid-insulator transition.

Due to the p-wave nature of the superfluid phase, the synthetic field b_y induces a pseudospin polarization $\langle \hat{S}_y \rangle = \chi b_y$, where $\hat{S} = \frac{1}{2N_d}\sum_i \boldsymbol{\sigma}_{ab}\hat{g}^\dagger_{ia}\hat{g}_{ib}$. The suscep-

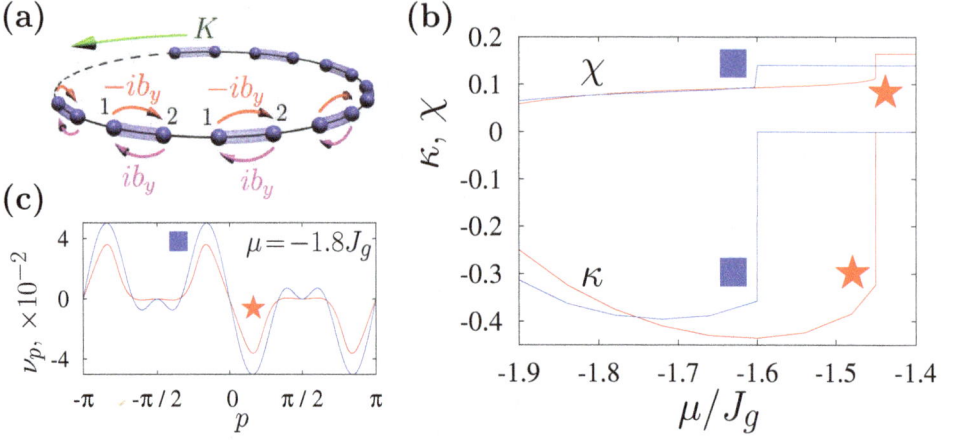

(a)

(b)

(c)

Fig. 6.   **(a)** Model (13) with a laser-assisted intra-dimer tunneling $\pm i\, b_y$ simulating a magnetic field $\boldsymbol{B} = b_y \boldsymbol{e}_y$. A supercurrent $\langle \hat{K} \rangle \sim b_y$ is induced in the topological superfluid phase. **(b)** Magneto-electric response $\kappa = \langle \hat{K} \rangle / b_y$ and longitudinal susceptibility $\chi = \langle \hat{S}^y \rangle / b_y$ as functions of the chemical potential μ. **(c)** Momentum distribution asymmetry ν_p inside the topological superfluid phase for $b_y = 0.3 J_g$ and $\mu = -1.8 J_g$. In (b) and (c), a red star (blue square) corresponds to $u_{gg}/J_g = 2.3$ (2.7) [cf Fig. 4].

tibility χ and magneto-electric coefficient κ can be related in the weak-coupling dilute limit $u_{gg} \ll J_g$, $n^g \ll 1$ at $T = 0$[41]. Indeed, similar calculations that led to Eq. (14) yield $\kappa = -4\chi = -\frac{1}{2 J_g N_d} \sum_k \left[1 - \frac{\epsilon_k - \mu}{E_k} \right] \cos \frac{k}{2}$ with ϵ_k and E_k defined after Eq. (14). Figure 6(b) shows χ and κ as functions of the chemical potential μ across the topological superfluid-insulator transition. In cold-atom experiments it is possible to measure χ, for example by a Ramsey-type protocol[66]: Assuming that the system is in its ground state $|\psi_0\rangle$ (with $b_y \neq 0$), at $t = 0$ we quench the Hamiltonian from (13) to $\hat{H}_x = J_g \hat{S}_x$ e.g. by making the intra-dimer g-atom tunneling dominant and let the system evolve for a time $t_0 = \frac{\pi}{2 J_g}$. As a result, the state becomes $|\psi\rangle = e^{-i\frac{\pi}{2}\hat{S}_x}|\psi_0\rangle$. Now we measure the difference in populations on two sites of a dimer, i.e. $\langle \psi | \hat{S}_z | \psi \rangle$. Because $e^{i\frac{\pi}{2}\hat{S}_x} \hat{S}_z e^{-i\frac{\pi}{2}\hat{S}_x} = \hat{S}_y$, the above protocol yields $\langle \psi_0 | \hat{S}_y | \psi_0 \rangle$ and can be used to obtain χ.

   Another physical effect induced by $\delta \hat{H}_{\text{ef}}$ is an asymmetry of the g-atom momentum distribution which can be detected in time-of-flight experiments[67]. Because these measurements involve crystal momentum p in the BZ of a *single-site* unit cell, we need to compute

$$\nu_p = \frac{J_g}{b_y} \langle \hat{g}_p^\dagger \hat{g}_p - \hat{g}_{-p}^\dagger \hat{g}_{-p} \rangle \tag{20}$$

with $\hat{g}_p \equiv \frac{1}{\sqrt{2 N_d}} \sum_x e^{-ipx} \hat{g}_x = \frac{1}{\sqrt{2 N_d}} \sum_{ia} e^{-2ipx_i - ip(a-1)} \hat{g}_{ia} = \frac{1}{\sqrt{2}} \left(g_{2p,1} + e^{-ip} g_{2p,2} \right)$. Figure 6(c) shows ν_p computed in the topological superfluid phase of Fig. 4(a). For comparison, in a non-superfluid system, $\nu_p \sim \delta_{k,k_F} - \delta_{k,-k_F}$ (k_F is the Fermi momentum). Superfluid correlations destroy Fermi points and lead to a finite asymmetry even away from $k = \pm k_F$.

3.4. *Preparation of the ^{87}Sr lattice clock*

The system in Fig. 3(a) and Eq. (12) can be realized with AEAs, such as fermionic ^{87}Sr, using the sequence shown in Fig. 7: **Step (0)** We start with a nuclear-spin polarized g-atom band insulator in a deep magic-wave lattice (in which e and g atoms experience equal light shifts and therefore same trapping potential[68]) with suppressed tunneling. **Step (1)** The system is irradiated by a laser that adiabatically applies a staggered synthetic gauge field[69,70] with wavevector $k = \pi$, Rabi frequency Ω and a detuning from the e-g transition δ. In the e-g basis, the single-atom Hamiltonian at the lattice site j is $\hat{H}_j = \frac{1}{2}[(-1)^j \Omega \sigma^x - \delta \sigma^z]$. Here the Pauli matrices $\boldsymbol{\sigma}$ act on the local g-e basis. As the detuning is decreased to zero, the Rabi frequency is simultaneously ramped up, thus adiabatically preparing atoms in their local ground state $\frac{1}{\sqrt{2}}|e \pm g\rangle_j$. **Step (2)** The laser wavevector is quenched from $k = \pi$ to 2π, making half of the local e-g mixtures excited states. **Step (3)** δ is adiabatically increased, while Ω is decreased. As a result, excited (ground state) coherent e-g superpositions are transferred to e (g) states. **Step (4)** A laser is used to directly transfer ground state e-atoms [$|e\,0\rangle$] to g-atoms [$|g\,0\rangle$] and $|g\,0\rangle$ to excited e-atoms [$|e\,1\rangle$], where $|e(g)\,n\rangle$ indicates an e (g) atom in n-th lattice band. **Step (5)** We adiabatically enable hoppings by decreasing the magic-wavelength lattice depth. Simultaneously, we create a dimerized e-superlattice by ramping up a potential experienced only by e-atoms[71] at twice the periodicity of the magic lattice, and transfer the states $|e\,1\rangle$ to excited antisymmetric motional states in each double-well, in order to satisfy the requirement $|J_e| > |J_g|$.

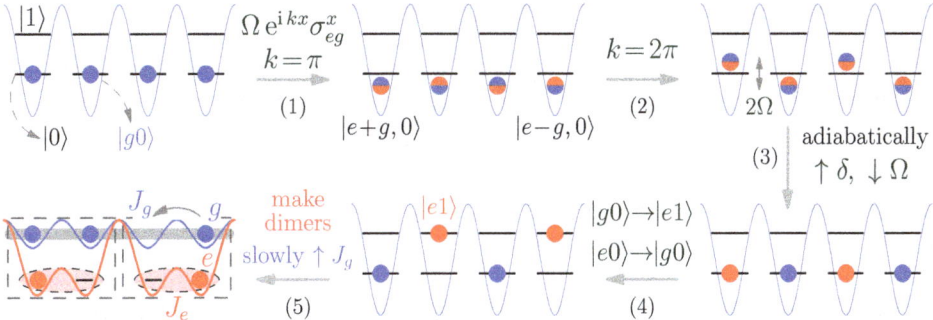

Fig. 7. A protocol to prepare the model in Fig. 3(a) and Eq. (12). Blue (red) color indicates g (e) atoms. Atoms are nuclear-spin polarized and hoppings are quenched until step (5). $|e(g)\,n\rangle$ means a state with one e (g) atom in the n-th spatial level ($n = 0$ means ground state). At steps (2) and (3), red-blue (blue-red) circles [from bottom to top] correspond to $|e \pm g\rangle$ states detuned from the lowest-band energy by $\pm \Omega$, respectively.

After the last step, g-atoms form a band-insulator with $n^g = 1$ (two atoms per dimer). Their filling can be controlled spectroscopically by removing atoms from k-states near band edges with a laser which drives a narrow transition whose detuning is adiabatically changed to scan the conduction band and access atoms deeper in the Fermi sea. This can be visualized as an adiabatic injection of holes in the presence of e-atom background. The newly added holes form Cooper pairs, thus building up a superfluid state.

4. Conclusions and outlook

Topological superfluidity in Fermi liquids is intimately related to the coupling between particles' spin and orbital motion. Unfortunately, in most systems this crucial ingredient is absent or too weak to yield a measurable topological structure of superfluid phases. In the present work we discussed a mechanism that bypasses this "rule" and allows a *coexistence of a strong spin-orbit coupling and pairing correlations*. The key ingredient in our theory is the lattice modulations that host localized degrees of freedom and play a dual role. On the one hand, quantum fluctuations of localized fermions stabilize superfluid states in the itinerant channel, *even when the bare interactions are repulsive*. On the other hand, the modulations enlarge the lattice unit cell and lead to an emergent *odd in momentum* spin-orbit coupling in the conduction band. A combination of these effects always results in a topologically non-trivial superfluid state in a number-conserving system with potential emergence of Majorana modes. We illustrated the above mechanism by studying a system of spinless fermions in a quasi-1D lattice with a dimerized structure and showed how one can observe this physics in a quantum simulator with alkaline-earth atoms with a variety of probes, including momentum-resolved spectroscopy and an analog of the spin-galvanic effect, i.e. a magneto-electric phenomenon that can be used to detect a superfluid phase with broken inversion symmetry.

The analysis presented above can be easily extended beyond one spatial dimension. In particular, in a 2D system where g-atoms propagate in a square lattice, and e-atoms are localized inside square plaquettes, similar arguments show that quantum fluctuations of the e-atoms stabilize a $p_x + i p_y$ superfluid state of the g-species. One can use fermion parity switches to establish the topological character of this superfluid phase[4,5]. As in the 1D case, our criterion combines the many-body ground-state energies of the system with N, $N-1$ and $N+1$ fermions, computed for both periodic and anti-periodic boundary conditions, into a topological invariant with the interpretation of an inverse compressibility that keeps track of switches in fermionic parity as the phase ϕ specifying twisted boundary conditions evolves from $\phi = 0$ (periodic boundary conditions) to $\phi = 4\pi$, passing through $\phi = 2\pi$ (antiperiodic boundary conditions). The behavior of this inverse compressibility in the thermodynamic limit dictates the topologically trivial or non-trivial character of the superfluid phase in question.

Concern with the role of particle (non-)conservation in mean-field theories of fermion superfluidity is as old as the theory itself[72,73]. We feel prompted to revisit this issue by recent *experimental* efforts to detect and control Majorana modes. Their presence is considered a key manifestation of topological fermion superfluidity. These quasi-particles at zero-energy emerge from the interplay between the existence of a topologically non-trivial vacuum and a, typically, symmetry-protected physical boundary (or defect). In recent literature, this connection goes under the name of bulk-boundary correspondence. Because of the expected resilience against decoherence and non-Abelian braiding properties, Majorana modes, or simply Majorana fermions, are key components of many blueprints of quantum-information processing devices. Given that either electrons or fermionic atoms are in fact locally conserved, it is imperative to investigate the conditions for emergence of Majorana fermions, and to establish procedures for their experimental detection beyond mean-field theory. Our previous work[4,5] suggests the idea that quantum-controlling Majorana fermions is so deeply rooted in the mean-field picture that it might not have a natural counterpart in more realistic particle-number conserving frameworks. Briefly stated, if a zero-energy mode of a superfluid system creates a superposition of states that differ in particle number/electric charge, it may not be possible to manipulate this mode without exchanging particles with an environment, as opposed to exchanging, say, energy only. Then, whether it is possible in practice to exchange *coherently* (charged) particles with a reservoir big enough to grant the mean-field picture of the (sub)system of interest becomes a big open question.

References

1. L. Landau, E. Lifshitz and L. Pitaevskii, *Statistical Physics*, no. pt. 2 (Pergamon Press, 1980).
2. X.-L. Qi and S.-C. Zhang, Topological insulators and superconductors, *Rev. Mod. Phys.* **83**, 1057 (2011).
3. C. Nayak, S. H. Simon, A. Stern, M. Freedman and S. Das Sarma, Non-abelian anyons and topological quantum computation, *Rev. Mod. Phys.* **80**, 1083 (2008).
4. G. Ortiz, J. Dukelsky, E. Cobanera, C. Esebbag and C. Beenakker, Many-body characterization of particle-conserving topological superfluids, *Phys. Rev. Lett.* **113**, 267002 (2014).
5. G. Ortiz and E. Cobanera, What is a particle-conserving topological superfluid? the fate of majorana modes beyond mean-field theory, *Annals of Physics* **372**, 357 (2016).
6. B. Zeng, X. Chen, D.-L. Zhou and X.-G. Wen, *arXiv:1508.02595*.
7. Z. Nussinov and G. Ortiz, Sufficient symmetry conditions for topological quantum order, *Proceedings of the National Academy of Sciences* **106**, 16944 (2009).
8. L. Isaev, A. Kaufman, G. Ortiz and A. M. Rey, *arXiv:1710.02768*.
9. J. Blaizot and G. Ripka, *Quantum Theory of Finite Systems* (Cambridge, MA, 1986).
10. J. Von Delft and D. Ralph, Spectroscopy of discrete energy levels in ultrasmall metallic grains, *Physics Report* **345**, 61 (2001).
11. G. Ortiz and J. Dukelsky, Bcs-to-bec crossover from the exact bcs solution, *Phys. Rev. A* **72**, 043611 (2005).
12. F. Iemini, L. Mazza, D. Rossini, R. Fazio and S. Diehl, Localized majorana-like modes

in a number-conserving setting: An exactly solvable model, *Phys. Rev. Lett.* **115**, 156402 (2015).

13. N. Lang and H. P. Büchler, Topological states in a microscopic model of interacting fermions, *Phys. Rev. B* **92**, 041118 (2015).

14. A. Kitaev, Unpaired majorana fermions in quantum wires, *Phys. Usp.* **44**, 131 (2001).

15. C. Beenakker, Search for majorana fermions in superconductors, *Annual Review of Condensed Matter Physics* **4**, 113 (2013).

16. A. Keselman, L. Fu, A. Stern and E. Berg, Inducing time-reversal-invariant topological superconductivity and fermion parity pumping in quantum wires, *Phys. Rev. Lett.* **111**, 116402 (2013).

17. C. W. J. Beenakker, J. M. Edge, J. P. Dahlhaus, D. I. Pikulin, S. Mi and M. Wimmer, Wigner-poisson statistics of topological transitions in a josephson junction, *Phys. Rev. Lett.* **111**, 037001 (2013).

18. J. D. Sau and E. Demler, Bound states at impurities as a probe of topological superconductivity in nanowires, *Phys. Rev. B* **88**, 205402 (2013).

19. A. Haim, A. Keselman, E. Berg and Y. Oreg, Time-reversal-invariant topological superconductivity induced by repulsive interactions in quantum wires, *Phys. Rev. B* **89**, 220504 (2014).

20. F. m. c. Crépin and B. Trauzettel, Parity measurement in topological josephson junctions, *Phys. Rev. Lett.* **112**, 077002 (2014).

21. C. W. J. Beenakker, *arXiv:1407.2131*.

22. A. Sakurai, Comments on superconductors with magnetic impurities, *Progress of Theoretical Physics* **44**, 1472 (1970).

23. A. V. Balatsky, I. Vekhter and J.-X. Zhu, Impurity-induced states in conventional and unconventional superconductors, *Rev. Mod. Phys.* **78**, 373 (2006).

24. W. Chang, V. E. Manucharyan, T. S. Jespersen, J. Nygård and C. M. Marcus, Tunneling spectroscopy of quasiparticle bound states in a spinful josephson junction, *Phys. Rev. Lett.* **110**, 217005 (2013).

25. E. Lee, L. Ma, D. Nath, C. Lee, Y. Wu and S. Rajan, Demonstration of 2d/3d p-mos2/n-sic junction *Device Research Conference - Conference Digest, DRC* 2014.

26. S. M. A. Rombouts, J. Dukelsky and G. Ortiz, Quantum phase diagram of the integrable $p_x + ip_y$ fermionic superfluid, *Phys. Rev. B* **82**, 224510 (2010).

27. S. Lerma H., S. M. A. Rombouts, J. Dukelsky and G. Ortiz, Integrable two-channel $p_x + ip_y$-wave model of a superfluid, *Phys. Rev. B* **84**, 100503 (2011).

28. K. Sengupta, I. Žutić, H.-J. Kwon, V. M. Yakovenko and S. Das Sarma, Midgap edge states and pairing symmetry of quasi-one-dimensional organic superconductors, *Phys. Rev. B* **63**, 144531 (2001).

29. S. Autti, V. V. Dmitriev, J. T. Mäkinen, A. A. Soldatov, G. E. Volovik, A. N. Yudin, V. V. Zavjalov and V. B. Eltsov, Observation of half-quantum vortices in topological superfluid ^{3}He, *Phys. Rev. Lett.* **117**, 255301 (2016).

30. M. Smidman, M. B. Salamon, H. Q. Yuan and D. F. Agterberg, Superconductivity and spin-orbit coupling in non-centrosymmetric materials: a review, *Reports on Progress in Physics* **80**, 036501 (2017).

31. M. Cazalilla and A. Rey, Ultracold fermi gases with emergent $su(n)$ symmetry, *Reports on Progress in Physics* **77**, 124401 (2014).

32. X. Zhang, M. Bishof, S. L. Bromley, C. V. Kraus, M. S. Safronova, P. Zoller, A. M. Rey and J. Ye, Spectroscopic observation of su(n)-symmetric interactions in sr orbital magnetism, *Science* **345**, 1467 (2014).

33. G. Cappellini, M. Mancini, G. Pagano, P. Lombardi, L. Livi, M. Siciliani de Cumis, P. Cancio, M. Pizzocaro, D. Calonico, F. Levi, C. Sias, J. Catani, M. Inguscio and

L. Fallani, Direct observation of coherent interorbital spin-exchange dynamics, *Phys. Rev. Lett.* **113**, 120402 (2014).

34. F. Scazza, C. Hofrichter, M. Höfer, P. C. De Groot, I. Bloch and S. Folling, Observation of two-orbital spin-exchange interactions with ultracold $su(n)$-symmetric fermions, *Nat. Phys.* **10**, 779 (2014).

35. J. Sebby-Strabley, M. Anderlini, P. S. Jessen and J. V. Porto, Lattice of double wells for manipulating pairs of cold atoms, *Phys. Rev. A* **73**, 033605 (2006).

36. P. J. Lee, M. Anderlini, B. L. Brown, J. Sebby-Strabley, W. D. Phillips and J. V. Porto, Sublattice addressing and spin-dependent motion of atoms in a double-well lattice, *Phys. Rev. Lett.* **99**, 020402 (2007).

37. S. Folling, S. Trotzky, P. Cheinet, M. Feld, R. Saers, A. Widera, T. Muller and I. Bloch, Direct observation of second-order atom tunnelling, *Nature* **448**, 1029 (2007).

38. S. Trotzky, P. Cheinet, S. Fölling, M. Feld, U. Schnorrberger, A. M. Rey, A. Polkovnikov, E. A. Demler, M. D. Lukin and I. Bloch, Time-resolved observation and control of superexchange interactions with ultracold atoms in optical lattices, *Science* **319**, 295 (2008).

39. J. T. Stewart, J. P. Gaebler and D. S. Jin, Using photoemission spectroscopy to probe a strongly interacting fermi gash, *Nature* **454**, 744 (2008).

40. T.-L. Dao, I. Carusotto and A. Georges, Probing quasiparticle states in strongly interacting atomic gases by momentum-resolved raman photoemission spectroscopy, *Phys. Rev. A* **80**, 023627 (2009).

41. L. P. Gor'kov and E. I. Rashba, Superconducting 2d system with lifted spin degeneracy: Mixed singlet-triplet state, *Phys. Rev. Lett.* **87**, 037004 (2001).

42. S. K. Yip, Two-dimensional superconductivity with strong spin-orbit interaction, *Phys. Rev. B* **65**, 144508 (2002).

43. T. Ojanen, Magnetoelectric effects in superconducting nanowires with rashba spin-orbit coupling, *Phys. Rev. Lett.* **109**, 226804 (2012).

44. M. Bishof, M. J. Martin, M. D. Swallows, C. Benko, Y. Lin, G. Quéméner, A. M. Rey and J. Ye, Inelastic collisions and density-dependent excitation suppression in a ^{87}sr optical lattice clock, *Phys. Rev. A* **84**, 052716 (2011).

45. N. D. Lemke, J. von Stecher, J. A. Sherman, A. M. Rey, C. W. Oates and A. D. Ludlow, p-wave cold collisions in an optical lattice clock, *Phys. Rev. Lett.* **107**, 103902 (2011).

46. M. Höfer, L. Riegger, F. Scazza, C. Hofrichter, D. R. Fernandes, M. M. Parish, J. Levinsen, I. Bloch and S. Fölling, Observation of an orbital interaction-induced feshbach resonance in ^{173}Yb, *Phys. Rev. Lett.* **115**, 265302 (2015).

47. G. Pagano, M. Mancini, G. Cappellini, L. Livi, C. Sias, J. Catani, M. Inguscio and L. Fallani, Strongly interacting gas of two-electron fermions at an orbital feshbach resonance, *Phys. Rev. Lett.* **115**, 265301 (2015).

48. C. A. Regal, C. Ticknor, J. L. Bohn and D. S. Jin, Tuning p-wave interactions in an ultracold fermi gas of atoms, *Phys. Rev. Lett.* **90**, 053201 (2003).

49. J. Zhang, E. G. M. van Kempen, T. Bourdel, L. Khaykovich, J. Cubizolles, F. Chevy, M. Teichmann, L. Tarruell, S. J. J. M. F. Kokkelmans and C. Salomon, p-wave feshbach resonances of ultracold li$_6$, *Phys. Rev. A* **70**, 030702 (2004).

50. J. P. Gaebler, J. T. Stewart, J. L. Bohn and D. S. Jin, p-wave feshbach molecules, *Phys. Rev. Lett.* **98**, 200403 (2007).

51. C. Chin, R. Grimm, P. Julienne and E. Tiesinga, Feshbach resonances in ultracold gases, *Rev. Mod. Phys.* **82**, 1225 (2010).

52. V. Galitski and I. B. Spielman, Spin-orbit coupling in quantum gases, *Nature* **494**, 49 (2013).

53. N. Goldman, G. Juzeliunas, P. Öhberg and I. B. Spielman, Light-induced gauge fields for ultracold atoms, *Reports on Progress in Physics* **77**, 126401 (2014).

54. A. Celi, P. Massignan, J. Ruseckas, N. Goldman, I. B. Spielman, G. Juzeliūnas and M. Lewenstein, Synthetic gauge fields in synthetic dimensions, *Phys. Rev. Lett.* **112**, 043001 (2014).

55. J. Dalibard, F. Gerbier, G. Juzeliūnas and P. Öhberg, *Colloquium*: Artificial gauge potentials for neutral atoms, *Rev. Mod. Phys.* **83**, 1523 (2011).

56. N. R. Cooper and G. V. Shlyapnikov, Stable topological superfluid phase of ultracold polar fermionic molecules, *Phys. Rev. Lett.* **103**, 155302 (2009).

57. C. V. Kraus, M. Dalmonte, M. A. Baranov, A. M. Läuchli and P. Zoller, Majorana edge states in atomic wires coupled by pair hopping, *Phys. Rev. Lett.* **111**, 173004 (2013).

58. A. Bühler, N. Lang, C. Kraus, G. Möller, S. Huber and H. Büchler, Majorana modes and p-wave superfluids for fermionic atoms in optical lattices, *Nature Communications* **5**, 4504 (2014).

59. B. Liu, X. Li, B. Wu and W. V. Liu, Chiral superfluidity with p-wave symmetry from an interacting s-wave atomic fermi gas, *Nature Communications* **5**, 5064 (2014).

60. B. Wang, Z. Zheng, H. Pu, X. Zou and G. Guo, Effective p-wave interaction and topological superfluids in s-wave quantum gases, *Phys. Rev. A* **93**, 031602 (2016).

61. F. Iemini, L. Mazza, L. Fallani, P. Zoller, R. Fazio and M. Dalmonte, Majorana quasiparticles protected by z_2 angular momentum conservation, *Phys. Rev. Lett.* **118**, 200404 (2017).

62. J. Eroles, G. Ortiz, A. V. Balatsky and A. R. Bishop, Inhomogeneity-induced superconductivity?, *EPL (Europhysics Letters)* **50**, 540 (2000).

63. W.-F. Tsai, H. Yao, A. Läuchli and S. A. Kivelson, Optimal inhomogeneity for superconductivity: Finite-size studies, *Phys. Rev. B* **77**, 214502 (2008).

64. A. M. Rey, R. Sensarma, S. Folling, M. Greiner, E. Demler and M. D. Lukin, Controlled preparation and detection of d-wave superfluidity in two-dimensional optical superlattices, *EPL (Europhysics Letters)* **87**, 60001 (2009).

65. L. Isaev, G. Ortiz and C. D. Batista, Superconductivity in strongly repulsive fermions: The role of kinetic-energy frustration, *Phys. Rev. Lett.* **105**, 187002 (2010).

66. S. L. Bromley, S. Kolkowitz, T. Bothwell, D. Kedar, A. Safavi-Naini, M. L. Wall, C. Salomon, A. M. Rey and J. Ye, *arXiv:1708.02704*.

67. M. Mancini, G. Pagano, G. Cappellini, L. Livi, M. Rider, J. Catani, C. Sias, P. Zoller, M. Inguscio, M. Dalmonte and L. Fallani, Observation of chiral edge states with neutral fermions in synthetic hall ribbons, *Science* **349**, 1510 (2015).

68. J. Ye, H. J. Kimble and H. Katori, Quantum state engineering and precision metrology using state-insensitive light traps, *Science* **320**, 1734 (2008).

69. N. R. Cooper and A. M. Rey, Adiabatic control of atomic dressed states for transport and sensing, *Phys. Rev. A* **92**, 021401 (2015).

70. M. L. Wall, A. P. Koller, S. Li, X. Zhang, N. R. Cooper, J. Ye and A. M. Rey, Synthetic spin-orbit coupling in an optical lattice clock, *Phys. Rev. Lett.* **116**, 035301 (2016).

71. M. S. Safronova, Z. Zuhrianda, U. I. Safronova and C. W. Clark, *arXiv:1507.06570*.

72. Y. Nambu, Quasi-particles and gauge invariance in the theory of superconductivity, *Physical Review* **117**, 648 (1960).

73. H. Lipkin, Collective motion in many-particle systems. part 1. the violation of conservation laws, *Annals of Physics* **9**, 272 (1960).

Clean and dirty bosons in 1D lattices

Thierry Giamarchi

University of Geneva
24 Quai Ernest Ansermet
1211 Geneva
Switzerland
Thierry.Giamarchi@unige.ch

These notes are a brief tour in the realm of quantum interacting one dimensional systems, and their connection to topological phase transitions. Two examples are examined. The first one is the superfluid-Mott transition for interacting 1D bosons. Via bosonization such systems are directly described by the Berezinskii-Kosterlitz-Thouless (BKT) transition. Both the universal jump at the transition and the nature of topological excitations can be probed experimentally. The second example is disordered 1D interacting bosons, for which the superfluid - Bose glass transition occurring in such systems is also a variant of the BKT transition. Some theoretical aspects and experimental realizations are discussed.

Keywords: One dimensional quantum systems; topological transition; BKT transition; Mott transition; Bose glass.

1. Introduction

Understanding the physics of strongly correlated systems is one of the most important challenges of the 21st century. Due to interactions novel physics (such as e.g. superconductivity) and phases (novel excitations such as fractional charges, topological excitations etc.) can appear opening a whole world of possibilities going way beyond what would be possible in a single particle world. From a fundamental point of view the combination of an exponentially growing Hilbert space (with the number of particles) and of the entanglement of a fermionic or bosonic many-body wave-function makes it a strong challenge both from the analytical and numerical point of view. From a practical point view interactions open the possibility of materials with unique properties.

One important element in our understanding of this physics is the realization that, in addition to "conventional" collective states that can be described by local order parameters (e.g. superconductivity, quantum magnetism, etc.) non-local excitations, topological in nature, can and do play a central role. In that sense the vortex excitations in the two dimensional classical XY model, leading to the celebrated Berezinskii-Kosterlitz-Thouless (BKT) transition[1-3], and the excitations existing in a spin 1 quantum chain, discovered by Haldane[4] were key works showing the importance of such non-local excitations with far reaching consequences in many domains ranging from hall effects to topological insulators and superconductors[5].

Among the whole cornucopia of such systems, one and quasi-one dimensional quantum systems play a very special and important role[6]. First, interactions play a major role in one dimension, since particle cannot "avoid" each other, and thus the physics properties are extremely different from the ones stemming out of an essentially single particle description such as Landau quasi-particles for fermions or Bogoliubov excitations for bosons. Actually the physics is entirely made out of collective excitations, leading to physical properties known as Tomonaga-Luttinger liquid (TLL) (another of Haldane major's contributions)[7,8]. In addition, because of the collective nature of the excitations, in one dimension the excitations can fractionalize by splitting into collective modes[6]. The most famous example of such phenomenon being the spins-charge separations occurring with one dimensional spinful fermions. In addition, the topological nature of the excitations is the norm and not the exception. One dimensional systems are thus prime candidate to realize and understand novel physics driven by the interactions. On the experimental side the last 15 years or so have seen a full explosion of experimental quantum one dimensional systems both in the condensed matter context[9] and in cold atoms[10,11]. These systems have allowed both to test for past predictions and to unravel new physics.

In these notes I will present two interesting cases of quantum phase transitions occurring with systems of interacting one dimensional bosons. These example are chosen as representative of the unusual one dimensional physics and its link with topological phase transitions and order parameters. They have been selected mostly based on my personal interest and research, and the notes do not have any pretention at being a full review of the field or at completeness. I will put a subset of references, more details and references on one dimension can be found in the literature[6,11,12].

The first case is the transition due to the presence of a periodic lattice. This is the Mott-superfluid transition. It allows for a remarkable test of the predictions of the BKT transition. Cold atomic systems have given an experimental realization with an unprecedented level of control, in particular on the interactions. In addition the Mott-superfluid transition is one of the simplest example on which one can illustrate and measure the topological nature of the order. The second example is the case of disordered one dimensional interacting bosons. In that case the transition is from the superfluid to a localized phase due to disorder, the Bose glass. This transition is a variant of a BKT phase transition.

The plan of the notes is as follows. In Sec. 2 I give a brief reminder of the connection between the BKT transition originally developed for classical 2D systems and one dimensional quantum systems. This section introduces the bosonization technique that will be used to make this connection. Sec. 3 discusses the case of the periodic lattice and the superfluid-Mott transition. Sec. 4 addresses the problem of disordered interacting bosons and the superfluid-Bose glass transition. Conclusions and perspectives are given in Sec. 5.

2. BKT transition and quantum systems

Let me give first a brief reminder on how to apply the BKT concepts to 1D quantum systems, in particular to quantum interacting bosons. To do so a key paper is the beautiful paper by Haldane[8] explaining how to "bosonize" bosons. As the paradox in the very name "bosonization" indicates, the general idea was that such a technique is to transform fermionic excitations into bosonic ones and that in order to use it you necessarily needed something fermionic. This ensured that there were Fermi points and a dispersion that you could linearize around such points. On the contrary, the key message of the paper[8] was that the essential ingredient of the technique is to reexpress the 1D excitations, which normally are expressed in the single particle basis, in term of *collective excitations* which are clearly the good excitations for a 1D systems. Thus the technique could be used for any kind of particles and in particular bosons. It also made transparent that interactions and statistics are intricately linked in one dimension.

The key ingredient is rewrite the density and single particle excitations in terms of two collective modes $\phi(x)$ and $\theta(x)$ via the following expressions

$$\rho(x) = \left(\rho_0 - \frac{1}{\pi}\nabla\phi(x)\right)\sum_p e^{i2p(\pi\rho_0 x - \phi(x))} \tag{1}$$

where p is a relative integer and

$$\psi^\dagger(x) = [\rho(x)]^{1/2}e^{-i\theta(x)} \tag{2}$$

The $\theta(x)$ phase is the standard superfluid phase for bosons and would be ordered in the case of a true superfluid. The phase ϕ can be interpreted essentially as the displacement field compared to the perfect positions of the particles in a crystal with a lattice spacing $a = \rho_0^{-1}$ (the average density of the particles). In the same way if the bosons were forming a perfect crystal this field would be fully ordered. Contrarily to the original single particle $\psi^\dagger(x)$ and density operators $\rho(x)$ which are highly discontinuous at the lengthscale of the interparticle distance a, these collective variables are smooth fields. They thus allow for an "hydrodynamic" description, and thus to a considerable simplification of the Hamiltonian, while at the same time allowing to describe all phenomena, even at scales smaller than a. Note that the representation of the density can be generalized to crystals in higher dimensions[13].

Because crystalline and superfluid orders are naturally conjugate, one can expect the fields $\phi(x)$ and $\theta(x)$ to obey canonical conjugation relations. Indeed using the commutation relations of the original bosons operators one can show that[6]:

$$\left[\frac{1}{\pi}\nabla\phi(x), \theta(x)\right] = -i\delta(x - x') \tag{3}$$

The momentum conjugate to the field $\phi(x)$ is thus

$$\Pi_\phi(x) = \frac{1}{\pi}\nabla\theta(x) \tag{4}$$

When expressed in terms of these fields the Hamilonian of a system of interacting bosons takes the form (at low energies)

$$H = \frac{1}{2\pi} \int dx \left[(uK)(\nabla\theta(x))^2 + \frac{u}{K}(\nabla\phi(x))^2 \right] \tag{5}$$

All the parameters of the original Hamiltonian have been included in the two independent parameters u and K, the so-called TLL parameters. u is the velocity of density excitations in the system (velocity of sound), and is of course finite if interactions are present. K is a dimensionless parameter, depending on the interactions which controls the decay of the various correlation functions. For example for the Lieb-Lininger model [14], of bosons with a contact interaction, $K = \infty$ of the interaction is zero and $K = 1$ if the bosons have hard core repulsion [6,12]. Longer range interactions between the bosons allows to reach value of K smaller than one.

The representation of the density is particularly useful to take into account the effect of a periodic lattice. I will confine myself here to the case where the lattice is commensurate with the boson density and refer the reader to the literature [6,12] for more general cases. Using the bosonization representation (1) one can write the coupling to a periodic lattice $V(x) = V_0 \cos(Qx)$ as

$$H_l = V_0 \int dx \, \cos(Qx)\rho(x) = V_0\rho_0 \sum_p \int dx \, \cos(Qx)e^{i2p(\pi\rho_0 x - \phi(x))} \tag{6}$$

If the lattice is commensurate with the number of bosons, for example if there is a boson per site, then $Q = 2\pi\rho_0$. In that case all oscillating terms vanish since the field $\phi(x)$ is slowly varying but terms with $p = 1$ remain leading to

$$H_l = V_0\rho_0 \int dx \, \cos(2\phi(x)) \tag{7}$$

in addition to the Hamiltonian (5). This is the famous sine-Gordon model.

One can thus immediately see that a phase transition is possible due to the lattice. The Hamiltonian (5) would like to let the field $\phi(x)$ have fluctuations, leading to a decay of order in $\phi(x)$ and thus a "liquid" (superfluid) state for the bosons. One the contrary the cosine term (7) wants to lock $\phi(x)$ in the minima of the cosine leading to long range order in $\phi(x)$ and thus to crystalline order. The competition between these two terms thus describes the Mott-Superfluid transition [8] of 1D interacting bosons in a lattice. The bosonized Hamiltonian allows also to compute many observable and I again refer the reader to [6,12] for more details and references on that point.

The nature of the transition can be easily understood by going to the action which reads

$$S = \frac{K}{2\pi} \int dx d\tau \left[\frac{1}{u}(\partial_\tau\theta(x,\tau))^2 + u(\partial_x\theta(x,\tau))^2 \right]^2 - V_0\rho_0 \int dx d\tau \, \cos(2\phi(x,\tau)) \tag{8}$$

where τ is the imaginary time.

The quadratic part has been purposefully written in terms of the phase θ to show that, pending the trivial rescaling of space and time by the velocity u, it is essentially the quadratic part of the energy that one would expect for a two dimensional XY model, for which θ would play the role of the spin angle. In the usual way x and τ play the role of the two spatial dimensions for the equivalent classical system. $K/(2\pi)$ corresponds to the inverse classical temperature. The cosine term is particularly interesting. Indeed given the commutation relation (3) one has

$$\phi(x,\tau) = \pi \int_{-\infty}^{x} dx' \Pi_\theta(x',\tau) \tag{9}$$

where Π_θ is the momentum conjugate to the field θ. One thus sees two things:

(1) The field ϕ is actually a non local field with respect to the phase θ. One can thus expect, as we will discuss below, that this field can create topological excitations. The fact that it is non-local means also that some non-local order parameter can emerge from the measurement of such a field. Such order parameter would be inaccessible by local measurements in a classical system. We will come back to both these points below.

(2) Because the field is expressed in terms of the momentum of θ is can create vortices in this field[6].

Indeed as is well known

$$e^{iaP}|x\rangle = |x + a\rangle \tag{10}$$

which means that the application of $e^{i2\phi(x)}$ on a state with well defined values of the field $\theta(x,\tau)$ shifts this field by 2π across an infinite string ending at the point (x,τ), as depicted in Fig. 1. This singularity can be unwrapped to obtain the more

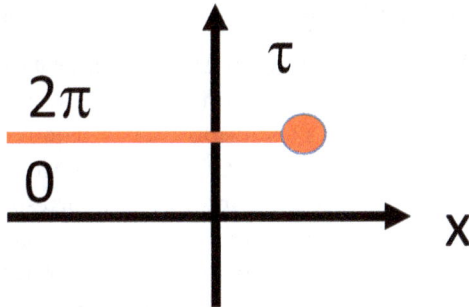

Fig. 1. Topological excitations: the operator $e^{i2\phi(x,\tau)}$ is a non local operator for the field θ. Applied to an eigenstate of θ it creates a discontinuity of 2π along a string ending at the point of coordinates (x,τ). Unwrapping this discontinuity allows to see that this operator is creating a conventional vortex in θ (and likewise $e^{i2\phi(x,\tau)}$ is the antivortex operator.

conventional form of the vortex. It is thus possible to map[15,16] the one dimensional sine-Gordon model (8) onto the two dimensional classical XY model. The main elements of the mapping are described in Table 1.

Table 1.   Relation between the 2D classical XY model and the 1D quantum sine-Gordon model.

2D classical XY model	1D quantum sine-Gordon
angle of the spins	field θ
finite size along y	inverse temperature β
inverse classical temperature β_{cl}	TLL parameter K
vortex fugacity	prefactor g of the cosine term

1D quantum systems are thus perfect physical realizations to hunt for the BKT transition. The control on the transition is exerted by varying parameters such as the interactions, which changes K, with the advantages that the quadratic part and the fugacity of vortices can usually be fully *independently* be controlled, contrarily to e.g. the case of the XY model where they depend on the single parameter J (the magnetic exchange). For example for (8) the periodic lattice strength controls the fugacity of the vortices. The drawback is that now the temperature is the enemy since it corresponds to a finite size effect of period β along the time direction. For one dimensional quantum bosons, the BKT transition corresponds thus to the Mott insulator-superfluid transition and we will examine it in more details in the next section. Similar ideas have been applied in various context connected with the BKT transition in particular in the context of superconducting films[17]. Another advantage is that for the quantum problem the non-local fields such as ϕ correspond actually to some *local* observable (for example the density for the field ϕ). It is thus possible, although non necessarily easy to measure such non local objects by looking at real physical quantities.

3. Mott transition with 1D bosons

Let us now turn to the Mott transition of 1D interacting bosons. This transition separates two phases. In one phase the cosine is relevant and thus ϕ is locked in *one* of the minima of the cosine. This means that ϕ acquires a fixed value (note: ϕ itself is locked and not just $\cos(2\phi)$). Of course this value can be anything of the form $2\phi = 2\pi q$ where q is a relative integer, and corresponds to the breaking of the translational symmetry of the system which has become discrete because of the existence of the periodic lattice. The fact that the translational symmetry is now discrete is what allows to get a phase transition even in one dimension (at zero temperature).

ϕ being ordered means that the density is ordered, and with the value that we have taken it is easy to see that it corresponds to an average of one boson per site. Similar transitions do occur for other commensurate values[18] and I refer the reader to reviews in the literature[6,12] for more details. Because of the cosine, the spectrum of (8) is now massive, and the various correlation functions decay exponentially. This corresponds to the high temperature phase of the BKT transition, in which vortices are relevant and destroy exponentially the order in the phase θ. For a quantum system it is easy to understand that since the field ϕ is ordered, its conjugate, is maximally disordered.

For large K, on the contrary, there might be enough quantum fluctuations in ϕ (since θ would fluctuate less) coming from the quadratic part to wipe out the cosine, as shown in Fig. 2. In that case the spectrum remains massless and the

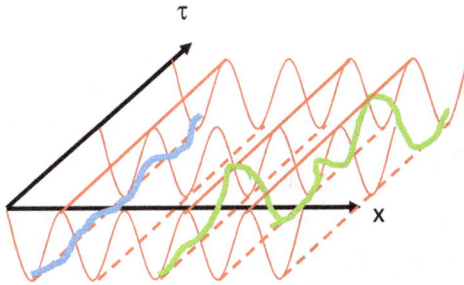

Fig. 2. The space-time trajectories of the bosons in the two phases: the periodic potential in space x and constant in imaginary time τ is indicated by the thin red lines. At small K the trajectory of a boson in the Mott phase stays in one of the minima of the potential (left thick blue line). On the contrary in the superfluid phase (large K) the bosons delocalize in all the minima of the potential (right thick green line). In the first case because the (non-local) field ϕ is ordered (see text) one can have non local and topological excitations.

phase θ has long range correlations only dictated by the (renormalized) quadratic part. The correlations of the form

$$\langle e^{i\theta(x,\tau)} e^{-i\theta(0,0)} \rangle \propto \left(\frac{a}{r}\right)^{\frac{1}{2K^*}} \tag{11}$$

where a is a short distance cutoff of the order of the lattice spacing and K^* the coefficient K with the renormalization coming from the presence of the (irrelevant) cosine. Of course using (2) this is directly related to the single particle correlation function which can be directly measured by e.g. time of flight measurements in cold atoms[10]. Using the standard BKT flow[3] the transition occurs on a certain separatrix in the (K, g) plane where g is coefficient of the cosine in (8). Exactly on the separatrix between the Mott and the superfluid phase the flow goes to $g \to 0$ and $K \to K^* = 2$. There is thus a universal value for the exponents of the correlations right at the transition. On the physical side, a remarkable fact is that, contrarily to what happens in higher dimension, an *arbitrarily* small periodic potential provokes the existence of a Mott insulator provided that the interactions are *repulsive enough* between the bosons (K value is such that $K < 2$). Of course the gap itself depends on V_0 which means that a very small periodic potential leads to a very small gap.

Nevertheless tracking in the (K, V_0) plane the position of the transition should give the value $K = 2$ for the boundary when $V_0 \to 0$. This provides an excellent way to test for the universal value of the BKT transition. Note[6] that the value of K is directly related to the superfluid stiffness $\mathcal{D}$ which is given by $\mathcal{D} = uK$. This means that there is a universal jump of the superfluid stiffness (suitably renormalized by the velocity) at the superfluid-Mott insulator transition.

Cold atomic systems provide an excellent system to test for this physics[19]. I will not discuss here the beautiful experiments in that field that directly tested for BKT in 2D systems of bosons (classical because of the finite temperature)[20], and focuss one the ones on 1D quantum systems. Pioneering experiments were done in the group of H.C. Nagerl[21], where bosons with a delta function repulsion of strength γ were put in a weak periodic lattice. Tracking the appearance of a gap in the spectrum showed that as the repulsion is increased, the critical value of the periodic potential needed to get the Mott insulator was decreasing. $V_0 \to 0$ led to a phase boundary tending to a *finite* value of γ. However determining the precise value of γ was difficult in this experiment. More recent experiments performed in LENS[22] allowed for a much more precise determination of the value of the critical interaction as shown in Fig. 3. For such a system of interacting bosons in the

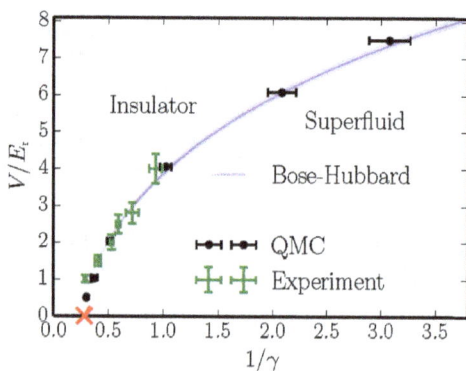

Fig. 3.   Phase diagram of bosons with a contact interaction γ put in periodic lattice of strength V (measured in units of the recoil energy E_r). Quantum monte carlo (QMC) data is compared with experiments in a system of ^{39}K atoms. The boundary separating the superfluid from the Mott insulator is computed and measured. The agreement between the calculation and the experiment is excellent. The transition line extrapolates to a *finite* value of the interaction when the lattice potential tends to zero showing that even an infinitesimal lattice produces a Mott insulator in 1D if the interactions are repulsive enough. The value of $K_c = 2$ predicted by the field theory, directly related to the microscopic value $\gamma_c = 0.297$ by the exact calculation of the TLL parameters for the Lieb-lininger model, shows remarkable agreement with the measured and numerically computed critical value. This system thus allows to test the universal value of the jump of the parameter K predicted by the BKT theory. After[22].

continuum with a contact interaction[14] the exact relation between the microscopic contact interaction γ and the TLL parameter K can be computed[12] (using another trick due also to F.D.M. Haldane[23]). The critical value of the interaction indeed corresponds well to $K = 2$ providing thus a remarkable experimental verification, coming from the 1D quantum system, of the universal jump at the BKT transition.

A non trivial, and perhaps less obvious aspect of such a transition is the existence of a non-local order parameter. Indeed as we discussed the field that orders at the transition is ϕ itself, and not just a periodic function of ϕ. The fact that there are no finite action excitations that allow the field to tunnel between the various minima[6] results from a theorem on the existence of instantons (see[24] for the theorem itself). This means that any function of ϕ will show order. This is especially important since the existence of the cosine in the Hamiltonian determines which type of excitation can in general be present. For example for the problem considered the $\cos(2\phi)$ only allow excitations such that $\phi(x \to -\infty)$ and $\phi(x \to \infty)$ tends to minima of the cosine, under penalty of paying an infinite action cost. One can thus classify the excitations based on the robust topological number $\delta = (\phi(x \to -\infty) - \phi(x \to +\infty))/\pi$. We see that in one dimension having such topological constraint is the norm rather than the exception. Of course the nature of the topological charge of the excitation depends on the problem at hand. For the case of the interacting bosons in the periodic potential the $\cos(2\phi)$ imposes that δ is a relative integer. Using the relation (1) we see that the total "charge" of such an excitation is

$$Q = -\frac{1}{\pi} \int_{-\infty}^{\infty} dx \, \nabla \phi(x) = \delta \tag{12}$$

One can thus interpret the excitations as saying that if there is a soliton of ϕ, e.g. between $\phi = \pi$ and $\phi = 0$ this corresponds to the injection of a charge $\delta = 1$ at the position of the soliton. Excitations that preserve the same value of ϕ at infinity (for example if one imposes periodic boundary conditions) need thus to have solitons and antisolitons and correspond in real space to doubly occupied sites and holes. Other periodicity of the cosine can occur in other problems. In that case the excitations can even be fractional due to the fact that a part of the excitation is "pushed" to infinity due to the non-local character. This is for example the case of quantum spin chains for which a $\cos(4\phi)$ exists in the Hamiltonian, leading to elementary excitation of spin $\delta = \pm 1/2$. Such excitations, known as spinons[6] have a smaller quantum number that what should have been the most elementary *local* spin excitation (namely flipping a spin from $S^z = -1/2$ to $S^z = 1/2$). such excitations can exist in 1D due to the collective nature of the excitation, allowing to push some part of it to infinity.

Going back to the Mott-superfluid transition, the ordering of ϕ means that one can in principle probe for any order of the form $\langle \cos(z\phi) \rangle$. If z is a multiple of two, this would simply correspond to local excitations of integer charge as we already discussed above. This simply means the obvious statement that in a Mott insulator the density is ordered (with an average of one particle per site). But one can also take fractional value such as $z = 1$ or even fractional values. For $z = 1$ an order parameter could be (going back to a lattice description and setting the origin of the chain at the site $j = 0$)[25]

$$\langle e^{-i\phi(x=aj_0)} \rangle = \langle e^{\pi \sum_{j=0}^{j_0} n_j} \rangle \tag{13}$$

where n_j is the density on site j and j_0 any arbitrary site. The non-local aspect of such an order parameter is obvious in (13) and as such cannot usually be measured by a local probe, such as X-ray or neutrons. In order to compute it, one needs the value of the density at each site of the lattice for *one* realization of the system, and not just the average value of such quantity. So for a while such order parameters were only existing in the realm of theory or numerical calculation. However cold atomic systems have provided an experimental alternative allowing for such a measurements. Indeed systems such as atom microscopes allow to obtain for a one shot measurement of the system the density at every site, allowing to obtain the quantity (13) for one realization of the system. Repeating the experiment many times give the quantum average. Measurement of the order parameter (13) was performed in the Mott phase[26] providing a very good experimental confirmation of the existent of such non-trivial type of order. Since then other non-local order parameters such as e.g. the string order parameter for a spin one chain have also been measured in cold atomic systems. More sophisticated topological phase transitions, involving two conjugate sets of topological order parameters as in an XY model with discrete symmetry breaking[27] have been even recently been identified and observed in quantum spin systems[28].

4. Disordered systems; Bose glass

Let us now move to disordered systems. The question of the combined effects of disorder and interactions is a long standing and particularly important problem. In the absence of interactions, disorder leads to the celebrated Anderson localization[29]. The eigenstates of the problem decay exponentially with distance and the system thus become an insulator due to the disorder. In one dimension all states are localized, even by an infinitesimal disorder, and the localization length is essentially twice the mean free path. Interactions can potentially drastically modify this behavior. It is not the place here to give a full review of this important class of problem and I will thus concentrate in what follows on one dimensional bosons again.

For bosons this competition is particularly drastic. First, contrarily to the case of fermions, the non-interacting limit is extremely singular for bosons since they can all go in the same quantum state. It is easy to check that since one single boson wavefunction would be localized in a finite region of space, for N bosons they would all go in this region, leading to a macroscopic number of bosons in a finite region of space, and thus infinite density. Such a state is obviously totally unstable to any infinitesimal repulsion. In a similar way, in the absence of disorder we know that the non-interacting point is singular. Its spectrum is quadratic and if offers no phase rigidity. Any repulsive interaction leads to excitation with an energy linear with momentum and thus allows for phase rigidity and superfluidity (with only quasi-long range order in one dimension). If we expect a superfluid to be robust to impurities, we can thus expect a strong competition between the superfluid tendency and the localization due to disorder.

As for the periodic system, the method of bosonizing bosons allows to solve this problem in one dimension[30]. The coupling to an external potential $V(x)$ becomes

$$H_d = \int dx\, V(x)\rho(x) = \int dx\, V(x)(e^{2\pi\rho_0 x - 2\phi(x)} + \text{h.c.}) \tag{14}$$

If $V(x)$ is periodic, only one Fourier component exists $V(x) = V_0 \cos(Qx)$ and the potential is effective only if $Q = \pm 2\pi\rho_0$. This is the case of the Mott potential discussed in the previous section. For a disordered potential all Fourier component exist. If we recenter the Fourier decomposition around the component $Q = 2\pi\rho_0$ the disorder potential can be written as

$$V(x) = \sum_k e^{ikx} V_k \simeq \sum_{\delta k} e^{i(2\pi\rho_0 x + \delta x)} V_{2\pi\rho_0 + \delta k} = e^{i2\pi\rho_0 x} \xi(x) \tag{15}$$

where $\xi(x)$ is also a random potential. One can thus rewrite (14) as

$$H_d = \int dx\, \xi(x) e^{2\phi(x)} + \text{h.c.} \tag{16}$$

$\xi(x)$ is in general a complex field. The case of the perfectly periodic potential (with the commensurability condition) would correspond to $\xi(x)$ being a constant. For a disorder potential $\xi(x)$ has only short range correlations in space (but is infinitely correlated in time).

The quadratic part of the action allows to take into account all the interactions between the bosons, as long as the disorder term (16) is weak enough that the bosonization mapping is valid. This assumes that the field $\phi(x)$ is smooth enough at the scale of the mean interparticle spacing. This description will thus be very well adapted to the case of weak disorder and moderate or strong interactions among the bosons. It can also be used for other one-dimensional or quasi-one dimensional quantum fluids (fermions, spins, ladders, coupled chains etc.). I refer the reader to[6] for more details on that point.

The effect of disorder on an interacting system of bosons can thus be understood in similar way than for the commensurate lattice potential. It is a competition between the fluctuations of the field imposed by the quadratic term, and the pinning term (16) which wants the field to be locked to classical value. If one decomposes the random field in amplitude and phase

$$\xi(x) = |\xi(x)| e^{-i\zeta(x)} \tag{17}$$

the disorder term wants essentially 2ϕ to follow the random phase $\zeta(x)$. The competition between the quadratic term and the disorder term can be studied by writing renormalization equations[30], in a spirit similar to the BKT equations[3]. There are some specific technical complications that I will not discuss here. The resulting equations are

$$\frac{dK}{dl} = -\text{Cst} D \tag{18}$$

$$\frac{dD}{dl} = (3 - 2K)D \tag{19}$$

where Cst is a constant and D is the strength of the disorder, as determined for example by the average of the correlations of the random field

$$\overline{\xi}(x)\xi^*(x') = D\delta(x - x') \tag{20}$$

The absence of spatial correlations of the disorder term have changed the dimension of the disorder operator to $3 - 2K$ instead of the $4 - 2K$ for the periodic potential. This modifies the critical value of the interactions at which the transition happens and the universal jump of the TLL parameter which now becomes $K^* = 3/2$ at the transition[6,30].

At the transition all correlations are controlled by this universal exponent. The system jumps from a superfluid phase, in which the disorder is essentially irrelevant, to a phase in which the disorder is dominant. Of course the nature of the phase in which the cosine is relevant is now very different than the one for the simple periodic case. Its nature cannot be directly determined from the RG. It can either be guessed on physical grounds, or by remarking that the point for which $K = 1$ can be mapped back on free fermions[31], which then will be localized due to the presence of the disorder[30]. One can also use more formal techniques such as a replica-based variational ansatz which allows to compute (approximately) the propagators in the localized phase[32]. Finally one can combine the RG with instanton calculations in the strong coupling regime[33] to extract some of the transport properties in the localized phase. In the end and despite all efforts we still have only a partial understanding of this localized phase of interacting particles. As a simple question whether this phase which is names "glass" has really glassy properties is still under debate. In higher dimensions than one, the analysis cannot be pursued with the same degree of control, since the interactions are more difficult to take into account, but scaling arguments show that a similar superfluid-Bose glass transition should still occur[34].

A summary of the phase diagram for one dimensional interacting bosons in the presence of disorder is indicated in Fig. 4.

Several points on this transition are worth mentioning.

(1) The superfluid phase "screens" the disorder and the transition towards the localized phase is essentially of a (modified) BKT type. This means in particular a jump of the TLL parameter $K^* = 3/2$, and thus of the superfluid density $D = uK$ at the transition, and *universal* exponents for the superfluid correlations at the transition.

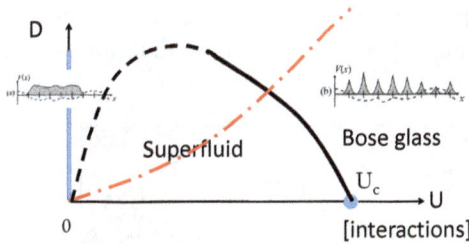

Fig. 4. Schematic phase diagram for disordered bosons as a function of the repulsion U between bosons and the strength of the disorder D. The non-interacting case is singular since a macroscopically large number of bosons go in a finite region of space. Such line is unstable to the presence of interactions. The red dot-dashed line indicates the region below which the bosonization is a faithful description of the system (see text) and the RG analysis can be fully trusted. The solid line indicates the boundary between the superfluid and Bose glass transition, as determined by the RG equations (see text). On this transition line the TLL parameter takes the *universal* value $K^* = 3/2$ leading to universal exponents for the correlation functions. There is a jump of the superfluid density at the transition. Because of the localization of the non-interacting bosons, the superfluid phase is re-entrant, as indicated schematically by the dashed line.

(2) In the localized phase the properties are more difficult to obtain. At $T = 0$ one can infer that the systems is compressible (the Bose glass is thus a compressible insulator, by opposition to the Mott insulator, which has a finite gap to excitations). Quantities such as the frequency dependent conductivity can be computed[32]. Finite temperature properties are more difficult to obtain. They can be computed in presence of a thermal bath[33].

These predictions have been the focuss of many tests, in particular using numerical methods. I refer the reader to the literature[6,12] more more details and references on these aspects. Note that one interesting aspect, that would need to be further explored is whether a string order parameter, similar to the one that exists for the Mott transition and has been discussed in the previous section would be useful to build for the disordered case.

Experimentally testing for the Bose glass phase is not totally easy due to the practical difficulty of realizing one dimensional disordered systems (especially bosonic). However in the recent years several remarkable realizations of disordered one dimensional bosonic systems have been realized. In the condensed matter realm the two systems that have been prominent in that respect have been the quantum spin systems[35-38] and Josephson junction arrays[39]. In the later a test of the RG predictions could also be performed leading to good agreement with the theory. Cold atoms have also provided remarkable realizations of disordered systems[40] either by using speckle potentials, other species or quasi-periodic potentials. Note that quasiperiodic potentials can be treated by the same RG method than discussed here[41,42] or numerical methods[43].

The later case could be realized with a control of the boson-boson repulsion thus providing an excellent system in which to test for the phase diagram[44], and compare with the theoretical predictions (slightly modified for quasi-periodic compared to pure disordered systems). The observed phase diagram is indicated in Fig. 5. The

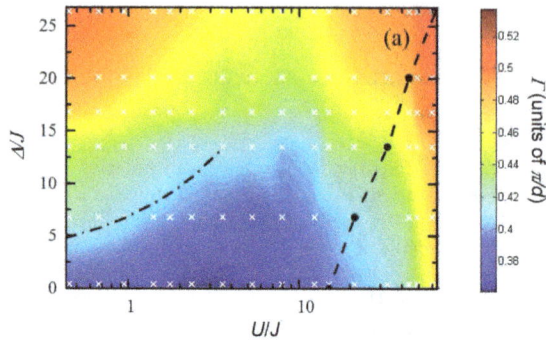

Fig. 5. Measurements of the width Γ of the momentum distribution of ^{39}K bosons in 1D tubes as a function of the their repulsion U and the strength of a quasi-periodic potential Δ. A superfluid phase is characterized by a narrow width, while a localized phase with exponentially decreasing superfluid correlations leads to a finite width of the order of the inverse localization length. One clearly sees a reentrant superlfuid-like region surrounded by non-superlfuid regions. Other measurements (not shown here) confirm that these regions are indeed localized. Some differences between this phase diagram and the "canonical" phase diagram of Fig. 4 come from the quasi-periodic vs disorder aspect of the potential (in particular the transition for non-interacting systems occur at a finite value of the potential) and the presence of the optical lattice that allows for a competing Mott phase at large interaction. (From [44])

phase diagram shows clearly a re-entrant phase in good agreement with what is to be expected from the theory point of view. Two main factors limit the comparison to theory

(1) The temperature, although low in absolute values is high due to the necessity to have a high optical lattice to ensure the quasi-periodic aspect of the potential. The finite temperature, which makes all correlation functions decay exponentially beyond a thermal length of order $1/T$ thus makes it more difficult to detect the localization in the system. A different experimental system in which the disorder is put via a speckle should thus address this point very efficiently.
(2) Because of the inherent lattice and the presence of a parabolic trapping potential the Bose glass is in competition with the Mott insulator phase. Possibility to put the system in a box or even better to remove the lattice as indicated in the previous point should thus kill two birds with the same stone.

Thus although excellent progress have been made, further experiments are clearly needed to probe further the various aspects of the Bose glass phase. The complementarity of the various new systems both in condensed matter and in cold atoms offer excellent perspectives though.

5. Conclusion and perspective

These notes presented a small tour of the importance of variants of the BKT transition in one dimensional quantum systems and how one can effectively use such systems to probe the physics of topological phase transitions and topological excitations. Because of the recent possibilities offered to realize in an extremely controlled way such systems, one-dimensional quantum problems have provided a remarkable playground for this physics. The bosonization technique for quantum fluids allow for a nice contact of this physics with the one of BKT transitions and topological excitations.

At the level of the simple sine-Gordon equation, which is the prototype model to describe the Mott-superfluid transition for 1D interacting quantum bosons in a lattice, experimental realization with cold atoms have provided a remarkable experimental way to test for the universal behavior of the BKT transition, and the topological nature of the excitations. There are of course many other directions that need to be explored. For example recent systems in quantum spin chains have provided realization of transitions with two competing topological excitations[28], such as the ones occurring in classical systems with symmetry breaking fields[27]. The sky is the limit and it is clear that the coming years will see many more use of 1D quantum systems.

Disordered systems are also opening a whole realm of possibilities. As we saw, both on the theoretical level and now also on the experimental side, there are ways to probe for interacting quantum disordered systems in a controlled way. The theory is on firm ground for $T = 0$. Still many quantities remain to be measured (such as e.g. the frequency dependent conductivity) or computed. Out of equilibrium properties also mostly remain to be explored. For finite temperatures the properties depend drastically on whether the system is in equilibrium with a thermostat In that case one can use the temperature as a controlled cutoff, and extract quantities from the RG or other techniques. In the absence of thermostat (which is what is now known under the generic name of many-body localization) a host of interesting properties is also showing[45,46], some of which have been probed experimentally[47]. How to compute in general such properties for interacting quantum particles remains a considerable challenge. Another interesting direction is to see whether some topological properties could be probed also for disordered system, as seem to be suggested by the bosonization solution.

Acknowledgments

It is a great pleasure and honor to write these notes in connection with the workshop on topological aspect of matter that honored M. Kosterlitz and F.D.M. Haldane. When starting my Phd, the two papers that had undoubtedly the most influence on my scientific growth were the paper by M. Kosterlitz on the renormalization of the 2D XY model[3], and the paper by F.D.M. Haldane on the bosonization of one dimensional quantum fluids[8]. These two papers have been at the heart of many aspects of my scientific life since then. These notes are a direct testimony of my admiration for these two wonderful pieces of physics and their authors.

The work presented in these notes has been the result of many scientific collaborations, too numerous to be all mentioned there, but I want to specially mention the late H.J. Schulz who introduced me to the above mentioned papers and more generally to the fascinating world of one dimensional quantum systems. The work presented in these notes has been supported in part by the Swiss National Science Foundation under Division II.

References

1. V. Berezinskii, *Soviet Physics JETP* **32**, p. 493 (1971).
2. J. M. Kosterlitz and D. J. Thouless, Ordering, metastability and phase transition in two-dimensional systems, *Journal of Physics C* **6**, p. 1181 (1973).
3. J. M. Kosterlitz, *J. Phys. C* **7**, p. 1046 (1974).
4. F. D. M. Haldane, *Physical Review Letters* **50**, p. 1153 (1983).
5. M. Z. Hasan and C. L. Kane, *Colloquium* : Topological insulators, *Rev. Mod. Phys.* **82**, 3045 (Nov 2010).
6. T. Giamarchi, *Quantum Physics in One Dimension*, International series of monographs on physics, Vol. 121 (Oxford University Press, Oxford, 2004).
7. F. D. M. Haldane, *J. Phys. C* **14**, p. 2585 (1981).
8. F. D. M. Haldane, *Physical Review Letters* **47**, p. 1840 (1981).
9. T. Giamarchi, Some experimental tests of tomonaga-luttinger liquids, *Int. J. Mod. Phys. B* **26**, p. 1244004 (2012).
10. I. Bloch, J. Dalibard and W. Zwerger, Many body physics with ultracold gases, *Reviews of Modern Physics* **80**, p. 885 (2008).
11. T. Giamarchi, One-dimensional physics in the 21st century, *C. R. Acad. Sci.* **17**, p. 322 (2016).
12. M. A. Cazallila, R. Citro, T. Giamarchi, E. Orignac and M. Rigol, One dimensional bosons: From condensed matter systems to ultracold gases, *Rev. Mod. Phys.* **83**, p. 1405 (Oct 2011).
13. T. Giamarchi and P. Le Doussal, *Physical Review B* **52**, p. 1242 (1995).
14. E. H. Lieb and W. Liniger, *Physical Review* **130**, p. 1605 (1963).
15. S. T. Chui and P. A. Lee, *Physical Review Letters* **35**, p. 315 (1975).
16. L. P. Kadanoff, *Journal of Physics A* **11**, p. 1399 (1978).
17. L. Benfatto, C. Castellani and T. Giamarchi, *Beresinskii-Kosterlitz-Thouless transition within the sine-Gordon approach: the role of the vortex-core energy*, in *Berezinskii-Kosterlitz-Thouless Transition*, ed. J. V. José (World Scientific, 2012).
18. T. Giamarchi, *Physica B* **230-232**, p. 975 (1997).
19. H. P. Buchler, G. Blatter and W. Zwerger, Commensurate-incommensurate transition of cold atoms in an optical lattice, *Phys. Rev. Lett.* **90**, p. 130401 (2003).
20. Z. Hadzibabic, P. Krüger, M. Cheneau, B. Battelier and J. Dalibard, Berezinskii-kosterlitz-thouless crossover in a trapped atomic gas, *Nature* **441**, p. 1118 (2006).
21. E. Haller, R. Hart, M. J. Mark, J. G. Danzl, L. Reichsöllner, M. Gustavsson, M. Dalmonte, G. Pupillo and H.-C. Nägerl, Pinning quantum phase transition for a luttinger liquid of strongly interacting bosons, *Nature* **466**, p. 597 (2010).
22. G. Boéris, L. Gori, M. D. Hoogerland, A. Kumar, E. Lucioni, L. Tanzi, M. Inguscio, T. Giamarchi, C. D'Errico, G. Carleo, G. Modugno and L. Sanchez-Palencia, Mott

transition for strongly-interacting 1d bosons in a shallow periodic potential, *Physical Review A* **93**, p. 011601(R) (2016).

23. F. D. M. Haldane, General relation of correlation exponents and spectral properties of one-dimensional fermi systems: application to the anisotropic $s = \frac{1}{2}$ heisenberg chain, *Physical Review Letters* **45**, p. 1358 (1980).

24. R. Rajaraman, *Solitons and Instantons: An Introduction to solitons and Instantons in Quantum Field Theory* (North Holland, Amsterdam, 1982).

25. E. Berg, E. G. Dalla Torre, T. Giamarchi and E. Altman, Rise and fall of hidden string order of lattice bosons, *Phys. Rev. B* **77**, p. 245119 (2008).

26. M. Endres, M. Cheneau, T. Fukuhara, C. Weitenberg, P. Schauß, C. Gross, L. Mazza, M. C. Bañuls, L. Pollet, I. Bloch and S. Kuhr, Observation of correlated particle-hole pairs and string order in low-dimensional mott insulators, *Science* **334**, 200 (2011).

27. J. V. José, L. P. Kadanoff, S. Kirkpatrick and D. R. Nelson, *Physical Review B* **16**, p. 1217 (1977).

28. Q. Faure, S. Takayoshi, S. Petit, V. Simonet, S. Raymond, L.-P. Regnault, M. Boehm, J. S. White, M. Månsson, C. Rüegg, P. Lejay, B. Canals, T. Lorenz, S. C. Furuya, T. Giamarchi and B. Grenier, Topological quantum phase transition in the ising-like antiferromagnetic spin chain BaCo2V2O8 (2017), arXiv:1706.05848.

29. P. W. Anderson, Absence of diffusion in certain random lattices, *Physical Review* **109**, p. 1492 (1958).

30. T. Giamarchi and H. J. Schulz, Anderson localization and interactions in one-dimensional metals, *Physical Review B* **37**, p. 325 (1988).

31. M. Girardeau, *Journal of Mathematical Physics* **1**, p. 516 (1960).

32. T. Giamarchi and P. Le Doussal, *Physical Review B* **53**, p. 15206 (1996).

33. T. Nattermann, T. Giamarchi and P. Le Doussal, *Physical Review Letters* **91**, p. 56603 (2003).

34. M. P. A. Fisher, P. B. Weichman, G. Grinstein and D. S. Fisher, *Phys. Rev. B* **40**, p. 546 (1989).

35. T. Giamarchi, C. Ruegg and O. Tchernyshyov, Bose-einstein condensation in magnetic insulators, *Nature Physics* **4**, p. 198 (2008).

36. F. Yamada, H. Tanaka, T. Ono and H. Nojiri, Transition from bose gass to a condensate of triplons in $tl_{1-x}k_x cucl_3$, *Physical Review B* **83**, p. 020409 (2011).

37. T. Hong, A. Zheludev, H. Manaka and L.-P. Regnault, Evidence of a magnetic bose glass in $(ch_3)_2 chnh_3 cu(cl_{0.95} br_{0.05})_3$ from neutron diffraction, *Physical Review B* **81**, p. 060410 (Feb 2010).

38. R. Yu, L. Yin, N. S. Sullivan, J. S. Xia, C. Huan, A. Paduan-Filho, N. F. Oliveira Jr., S. Haas, A. Steppke, C. F. Miclea, F. Weickert, R. Movshovich, E.-D. Mun, B. S. Scott, V. S. Zapf and T. Roscilde, Bose glass in dtn, *Nature* **489**, p. 379 (2012).

39. K. Cedergren, R. Ackroyd, S. Kafanov, N. Vogt, A. Shnirman and T. Duty, Insulating josephson junction chains as pinned luttinger liquids, *Physical Review Letters* **119**, p. 167701 (2017).

40. L. Sanchez-Palencia and M. Lewenstein, Disordered quantum gases under control, *Nat. Phys.* **6**, 87 (Feb 2010).

41. J. Vidal, D. Mouhanna and T. Giamarchi, *Physical Review Letters* **83**, p. 3908 (1999).

42. K. Hida, *Journal of the Physical Society of Japan* **69**, p. 311 (2000).

43. G. Roux, T. Barthel, I. P. McCulloch, C. Kollath, U. Schollwöck and T. Giamarchi, Quasiperiodic bose-hubbard model and localization in one-dimensional cold atomic gases, *Phys. Rev. A* **78**, p. 023628 (Aug 2008).

44. C. D'Errico, E. Lucioni, L. Tanzi, L. Gori, G. Roux, I. P. McCulloch, T. Giamarchi, M. Inguscio and G. Modugno, Observation of a disordered bosonic insulator from weak to strong interactions, *Phys. Rev. Lett.* **113**, p. 095301 (Aug 2014).
45. D. M. Basko, I. L. Aleiner and B. L. Altshuler, *Ann. Phys.* **321**, p. 1226 (2006).
46. P. Michal, V. L. Altshuler, B. and V. Shlyapnikov, G. Delocalization of weakly interacting bosons in a 1d quasiperiodic potential, *Phys. Rev. Lett.* **113**, p. 045304 (Jul 2014).
47. M. Schreiber, S. S. Hodgman, P. Bordia, H. P. Lüschen, M. H. Fischer, R. Vosk, E. Altman, U. Schneider and I. Bloch, *Science* **349**, p. 842 (2015).

Realizing quantum materials with Helium:
Helium films at ultralow temperatures, from strongly correlated atomically layered films to topological superfluidity

J. Saunders

Department of Physics, Royal Holloway University of London,
Egham, Surrey TW20 0EX, UK
E-mail: j.saunders@rhul.ac.uk
www.royalholloway.ac.uk

This article provides an overview, primarily from an experimental perspective, of recent progress and future prospects in using helium to realize a range of quantum materials of generic interest, by "top-down" and "bottom-up" nanotechnology. We can grow model systems to realise new quantum states of matter, and explore key issues in condensed matter physics. In the language of cold atomic gases, two dimensional and confined ^{3}He and ^{4}He provide "quantum simulators", with the potential to uncover new emergent quantum states. These include: strictly 2D Fermi system with Mott-Hubbard transition; interacting coupled 2D fermion-boson system; heavy fermion quantum criticality; ideal 2D frustrated ferromagnetism; 2D quantum spin liquid; intertwined superfluid and density wave order with emergent large symmetry; topological mesoscopic superfluidity (new materials and emergent excitations).

Keywords: Strongly correlated fermions; two dimensional; Kosterlitz-Thouless superfluid transition: supersolid; intertwined order; quantum spin liquid; topological superconductivity and superfluidity; chiral superfluid; Majorana fermions.

1. Introduction

Helium is unique: for both isotopes the combination of relatively weak van der Waals atomic attractions and strong zero point motion, due to small mass, leads to the stability of the bulk liquid phases down to absolute zero. These model systems of strongly correlated bosons (^{4}He) and fermions (^{3}He) have played a central role in the development of concepts in condensed matter physics. Superfluid ^{4}He demonstrated the first BEC, albeit with condensate fraction strongly depleted by interactions, and the first macroscopic quantum state. Liquid ^{3}He led to the development of Landau Fermi liquid theory—the standard model of strongly correlated fermions—with the striking *prediction* of collisionless zero sound, a collective-mode of the Fermi surface that can be driven and detected ultrasonically.

The topic of this workshop is topological phase transitions and topological quantum matter. Topological quantum matter has been classified[1] and has developed into a concept of wide applicability[2,3] with the discovery of topological insulators, and proposals of potential topological superconductors. Thus as an established topological superfluid ^{3}He is clearly a system of significant contemporary importance[4,5].

Superfluid ^{3}He was the first discovered unconventional superconductor/ superfluid (Nobel Prize 1996, 2003)[6-9]. It has $L = 1$ (p-wave) and $S = 1$ pairing, with a 9 (complex) component tensor order parameter allowing multiple superfluid phases. Normal liquid ^{3}He is the paradigm Landau Fermi liquid, underpinning the "standard model" of strongly correlated quantum matter; there is no lattice and the Fermi surface is a perfect sphere. The high degree of symmetry, $SO(3)_S \times SO(3)_L \times U(1) \times T \times P$, of the normal state is a key simplifying factor in revealing the broken symmetries of the emergent topological superfluid phases. It is easier to study anisotropic superfluidity if the Fermi surface is isotropic. There are two main stable phases in bulk, which break the symmetry of the normal state in different ways. There is clear evidence or pairing by spin fluctuations[10], extensively sought and discussed in heavy fermion superconductivity. Furthermore the importance of topology in momentum space, and the emergence of surface and edge excitations in these superfluids was discussed extensively in the relatively early literature[11-16].

The 2016 Nobel Prize reminds us of the power of ^{4}He to experimentally test theoretical concepts, in this case through the observation of the topological phase transition in superfluid ^{4}He films, as discussed elsewhere in this volume. (The fact that this was almost 40 years after the first detection of superfluid ^{4}He film flow graphically illustrates the power of interplay between theory and experiment). The ^{4}He film was grown on mylar, which is a heterogeneous substrate (the binding potential is non-uniform over the surface). The key features are that, beyond a ^{4}He so-called "dead-layer" of localised atoms a uniform mobile film is created, the density of which can be tuned continuously. The vortex interactions are precisely of the required logarithmic form (avoiding screening effects that plagued the observation in a 2D superconductor). And the experimental method of using a high Q torsional oscillator to observe the decoupling of the film from the substrate at the superfluid transition yield both the real and imaginary parts of the response function to validate the finite-frequency theory. This provides one topical template to highlight the power of helium to test key ideas in condensed matter physics.

This article mainly discusses the work of our group, as reported at this symposium. It is not a comprehensive review. It is intended to provide a broadly accessible overview of our research on helium films.

2. Bottom up: Strong correlation effects in two dimensional helium

The approach is to use graphite as a substrate. Atoms are physisorbed onto this substrate in a controlled way. The resultant film is "self-assembled", driven by a combination of atomic interactions with graphite, helium interatomic interactions, and zero point energy.

Graphite is atomically flat so that helium films grown on its surface are atomically layered. This is established both experimentally and theoretically. The binding energy of a ^{3}He atom to graphite is around $150\,$K, while that of ^{4}He is higher due to the larger mass and hence smaller zero point energy[17].

As the helium film is grown the complete first layer forms a compressed incommensurate solid on a triangular lattice. Atoms added to the second layer are subject to a periodic potential due to these first layer atoms. Typically films up to eight atomic layers have been studied. A key tool in the study of 2D ^{3}He is to pre-plate the graphite surface to effectively create a new substrate. This exploits the fact that ^{3}He is more weakly bound than any other species. Before growing a ^{3}He film, the substrate can be pre-plated, for example with: a solid monolayer of ^{4}He; a solid bilayer of ^{4}He; a solid HD bilayer. We find that each composite substrate allows us to use the subsequently grown ^{3}He film to model different physics. Thicker ^{4}He films consist of a number of solid layers next to the substrate with a superfluid ^{4}He overlayer. When ^{3}He is adsorbed onto this surface, the superfluid ^{4}He "substrate" plays an active role due to fermion-boson coupling.

These atomically layered helium films have allowed the study of a wide variety of phenomena:

- the study of 2D Fermi systems to test of the applicability of Landau Fermi liquid theory in 2D;
- the density-tuned Mott transition of a ^{3}He monolayer;
- Kondo-breakdown quantum criticality in a ^{3}He bi-layer;
- rustrated magnetism in 2D; Heisenberg ferromagnet in 2D;
- quantum spin liquid in a ^{3}He solid monolayer;
- Kosterlitz-Thouless transition in ^{4}He films;
- intertwined superfluid and density wave order in ^{4}He films – a 2D supersolid.

In practice exfoliated graphite is used as the adsorbate, to provide sufficient total surface area for measuring the properties of the adsorbed helium film. Typical specific surface areas range from $1-20\,$m^2/gm, depending on preparation. The potential influence of finite platelet size, interconnectivity and orientation of platelets, residual surface disorder are taken into account in the analysis. In particular, the detection of superfluidity of a 2D Fermi system, that is of a ^{3}He fluid monolayer, requires an adequately small level of surface disorder, to which p-wave superfluidity would be particularly sensitive.

2.1. *Mott-Hubbard transition*

A monolayer of ^{3}He on graphite preplated by a bilayer of HD shows, from heat capacity and magnetization measurements, a clear density-tuned effective mass-diverging Mott transition, at which the Landau Fermi liquid parameter F_0^a saturates[18]. This confirms the almost-localised fermion picture of strongly correlated fluid ^{3}He, in which the strongly repulsive hard core repulsion is dominant[19]. This system also allowed the measurement of non-analytic finite T corrections to the heat capacity,

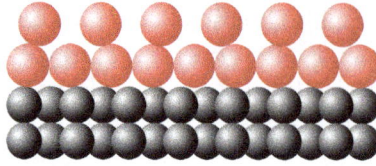

Fig. 1.   Cartoon of an atomically layered helium film on the atomically flat surface of graphite. Helium atoms are represented by hard spheres. Shown is a two layer film. The complete first layer forms a compressed solid on a triangular lattice. In such films the total areal density (coverage) can be adjusted with precision.

which support the theoretical claim that, although anomalous, Fermi liquids can survive in 2D[20], in contrast to earlier ideas[21].

A related case is the second layer of ^{3}He on graphite (including the case where the first layer is replaced by ^{4}He)[22]. Our recent NMR study[23] shows a relatively wide density range over which there is a quantum coexistence of fluid and solid, with no evidence for a hole-doped Mott insulator and associated Fermi surface reconstruction. In this case we argue that the ^{3}He experiences a density tuned Wigner-Mott-Hubbard transition. However, unresolved issues remain with behaviour at the very lowest temperatures $T \ll T_F^*$, in terms of unexplained temperature dependences of the heat capacity and magnetization. Whether this behaviour is intrinsic, arising from fermionic correlations (for example formation of a fermionic flat-band)[24], or extrinsic, arising from residual surface disorder, may be tested by experiments using substrates of improved quality.

In both cases discussed above the Mott insulator is a strong candidate for the long-sought quantum spin liquid material, as discussed later.

2.2. *Heavy fermion physics and quantum criticality in a ^{3}He bilayer*

When, on the other hand, ^{3}He is grown on graphite plated by a bilayer of solid ^{4}He, yet new strongly correlated physics emerges. The fluid ^{3}He bi-layer behaves as a heavy fermion system[25]. The lower layer (L1) plays the role of the f-fermions and the second layer (L2) is analogous to the mobile conduction electrons. The two layers are hybridized by a Kondo interaction: in this case exchange of atoms between the two layers. This is tuned by the density of the upper layer. A density-tuned quantum critical point (QCP) is found at which the effective mass diverges. This appears to fall into the class of orbital-selective Mott transition[26], with a Kondo-breakdown QCP[27-31], at which the effective mass diverges. Beyond this QCP, layer L1 is localised and layer L2 is itinerant, consisting of weakly-interacting 2D fermions. The frustrated magnetism of atomic ring exchange plays a role in L1. Approach to the QCP is intercepted by a magnetic instability, which it is believed is triggered when the ferromagnetic intralayer exchange in L1, dominates the interlayer Kondo coupling[32,33].

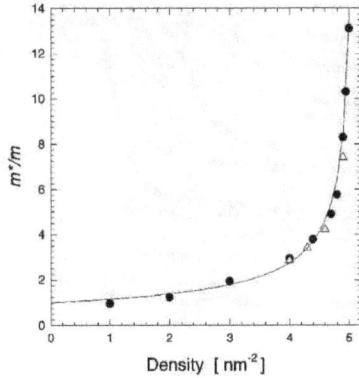

Fig. 2. Effective mass divergence at a density driven Mott transition, in a 2D ^{3}He monolayer. The substrate is graphite, plated by a bilayer of HD[18]. The tuning parameter is the ^{3}He surface density. The effective mass is determined from the heat capacity. The effective mass inferred from magnetization data are consistent, within the framework of the almost localized fermion model, for which $(1 + F_0^a)$ saturates at a finite value of approximately 1/4. 2D ^{3}He is almost localised *not* almost ferromagnetic. The Mott insulator is a strong candidate to be the elusive quantum spin liquid.

2.3. Superfluidity of atomically layered ^{4}He films

If a ^{4}He film is grown on graphite, superfluidity is observed in the third and sub-sequent layers[34]. The first two layers are solid ^{4}He. Anomalous properties in the second layer, interpreted as a 2D supersolid are discussed in Section 2.6. A related system is ^{4}He on graphite plated by HD[35]. Superfluidity in these atomically layered films shows a Kosterlitz-Thouless (KT) transition. The KT transition when measured using a torsional oscillator at finite frequency has a somewhat rounded "jump" in the superfluid density accompanied by a peak in dissipation. These are inferred from the period shift and quality factor, which determine the real and imaginary parts of the vortex response function, respectively. A parametric Cole-Cole plot of the real *vs.* imaginary parts of the superfluid response function provides a characteristic fingerprint of the transition. Indeed the data for a range of ^{4}He coverages, and for a variety of surface preplatings collapse well onto a "universal" contour[36], albeit with small quantitative discrepancies relative to the theoretically predicted contour.

There is a further important feature of the growth of these atomically layered ^{4}He films. As each layer forms, it undergoes spinodal decomposition, at layer densities below around $4\,\mathrm{nm}^{-2}$. In other words each 2D fluid layer will self-condense (liquid-gas instability in 2D)[37,38] forming puddles with a self-bound density $4\,\mathrm{nm}^{-2}$. This effect is inhibited in ^{4}He films on heterogeneous substrates. We draw the reader's attention to an unexplained observation that we find particularly striking. This is the suppression of superfluidity approaching the spinodal point from higher densities, observed both with and without HD pre-plating, that suggests an intrinsic

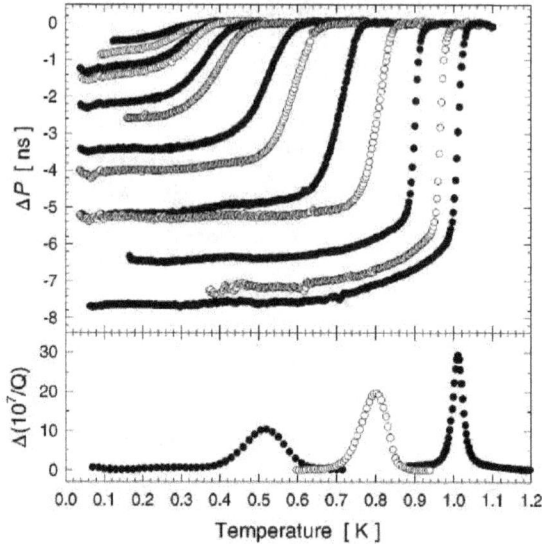

Fig. 3. Kosterlitz-Thouless superfluid transitions of ^{4}He films on graphite plated by a trilayer of HD[35]. ^{4}He coverages range from 7.0 to 13.1 nm^{-2}. Similar results are found with HD bilayer preplating. A parametric plot of the real vs. imaginary parts of the superfluid response, for both preplatings, collapse onto a "universal" curve[36].

mechanism for the suppression of 2D superfluidity, as yet theoretically unexplained. We conjecture that this may either arise from the strong periodic potential of the first solid ^{4}He layer atoms, or from the influence of the proximity of the spinodal point on the vortices and their dynamics.

2.4. ^{3}He 2D Fermi liquid

Nevertheless, the three layer film (superfluid ^{4}He monolayer sitting on a solid ^{4}He bilayer), and the four layer film (superfluid ^{4}He bilayer sitting on a solid ^{4}He) are relatively well understood. They appear to provide ideal "substrates" for the study 2D ^{3}He. The study of these, so-called, helium mixture films has been the the subject of extensive prior work using heterogeneous substrates[39,40]. The use of atomically layered ^{4}He films on graphite, as in the work descibed here, has significant advantages. These films are adsorbed on graphite, and so can be readily cooled into the microkelvin regime. Furthermore, the uniformity of the ^{4}He film is of crucial importance, since it guarantees a uniformity of the fermion-boson coupling, which is enhanced by the graphite substrate[41].

As ^{3}He is added to the ^{4}He film, a 2D ^{3}He Fermi system of tuneable density is built on the single particle surface ground state for motion normal to the surface. We find that on the 3-layer ^{4}He film, the ^{3}He system shows a number of instabilities for coverages below 1 nm^{-2}, with possible evidence for ^{3}He dimer formation in a very low density component[42]. At higher ^{3}He densities above 1 nm^{-2} the superfluidity of the ^{4}He film is suppressed and quenched for coverages above around $4\,\mathrm{nm}^{-2}$. On the other hand, for the 4-layer film the instability region at low ^{3}He coverages is significantly reduced, and a single Fermi system can be studied up to $4.5\,\mathrm{nm}^{-2}$, at which (in the most naïve picture) it becomes energetically favorable for two (ground and first excited) surface normal states to become occupied[43], see Fig. 4. In practice these two sub-bands will be hybridized versions of those expected in a non-interacting ideal Fermi gas picture.

Over the coverage range at which the ^{3}He forms a single 2D Fermi system, the conditions are satisfied for the interactions in this 2D system to be strictly two dimensional. This is in contrast to 2D cold atom systems studied thus far, where interactions are 3D (eg. the system can be tuned through a Feshbach resonance, absent for 2D interactions). Thus ^{3}He provides an interesting model system to explore interactions in 2D, of significant interest theoretically[44–49]. This is important in the wider context of: two dimensional/interface superconductivity in metals[50], incuding FeSe/STO[51], Li/graphene[52,53]; cold atoms, experiments on multiple quasi-2D layers[54,55].

The ^{3}He system allows a number of key questions to be addressed:

- Do Fermi liquids survive in 2D[21]?
- What is the nature of the interactions, in particular are there anomalies associated with interactions in 2D? [Here the ability to tune density over a wide range is important].
- Is there evidence of ^{3}He–^{3}He interactions mediated by the ^{4}He bosonic film? [These interactions, mediated by phonon/3rd sound-like excitations are expected to be (graphite) substrate enhanced. Fermion and boson modes should be hybridized[41]].
- Can we enter a low density regime in which Fermi gas theory applies[56,57], and Landau parameters can be calculated by a single, density dependent, scattering parameter?
  With regard to the last point, this is challenging because of the logarithimic dependence of the interaction parameter in 2D;

$$g = -1/\ln(2E_\mathrm{F}/E_\mathrm{b}) \quad \text{or} \quad g = -1/2\ln(\sqrt{2}k_\mathrm{F}a).$$

- Under what conditions might the 2D superfluidity of a system of fermions in 2D be realised[48,58]?

So far the experiments have been to measure heat capacity and nuclear magnetic susceptibility[42,43]. The enhancement of the Pauli susceptibility relative to the ideal Fermi gas value has been determined by the highly sensitive low frequency SQUID NMR method[59], so far to temperatures as low as $200\,\mu\mathrm{K}$. NMR selectively measures the susceptibility of the Fermi system; there are no background corrections. In

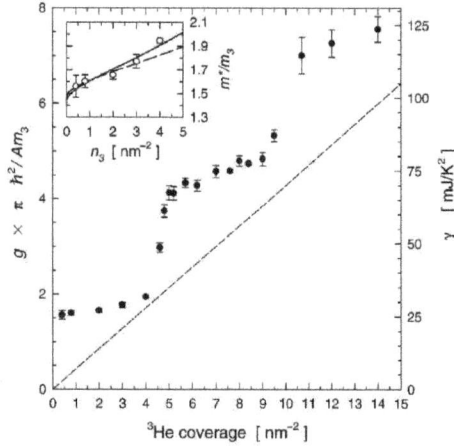

Fig. 4. Energy density of states, normalised by Fermi gas value, for ^{3}He on a four layer ^{4}He film on graphite. Below a ^{3}He coverage of 4 nm^{-2} the ^{3}He forms a 2D Fermi system built on the ground state for motion along the surface normal. Steps indicate the population of excited surface-normal states ("multiple sub-band occupancy").

Landau Fermi liquid theory it is given by

$$\frac{\chi}{\chi_0} = \frac{1 + \frac{1}{2}F_1^s}{1 + F_0^a} \frac{m_H}{m_3},$$

This is temperature independent for $T \ll T_F^{**}$, where $T_F^{**} = \frac{0.505(1+F_0^a)}{m^*/m} n_3$ K nm^2. The Fermi gas susceptibility, χ_0, is independent of ^{3}He density in 2D. Thus the nuclear magnetic susceptibility provides a good measure of correlation effects to arbitrary low temperatures and for arbitrary low surface density. The enhancement of the coefficient of the linear in T heat capacity is $\gamma/\gamma_0 = (1 + \frac{1}{2}F_1^s)(m_H/m_3)$. In this way F_0^a can be determined with precision. The relative density dependence of F_0^a and F_1^s found experimentally[60], can be compared with the predictions of microscopic theory[47], and we find that both s-wave and p-wave scattering are essential to explain results, such that back-scattering dominates.

At somewhat higher density, with "multiple sub-band occupancy", we have two coupled 2D Fermi systems, of potential relevance to coupled 2D layers in cuprate superconductors.

2.5. *The second layer of helium films on graphite: 2D supersolid and quantum spin liquid*

In this section we focus on to two quantum states: a 2D supersolid (realised in a ^{4}He atomic layer) and a quantum spin liquid (for which there is mounting evidence in a ^{3}He atomic layer), and we speculate on the potential relationship between them. As we shall see these two states are manifested at essentially the same helium coverage, occurring in the second atomic layer of helium on graphite, prior to the completion of that layer. All we do is replace bosons by fermions, with the same interactions, albeit different mass. We ask, might this fermion-boson "correspondence" be significant?

The realization of supersolid in cold atomic gases is a matter of active research[61,62]. The realization of the quantum spin liquid in layered magnetic materials has been a long-term quest, with many "candidates"[63,64]. It would be quite appealing if both of these highly entangled quantum ground states were found in two dimensional helium.

2.6. *Intertwined superfluid and density wave order: 2D supersolid*

The 2D supersolid phase that we identify[65,66] is quite different from the putative supersolid phases proposed in the literature in the context of bulk solid ^{4}He. We argue that the state of intertwined superfluid and density wave order, has a larger emergent symmetry[67]. It is a new quantum state of matter.

On the other hand the "classic" supersolid has a well-defined solid crystal structure with mobile vacancies constituting the superfluid component[68–71]. In the case of bulk solid ^{4}He it was proposed that these arise spontaneously[72]. Experiments using torsional oscillators to detect the mass decoupling associated with superfluidity of the solid, typically performed at kHz frequencies, must contend with a large visco-elastic response of the crystal[73]. There is no clear experimental evidence for superfluidity by this method. There is evidence for superfluidity at the cores of dislocations[74,75], as predicted theoretically[76]. All these proposals concern superfluidity associated with defects, coexisting with crystalline order.

On the other hand, in the ^{4}He layer, we propose a 2D supersolid as a new quantum material[65,66]. This exotic quantum state would have both the rigidity of a solid, but also paradoxically be able to flow without resistance. To account for the observations, we propose a new quantum state in which density wave order and superfluid order are fully quantum entangled, represented by

$$|\Psi\rangle = \exp\left(\alpha_0 b^\dagger_{\mathbf{q}=0} + \sum_{\mathbf{G}} \alpha_{\mathbf{G}} b^\dagger_{\mathbf{G}}\right) |0\rangle,$$

which describes a quasi-condensate at both zero momentum and finite momenta $\mathbf{G}_i$, the set of reciprocal lattice vectors of triangular lattice. A quasi-condensate at $\mathbf{q} = 0$, in the presence of density wave order, necessarily implies a quasi-condensate at the set of wavevectors $\mathbf{G}_i$.

We hypothesize that the state vector freely explores the Bloch hyper-sphere, and hence varying relative occupation of condensates, as well as varying degrees of "solid" and "superfluid" order. In other words, the Hamiltonian commutes with rotations transferring particles between the quasi-condensates. This is not a fragmented condensate[77]. It is a quantum material described by a macroscopic wave function (actually with power law correlations in space because we are in 2D), which is a Schrödinger-cat like state: the entire system can be both solid *and* superfluid.

Thus quantum mechanics resolves the central paradox of the supersolid, arising from our everyday intuition which tells us that the properties of a solid and a superfluid are contradictory. This is not a state of matter in which solid and superfluid simply coexist. It is an entangled quantum state in which density wave order and superfluid order are intertwined.

We briefly summarise the observations that lead to these proposals; a fully referenced discussion is provided elsewhere[65]. We measure the superfluid response of a bilayer ^{4}He film adsorbed on exfoliated graphite with a torsional oscillator. The first layer forms a compressed solid and, apart from some viscoelastic response, which can be corrected for, can be considered passive. Superfluid response occurs over a narrow range of film densities in the second layer close to its completion. The superfluid response shows no evidence of a KT transition (jump in superfluid density) and even shows no sharp onset. The superfluid density shows an anomalous temperature dependence; the leading order dependence in the $T \to 0$ limit is linear in T. We introduce the coverage/density dependent characteristic energy scale $\Delta(n)$, which governs both the temperature dependence of the normal fraction, and the $T = 0$ superfluid fraction:

$$\frac{\rho_s(T, n)}{\rho} = \frac{\Delta(n)}{T_0} f(T/\Delta(n)).$$

We can scale data to demonstrate quantum criticality. The quantum critical point, at which $\Delta(n)$ extrapolates to zero, is at layer completion. The normal density in the $T \to 0$ limit has an unusual linear in T dependence. This is accounted for by an *ansatz* for the elementary excitation spectrum, in the spirit of Landau. The Landau prescription for calculating ρ_n involves a momentum-weighted integral. The proposed spectrum has an extremely soft roton-like minimum, such that the roton gap is smaller than the minimum temperature explored so far experimentally (around $2\,\mathrm{mK}$). Nozières refers to the roton as the "ghost of the Bragg spot". The small roton gap also implies a strong peak in the structure factor via the Feynman-Cohen relation $E(q) = \hbar^2 q^2 / 2mS(q)$. Thus the inferred excitation spectrum implies density wave order. The *ansatz* for the quasi-condensate wavefunction has both superfluid and density wave order. The fact no Kosterlitz-Thouless transition is observed can be explained if the state is not such that the U(1) symmetries of both orders are separately broken. Rather the two orders a fully intertwined, the ground state has SU(N) symmetry, and vortices are not stable. This is a non-Abelian superfluid. It is important to note that these conclusions are drawn in the absence of knowledge of the system Hamiltonian. The starting point is not a model Hamiltonian that we simulate, but rather the emergent properties we experimentally observe.

The fact that this supersolid is observed in the second layer of helium on graphite is a manifestation of the highly quantum nature of the system. In bulk quantum solid helium[72], the atoms have high zero-point motion, the amplitude of which is a significant fraction of the lattice parameter. The overlap of neighbouring atomic wave-functions gives rise to atomic exchange, at a rate orders of magnitude smaller than the Debye frequency, and strongly dependent on lattice parameter (a function of pressure)[78], decreasing with molar volume as V_m^γ with $\gamma \approx 60$. Thus the atoms in solid helium move from site to site, and in solid ^{3}He this manifests as an exchange interaction between the nuclear spins. Now, in the region of the second layer of helium on graphite that is presently of concern to us, the rate of inter-site atomic motion is significantly higher than in bulk solid because of the much lower density. This motion also has been detected through thermodynamic studies of ^{3}He impurities introduced into the ^{4}He second layer[79,80].

2.7. Quantum spin liquid

We now turn to the potential realization of a quantum spin liquid phase in 2D solid ^{3}He. Here the magnetism arises from the ^{3}He nuclear spin, and is interrogated by NMR and heat capacity measurements[81–85]. In contrast to ^{4}He, different phases (solid, fluid) can easily be distinguished through signatures in these thermodynamic probes.

The key feature of 2D solid ^{3}He is that it forms a triangular lattice, and is therefore geometrically frustrated. Furthermore, the spin interactions are highly frustrated by competing atomic ring exchange[86]. Ring exchange of an odd number of particles is ferromagnetic (FM), even is antiferromagnetic (AFM). The ring exchange interactions are strong in 2D, and significantly higher than in 3D solid helium, because of both high in-plane zero point motion, low density and zero point motion out of plane. Thouless[87] first proposed the effective spin Hamiltonian, in terms of permutation operators:

$$P_2 = \tfrac{1}{2}(1 + \boldsymbol{\sigma}_1 \cdot \boldsymbol{\sigma}_2)$$

$$\mathcal{H} = \sum_n (-1)^n J_n P_n \qquad P_3 = \tfrac{1}{2}(1 + \boldsymbol{\sigma}_1 \cdot \boldsymbol{\sigma}_2 + \boldsymbol{\sigma}_2 \cdot \boldsymbol{\sigma}_3 + \boldsymbol{\sigma}_3 \cdot \boldsymbol{\sigma}_1)$$

$$P_4 \text{ includes terms like } (\boldsymbol{\sigma}_1 \cdot \boldsymbol{\sigma}_2)(\boldsymbol{\sigma}_3 \cdot \boldsymbol{\sigma}_4)$$

The effective Heisenberg Hamiltonian $J = J_2 - 2J_3$ is FM because three particle exchange dominates two particle exchange. This is a consequence of the fact that helium atoms are "hard spheres".

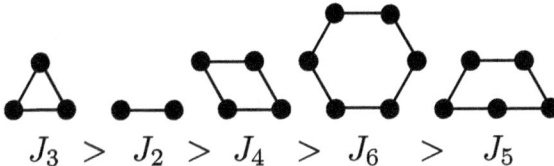

$$J_3 > J_2 > J_4 > J_6 > J_5$$

Fig. 5. Hierarchy of cyclic ring-exchange interactions in 2D ^{3}He on a triangular lattice[86,88].

For simplicity, and the purposes of illustration, we truncate at 4 particle exchange. We refer to this two parameter model as the $J - J_4$ model. In principle these exchange parameters can be inferred from experiment, since the effective exchange parameters which enter the magnetic susceptibility, heat capacity, and spin wave velocity to leading order are different and take the form:

Curie-Weiss constant $\quad J_\chi = -(J + 3J_4) \qquad M = \dfrac{c}{T - \theta} \quad \theta = 3J_\chi$

Spin wave velocity $\quad J_S = -(J + 4J_4)$

Heat capacity $\quad J_c^2 = (J + 5J_4/2)^2 + 2J_4^2 \qquad C = \dfrac{9}{4} N k_B \left(\dfrac{J_c^2}{T^2} \right).$

These exchange parameters are tuned by the total film coverage, as shown in measurements of magnetization and heat capacity on the same samples[83,89]. The second layer with a fluid overlayer behaves as a 2D Heisenberg magnet with weak frustration: its properties are fully consistent with the Mermin-Wagner theorem[90] (there is no finite$-T$ phase transition). The temperature dependent spontaneous magnetism (observed by NMR in a weak magnetic field) at finite T is accounted for by 2D spin-wave theory, with a small Zeeman gap. Perhaps this is the cleanest example of 2D ferromagnetism, with no complications from inter-layer coupling as in quasi-2D magnetic materials, and well characterised exhange.

Our recent results suggest that the solid second layer is driven to ferromagnetism, by the indirect RKKY interactions mediated by the fluid overlayer[91]. This is inferred from the evolution of the NMR lineshape (obtained using low frequency SQUID NMR), as it has been established tha the third (fluid) overlayer initially forms in low density self-condensed puddles[92].

We now turn to the three physical ^{3}He-on-graphite systems in which the quantum spin liquid (QSL) may manifest:

(i) The second layer of ^{3}He on graphite, where the first layer is ^{3}He. In this case the first layer of ^{3}He is a compressed solid on a triangular lattice [confirmed by neutron scattering][88], paramagnetic, with very weak exchange interaction with the second layer. The coupled magnetism of the first and second layer is a complication. However the fact that the first layer is a weakly interacting "spectator" of the putative QSL in the second layer, may prove to be advantageous.

(ii) A monolayer ^{3}He on graphite, preplated by a solid monolayer of ^{4}He. This system is very closely related to (i). However the paramagnetic ^{3}He first layer is replaced with non-magnetic ^{4}He. The density of the close packed ^{4}He first layer triangular lattice is about 5% higher than the ^{3}He first layer. Given this close correspondence we will refer to this system also as "the second layer of ^{3}He on graphite".

(iii) A monolayer of ^{3}He on graphite, preplated by a solid bilayer of HD[94,95].

In all of the above cases, the putative QSL is a 2D solid on a triangular lattice, at the border of a density tuned Mott-Hubbard transition. This is a spin $\frac{1}{2}$ system [^{3}He nuclear spin]. As well as the geometrical frustration of the triangular lattice, there is strong frustration due to competing atomic ring exchange interac-

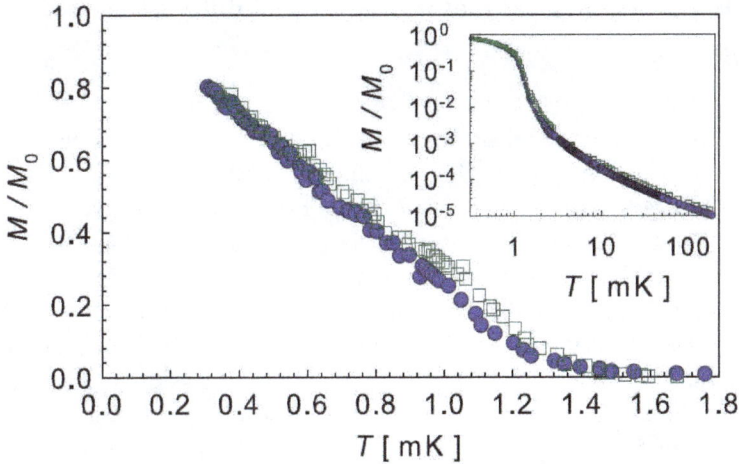

Fig. 6.   Magnetization of the second ^{3}He layer on graphite, in presence of a fluid overlayer, relative to its fully polarized saturation value at $T = 0$. Results are consistent with Mermin-Wagner theorem for a 2D ferromagnet. Measurements in two magnetic fields show system is well described by 2D spin-wave theory. Results show that spin-wave spectrum is governed by a different exchange constant from that determining high-T magnetization (see text), as a result weak frustration[93].

tions. The magnetization is directly, and selectively, measurable by NMR. The high temperature magnetism shows that the system has an antiferromagnetic character. Magnetization measurements on both system (ii) and system (iii) support a gapless spin liquid[96]. In our recent work in system (ii) we find that the low temperature magnetism is consistent with a Pauli susceptibility[23], as expected for a gapless spin liquid, with a characteristic energy scale of a few hundred μK.

Our current belief is that system (ii) and (iii) reflect a different balance between the periodic potential of the solid underlayer (HD bilayer or ^{4}He) on a triangular lattice and intralayer ^{3}He interactions. The HD bilayer is of significantly lower density than the ^{4}He first layer. It shows a Mott-Hubbard transition into a 4/7 or 7/12 triangular superlattice phase. The results for system (ii) are more consistent with a density wave instability in the ^{3}He layer. In system (iii) we have shown that exchange in the Mott insulator is much stronger than in system (ii)[94]. This is understood in terms of the lower density. Therefore a monolayer of ^{3}He on graphite, preplated by a solid bilayer of HD may be the most promising for demonstrating quantum spin liquid behaviour.

We can also use the information on ring exchange processes from the frustrated magnetism of ^{3}He system (ii) to estimate the characteristic energy scale governing atomic mobility in the ^{4}He (supersolid) case. Now all ring exchange processes add to translate a ^{4}He atom between sites, so we have $t = J_2 + 2J_3 + 4J_4 + \ldots$. Using the values calculated by PIMC we find $t \sim 200$ mK, comparable to the temperature at which a superfluid response is detected.

3. Top down: Topological mesoscopic superfluidity

3.1. *Motivation*

In contrast with the "self-assembled" films discussed in section 2, here we use nanofabrication methods to define a cavity. In the simplest case we have a thin slab geometry, into which helium is admitted through a fill line. This can be thought of as a film, of thickness precisely defined by the height of the cavity, with equivalent upper and lower surfaces. This strategy is particularly suitable for the study of superfluid ^{3}He, where the diameter of the Cooper pair ξ_0 at zero pressure is around 80 nm. [Stabilization of van der Waals films of such thickness is tricky in the face of competing effects of surface tension and gravity. Moreover it is not easy to precisely determine the film thickness of such a "bottom-up" thick film]. So far cavities of height D in the range 1000 to 100 nm have been studied, $k_F^{-1} \ll D < 10\xi_0$. While this is far away from the strictly 2D limit $k_F^{-1} \sim D \ll \xi_0$, it should allow that limit to be approached in a controlled way in future experiments. In our approach, the thickness of the "film" formed by helium in the cavity is well defined by the confining geometry. Moreover for fixed cavity height the effective confinement ξ_0/D is tuneable by pressure, since $\xi_0 = \hbar v_F/2\pi k_B T_c$.

Our motivation is the study of topological superfluid ^{3}He under confinement, as a model system for topological superconductivity. The topological classification of condensed matter systems has now become established as a key general principle to understand, predict and design new states of matter. There are several *candidate* topological superconductors, discussed elsewhere in this volume. Here we exploit superfluid ^{3}He, where the topological classification is firm. Under confinement the goal is to use it as a model system for the investigation of emergent surface excitations.

The gift of Nature is that the p-wave superfluid ^{3}He supports both a time reversal invariant (TRI) phase (^{3}He-B) and a chiral phase (^{3}He-A). The richness of the order parameter of superfluid ^{3}He also allows new potential phases (*new "materials"*), which should be stabilised by confinement. The most striking feature in topological quantum matter is the surface and edge excitations which emerge through bulk-surface/edge correspondence. In a topological superfluid these are (TRI superfluid) Majorana or (chiral superfluid) Weyl fermions. It is the understanding, detection, characterization and possible quantum control of these new particles, which is the key challenge. In our neutral topological "superconductor", the topological invariant is a property of the emergent superfluid order. Thus the "protection" of the surface/edge states is expected to be influenced by a subtle interplay between broken symmetry and topology[4,5].

This approach opens the way to the study of ^{3}He is studied in hybrid nanostructures, "topological mesoscopic superfluidity". ^{3}He allows a degree of control that is unprecedented. Confinement (for example in a cavity of height D) is the key new control parameter and is used to stabilise distinct superfluid phases or normal phase, and the interfaces between them, with exquisite control over interface quality, the elimination of disorder, and flexibility of geometry. The ^{3}He hybrid mesoscopic structure can built from a range of "materials", with ^{3}He tuned into a particular state by the scale and sculpture of confinement. Each superfluid phase has different symmetry, topology and surface/edge excitations, with further control over excitations at engineered intra-fluid interfaces. The helium-cavity surface interfaces are also of high quality: they are atomically flat and, as has been shown, the surface scattering is tuneable *in situ*, by a ^{4}He surface film[97–99]. The flexibility this approach offers provides a clear opportunity to reveal edge, surface and interface states in a model topological superfluid.

3.2. Topological superfluid ^{3}He under engineered nanoscale confinement: Methods

The first experiments[100] required a number of technical breakthroughs: to confine ^{3}He in a nanofabricated rectangular cavity of thickness comparable to the coherence length, including a precise characterisation of the geometry and surfaces of cavity; to cool the sample to well below the superfluid transition temperature; to "fingerprint" the superfluid order parameter by developing an NMR spectrometer of unprecedented sensitivity[101]; to measure superfluid density of such a small sample with a torsion pendulum.

The first generation of cells (fabricated by anodically bonding Hoya SD-2 glass and Silicon)[102], were of typical area 1 cm^2, with a cavity height of 700 nm. We subsequently developed all-silicon structures, direct wafer bonded, with cavity heights (in the range 50 to 300 nm) precisely defined by an array of posts, and modelled successfully by finite-element techniques. The ^{3}He confined within this cavity is cooled through the ^{3}He in the metallic fill line. This relies on the fact that the thermal conductivity of normal liquid ^{3}He at 1 mK is high, comparable to that of high quality copper. The fill line links the sample to a silver sinter heat exchanger mounted on a cold-plate, which is thermally linked tothe nuclear stage.

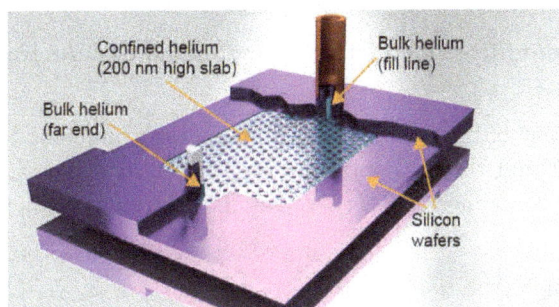

Fig. 7.   Artist's impression of an all-silicon nanofluidic cell. Cavity height is 200 nm, defined by array of silicon posts. Small volumes of bulk helium act as markers to compare with response of helium confined in cavity.

NMR plays a central role in the study of superfluid ^{3}He. The ^{3}He nuclear spins are the spin degrees of freedom of the Cooper pairs; NMR directly interrogates the Cooper pairs, and through the coherent dipole interaction is used to "fingerprint" the superfluid order parameter ("pair-wavefunction"), in a direct and non-invasive way[6,103]. In the absence of Meissner effect and skin depth effects (present in superconductors) it can be used to probe the entire sample. This is very different from NMR studies of other quantum materials, where the nuclei are essentially spectators, hyperfine coupled to the strongly correlated electron system, and the focus is on Knight shift and T_1. The power of NMR was graphically illustrated by the rapidity with which the order parameters of superfluid ^{3}He phases were established, following initial discovery[6,103,104]. This should be contrasted with unconventional metallic and heavy fermion superconductors, where the complexity of the Fermi surface adds further to the problem.

While NMR thus provides, in principle, the method to directly determine the influence of confinement on the order parameter, including the stabilization of new phase, the experimental challenge is signal sensitivity of the NMR spectrometer. To address this question we developed a SQUID NMR method[101]. The SQUID NMR technique allows measurements on a single well-characterized ^{3}He slab. Previously confinement was achieved by immersing a stack of typically 1000 mylar sheets separated by dispersed polyamide spheres[105,106]. Unfortunately, this inevitably results non-uniformity of confinement, which is undesirable since confinement is the key control parameter. The presence of significant signals from surrounding bulk liquid also can complicate the discrimination of the NMR response of the confined sample.

The flexibility of nanofabrication allows the introduction of small volumes of "bulk" superfluid ^{3}He, both near the fill line entrance to the cavity and at the far end of the cavity. These "bulk-markers", with well defined geometry, can be resolved from the cavity NMR signal, by simple NMR imaging techniques ("zeugmatography" – involving the applications of magnetic field gradients for spatial resolution). The "bulk-markers"define the bulk-liquid superfluid transition temperature, and show that temperature gradients across the cavity are insignificant.

3.3. *Profound modification of the superfluid phase diagram due to confinement*

The relative stability of the A and B phases is strongly influenced by confinement. This arises because all three components of the orbital triplet are present in the B-phase, and the component with $l_z = 0$ is most strongly suppressed by surface scattering.

$$\mathbf{\Delta}(\mathbf{p}) = \left[\Delta_\parallel(-\hat{p}_x + i\hat{p}_y)\left|\uparrow\uparrow\right\rangle + \Delta_\parallel(\hat{p}_x + i\hat{p}_y)\left|\downarrow\downarrow\right\rangle + \Delta_\perp\hat{p}_z(\left|\uparrow\downarrow\right. + \left.\downarrow\uparrow\right\rangle)\right].$$

On the other hand, in the A-phase, all pairs have the same orientation of their angular momentum,

$$\mathbf{\Delta}(\mathbf{p}) = \Delta(\hat{p}_x + i\hat{p}_y)\left(\left|\uparrow\uparrow\right\rangle + \left|\downarrow\downarrow\right\rangle\right).$$

Under confinement the orbital angular momentum orients perpendicular to the wall. As a result the A-phase, which in bulk is only stable at high pressures in zero magnetic field, is favoured at low pressures, where the effective confinement parameter D/ξ_0 is smallest. In a 700 nm cavity the A phase is stabilised at $T = 0$ from zero up to around 2 bar.[100]

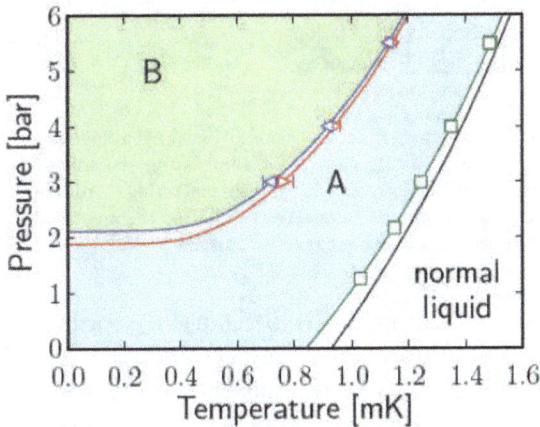

Fig. 8. The phase diagram of superfluid ^{3}He confined in a slab-like cavity of height 700 nm[100]. A-phase is stabilized at low pressure. Phase diagram shown is for diffusely scattering walls. In this case there is also a pressure dependent suppression of slab T_c relative to bulk (solid black line).

Consistent results for the A-B transition line were found in NMR experiments on 700 nm and 1100 nm cavities[100,107] and torsional oscillator experiments on a 1080 nm cavity[108]. The critical value $D/\xi(T_{AB})$ is pressure dependent and differs at all pressures from the prediction of weak coupling theory. The conclusion from these observations is that strong coupling corrections persist to zero pressure and are temperature dependent at each pressure. [Strong coupling refers to the dependence of the pairing interaction on the superfluid order parameter[109]].

As the cavity height is reduced,the expectation is that the A-phase will be stable over a progressively wider range of pressure. We expect that a re-entrant bubble (in the $p-T$ plane) of planar distorted B-phase will shrink and eventually disappear on reducing the cavity height: the details will depend on strong coupling effects and the influence of confinement. The superfluid phase under strong confinement $D \sim \xi_0$ is of great interest. These require control over the surface scatering conditions, which we now discuss.

3.3.1. Tuning the surface scattering

Since the discovery of superconductivity in heavy fermion metals and oxide materials the majority of emerging superconducting materials exhibit unconventional, non-s-wave, pairing. In contrast to s-wave superconductors, they are extremely sensitive to non-magnetic defect, and surface scattering. Superfluid ^{3}He is naturally defect-free (although there is a significant body of work investigating the influence of introducing disorder by immersing the superfluid in silica-aerogels). In

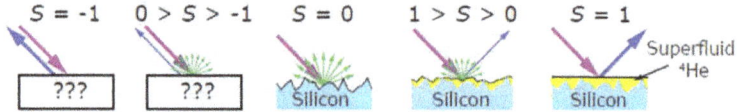

Fig. 9. Cartoon of different surface scattering conditions: from specular ($S = 1$) through fully diffuse ($S = 0$) to fully retro-reflecting $S = -1$. Eliminating the solid ^{3}He boundary layer, by plating the surface with solid ^{4}He, results in diffuse scattering. Adding further ^{4}He to create a superfluid layer results in specular scattering. Creating the maximally pair-breaking retro-reflecting surface will require a specially engineered surface profile.

our experiments under confinement, the distinguishing feature is that we just have surface scattering, which is *tuneable*; the surface scattering depends strongly on the helium surface boundary layer. Theoretically the calculation of surface suppression of components of the gap is made using quasiclassical theory; surface scattering is introduced phenomenologically[110]. The limits are retroreflection, (random) diffuse, specular: specularity can be varied continuously between these limits. We have demonstrated the ability to tune surface quasiparticle scattering from magnetic, to non-magnetic diffuse, to specular, by precise measurements of T_c and gap suppression using NMR on a single cavity of height 200 nm[99]. The chiral A-phase is found to be stable at all pressures studied, $0 - 5.5$ bar. Quasiclassical theory provides a consistent desciption of both the measured T_c suppression and gap suppression, in the non-magnetic scattering case.

With diffusively scattering surfaces p-wave superfluidity is suppressed for $D < \xi_0$. For perfectly specular surfaces both T_c and gap suppression are eliminated for order parameters with only $l_z = \pm 1$ components. These are the chiral A-phase and the TRI planar phase, which are degenerate in the weak coupling limit. Experiments are in progress, at the time of writing, to establish the phase diagram in 100 nm cavities. In part they rely on the ability, demonstrated in the experiments on a 200 nm cavity, to reproduce close to specular surface scattering conditions.

3.4. *Planar distorted B-phase*

In bulk the energy gap of the B-phase is isotropic. Under confinement the planar distorted a phase develops a strong gap anisotropy along $o = \pm z$, where z is the surface normal of the cavity . This results in a strong susceptibility anisotropy along w, a vector rotated with respect to o by a rotation matrix R, which rotates the spin relative to the orbital coordinates of the pair, changing only the dipole interaction. The Zeeman interaction is minimised for $w = \pm H$. This gives rise, in sufficiently strong magnetic fields, along $\mathbf{z}$, to two orientations with different dipole energies and hence different NMR frequency shift. For $\mathbf{o} = \mathbf{z}$ the dipole energy is minimised and the frequency shift is positive, while for $\mathbf{o} = -\mathbf{z}$ the frequency shift is negative. This latter metastable state is formed stochastically into the plnara distorted B-phase from A-phase. It can be eliminated by growing the B phase in zero magnetic field. Domain walls between these two orientations support surface states, including fermionic zero modes. We have proposed that they can be pinned at controlled sites on the surface and fused by manipulating the magnetic field[111]. This is potentially a way of manipulating Majorana zero modes. These domain walls are "soft", of characteristic width $\xi_H \sim 10\ \mu\text{m}$, the dipole length, and significantly wider than those referred to in the next section:

$$\xi_H = \xi_0 \sqrt{N_F \Delta_\parallel^2 / \Delta\chi H^2}.$$

Under confinement the gap is spatially-dependent, due to wall boundary conditions. By measuring the NMR frequency shift as a function of tipping angle it is possible to measure the following spatial averages of the gap across the cavity:

$$\bar{q} = \frac{\langle \Delta_\parallel(\mathbf{r})\Delta_\perp(\mathbf{r})\rangle}{\langle\Delta_\parallel^2(\mathbf{r})\rangle}, \quad \bar{Q}^2 = \frac{\langle\Delta_\perp^2(\mathbf{r})\rangle}{\langle\Delta_\parallel^2(\mathbf{r})\rangle}.$$

These can be compared with with predictions of the gap distortion from microscopic theory[111,112].

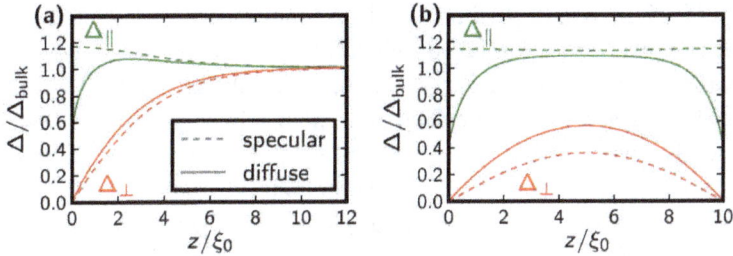

Fig. 10. Suppression of components of the B-phase gap for specular and diffuse scattering limits[113,114]. (a) Single surface, (b) Slab of height $10\ \xi_0$. Under such confinement the surfaces, and surface scattering play a dominant role. This provides a laboratory to investigate emergent surface excitations.

3.5. Spatially-modulated superfluid order

There is widespread interest in superfluid/superconducting states in which the order parameter spontaneously acquires a spatial modulation. Such FFLO states[115,116] have been widely discussed in spin-singlet superconductors, induced by a spin splitting of the Fermi surface by magnetic fields close to the Pauli limiting field, under conditions where orbital effects are inhibited. This conditions are potentially achieved in heavy fermion superconductors[117,118], or low dimensional organic superconductors[119]. A spatially modulated superfluid state has also been predicted

$$\xi(T)$$

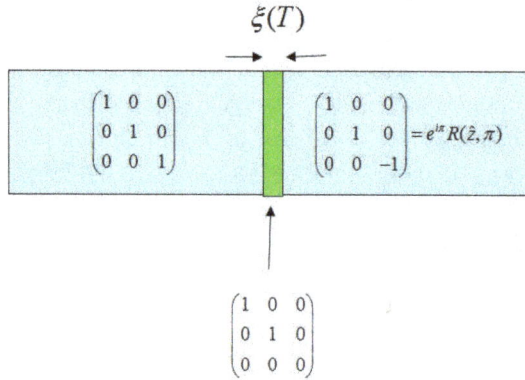

Fig. 11.   The order parameter of the B-phase supports a number of possible "hard" domain walls of thickness of order the coherence length[120]. A *simplified* version of one of these is shown. One component of the order parameter changes sign, and the domain wall itself is the planar phase. Under confinement such a domain walls reduce surface pair breaking and can acquire negative surface energy. This is predicted to lead to their proliferation and the formation of the stripe phase, a spatially-modulated superfluid[121].

in topological superfluid, ^{3}He-B confined in a thin cavity with slab geometry: the stripe phase[121]. Since the superfluid is a clean condensed matter system, totally free from impurities, conditions are favourable to observe a state that is *intrinsically* inhomogeneous. A distinguishing feature of this spatially modulated superfluid state is that it arises in a p-wave superfluid with spin triplet pairing, unlike the spin-singlet cases discussed above. As we have discussed, the superfluid order parameter necessarily varies across the slab due to the boundary conditions on the upper and lower surfaces. However, in addition, an in-plane inhomogeneity is predicted due to the spontaneous appearance of domain walls, on either side of which one component of the order parameter changes sign. These are "hard" domain walls, of thickness of order the superfluid coherence length between energetically equivalent states in the B-phase manifold. They were originally discussed in the context of bulk liquid[120]. In this proposal they can be stabilized under confinement, since the presence of the domain wall reduces surface pair breaking, and the domain wall acquires negative surface energy. Their proliferation potentially leads to the formation of a spatially modulated phase.

The simplest possible structure is a periodic array of regularly spaced and linear domain walls, referred to as the stripe phase. Superfluid ^{3}He has the advantage that NMR directly probes the spin degrees of freedom of the Cooper pairs. A stripe phase has $q = 0$, which has a clear signature in the tipping angle dependence of the NMR response. Its stability is predicted to be favoured by weak coupling, so an experiment was performed at zero pressure where strong coupling effects should be minimised. We therefore chose a 1.1 μm cavity, which has an A-B transition at zero pressure[107]. While the experiment rules out the stripe phase, there is NMR evidence for a spatially modulated superfluid of two-dimensional morphology, characterised by two wavevectors, rather than the single wavevector of the stripe phase. We refer to this as the polka dot phase. It is similar to states discussed in the context of FFLO[117]. In our context it may arise because of a lower nucleation barrier of "dots" compared to stripes, which are macroscopic in one dimension.

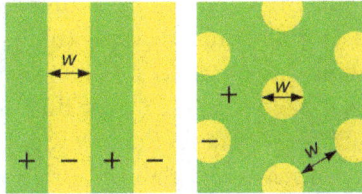

Fig. 12.   Two possible spatially modulations of planar distorted superfluid ^{3}He-B under confinement: stripe phase and "polka-dot".

3.6. *Nucleation of B-phase under confinement*

In bulk the nucleation of the B phase from the A phase, a first order phase transition, has been widely discussed[122]. Study of the A-B transition under confinement at low pressures, using a torsional pendulum, shows that it is easy to nucleate the B phase, and very little supercooling is observed[108]. We argued that this demonstrates a new intrinsic nucleation mechanism, in which the stripe phase and resonant tunnelling model (significant in models of the early universe) are proposed to play a role[108]. Using the new techniques of nanoscale confinement, this is subject to further test. We can engineer an isolated "mesa" of superfluid ^{3}He, cooled through normal "leads"(where the normal state is stabilized by high confinement). This setup provides an environment to answer fundamental questions on phase transitions in the superfluid of broad cosmological relevance.

4. Topological mesoscopic superfluidity: Future prospects

4.1. *Strong confinement*

The demonstration that, when the surface boundary layer includes a thin superfluid ^{4}He film, there is no suppression of T_c or the superfluid energy gap, and that surface scattering is therefore specular, opens the way to studies of superfluid ^{3}He under progressively stronger confinement. As the "film" is made thinner we enter the quasi-2D limit, in which size quantization along $\mathbf{z}$ plays a role and the Fermi sphere breaks up into Fermi discs, where the number of 2D mini-bands is $j = k_F D/\pi$. This opens up a wealth of new quantum states, associated with the integer number of bands, which in principle can be tuned by slab thickness, [13], incuding detection the thermal quantum Hall effect in chiral superfluid ^{3}He-A (if this is indeed the stable phase).

In our opinion, the quasi-2D regime requires a new theoretical approach on a number of grounds. Firstly the pairing interaction, and strong coupling effects, may be modified, possibly through a dependence of the spin-fluctuation spectrum on dimensionality. Secondly, quasi-classical theory as presently constituted is not adequate, under conditions of high confinement. This theory was developed to treat the influence of surfaces and interfaces within bulk superfluid ^{3}He. The notion of an arbitrary quasiparticle trajectory impinging on the confining surface clearly breaks down in quasi-2D. In this limit the quasi-2D minibands are subject to an effective disorder potential $v(x,y)$ that is determined by the fluctuations in confining cavity height $D+d(x,y)$, due to surface roughness or longer length scale variations in cavity height [123,124]. Since variations in cavity height can be measured, at least in principle, we have the unusual situation of a disorder potential that can be fully determined experimentally. The success of this approach has already been demonstrated in studies of the flow of an unsaturated normal ^{3}He film over a polished silver surface with fully characterised surface roughness [125,126]. The challenge now is to extend this approach to analyse quasi-2D superfluid ^{3}He. [It should be noted that it is necessary to eliminate experimentally other sources of disorder. Each surface should be an equipotential to eliminate fluctuating electric fields, and consequent electrostrictive effects].

The ultimate objective is a 2D p-wave superfluid. Because of the different classes of the topological defects in a p-wave superfluid, the topological phase transitions into this system are particularly rich [127,128].

4.2. Hybrid superfluid structures

In hybrid metallic nanostructures, normal metals and s-wave superconductors are combined to create new devices. Here a new approach will be the creation and investigation of a range of superfluid ^{3}He hybrid nanostructures to reveal edge, surface and interface states in a model topological superfluid. This will serve as a model of such mesoscopic devices based on unconventional, p-wave, superconductors, including topological superconductors. ^{3}He allows a degree of control that is unprecedented. The ^{3}He order parameter is a 3×3 matrix with complex components encoding the spin state as a function of position over the Fermi surface, the gap anisotropy and the topology. As we have seen, confinement is the key new control parameter and is used to stabilise distinct superfluid phases or normal phase, and the interfaces between them, with exquisite control over interface quality, the elimination of disorder, and flexibility of geometry.

4.2.1. *Order parameter sculpture*

The hybrid mesoscopic structure is built from a range of "materials", with ^{3}He tuned into a particular state by the scale and sculpture of confinement, or the creation of p-wave superfluid meta-materials. For example consider a cavity with a square array of posts separated by $\sim 10\xi_0$, for which the phase diagram has been calculated[129]. Near T_c the polar phase is stabilized. This geometry should allow the creation of polar phase in the absence of disorder, and even to engineer disorder in a controlled way. [It has recently been shown that the polar phase is stabilized in nematic aerogels, and furthermore stabilizes half-quantum-vortices[130]]. At lower temperatures a new variant of the B-phase with four-fold symmetry is predicted[129].

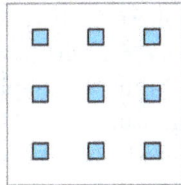

Fig. 13.   Periodic array of posts to stabilize polar phase, and a variant of B-phase with four-fold symmetry[129].

On the other hand in narrow channels the following sequence of phases is predicted[131]: chiral-A; chiral/polar with periodic domain walls; polar.

4.2.2. *Superfluid ^{3}He meta-materials*

It is also possible to imagine the creation of superfluid ^{3}He meta-materials, such as a regular array of islands, each hosting a macroscopic quantum state, with interconnecting channels. Of interest is a periodic array of islands, which are not equivalent in terms of their superfluid phase (A, B, polar, spatially modulated etc). Furthermore, the interconnecting channels, "bonds", can be tuned. A higher level structure in which two or more meta-materials are coupled together is even possible.

4.3. *Majorana and Majorana-Weyl excitations at surfaces, interfaces and edges*

Topological superfluids/superconductors necessarily host surface and edge excitations. The spectrum of these emergent surface excitations is calculated self consistently with the gap suppression[110,114,132]. The energy density of states of these mid-gap excitations depends on the surface scattering conditions. As we have seen a variety of p-wave superfluids are stabilized by confinement. Each superfluid phase has different symmetry, topology and surface/edge excitations.

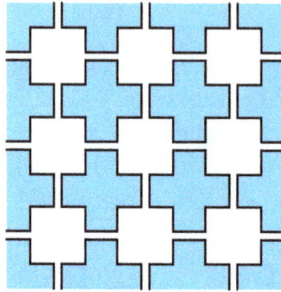

Fig. 14. Periodic array of inter-connected cavities to create superfluid ^{3}He meta-materials. Adjusting height of cavity will stabilize different phases. Not all cavities need be the same height.

The time-reversal-invariant B-phase with specular scattering has linearly dispersing Majorana surface excitations[133], with strong spin-orbit locking[134]. They carry a ground state spin current. Again, these spins are the ^{3}He nuclear spins, and there should therefore be clear signatures in NMR response[135]. The quasiclassical theory of how the spin-orbit locked surface excitations may influence the NMR response of the confined superfluid is currently under development[136]. Ultimately we might hope to detect a non-local response of surface Majorana excitations in confined superfluid ^{3}He-B using local NMR probes.

Furthermore the new topological mesoscopic ^{3}He structures offer the prospect of further control over excitations at engineered intra-fluid interfaces. To give an example, we envisage a structure with a step in cavity height at which we can stabilize an interface, either between superfluid and normal liquid (SN) or between two topologically distinct superfluid phases (SS'). Unlike an SN interface between a superconductor and normal metal this interface is within a single material (reminiscent of the high quality pn junction in silicon). By combining such ingredients we can create progressively more complex hybrid structures. The new feature of such

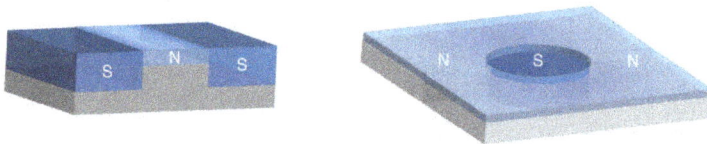

Fig. 15. Mesoscopic structures: SNS junction (SS'S is also possible); superfluid mesa, isolated by normal slabs.

junctions is that we are dealing with a p-wave superfluid. If B-phase, Majorana-like excitations will exist at each SN interface. The spectrum of Andreev bound states in the junction is controlled by L_N, and we can exploit the fact that at T_c the inelastic scattering length is of order 50 microns.

4.4. *Further probes of surface excitations*

Furthermore, the breaking of gauge and rotational symmetries in superfluid ^{3}He gives rise to gapped Anderson-Higgs modes, as well as gapless Nambu-Goldstone modes. These modes couple to longitudinal and transverse zero sound. Thus, uniquely, superfluid ^{3}He offers the prospect to study the coupling of bosonic collective modes of the superfluid order to fermionic degrees of freedom, in particular the predicted exotic surface states [137]. Moreover, since our topological superconductor is a liquid, we can insert probes into it. A new generation of mesoscopic devices, such as nanowires of diameter comparable to the superfluid coherence length is under development, with resonant frequencies comparable to the superfluid gap, to study the interplay of vibration and surface/edge excitations. In part this is motivated by closely related studies of superfluid ^{3}He using ions, trapped under the free surface [138].

5. Conclusion

The cold atoms ^{3}He and ^{4}He can be grown in atomically layered thin films, on graphite with a variety of pre-platings, or confined in precisely engineered nanofluidic geometries. By manipulating helium in this way it is possible to effectively create new "materials" with interesting classes of quantum ground states, some anticipated, others emergent. These new materials can be well-characterized, while disorder and impurity effects are small and controllable. In this way important classes of quantum materials can be realized using helium as a amodel system, exploiting its relative simplicity, and providing a robust confrontation between theory and experiment.

Acknowledgements

The recent experiments at Royal Holloway reported here were a collaboration with Frank Arnold, Andrew Casey, Brian Cowan, Petri Heikkinen, Lev Levitin, Jan Nyéki, Xavier Rojas, Alexander Waterworth, and Jeevak Parpia at Cornell (NSF DMR 17808341). Thanks to Brian Cowan, who helped with the preparation of this manuscript. Discussions with Matthias Eschrig, Piers Coleman, Andrew Ho, Derek Lee, James Sauls are gratefully acknowledged. The research on atomically layered films was most recently supported by EPSRC (UK) through EP/H048375/1. The work on topological superfluidity was most recently supported by EP/J022004/1. We also acknowledge support of the European Microkelvin Platform, and invaluable discussions with its members.

References

1. A. P. Schnyder, S. Ryu, A. Furusaki and A. W. W. Ludwig, Classification of topological insulators and superconductors in three spatial dimensions, *Physical Review B* **78**, 195125 (2008).

2. X.-L. Qi and S.-C. Zhang, Topological insulators and superconductors, *Reviews of Modern Physics* **83**, 1057 (2011).

3. M. Z. Hasan and C. L. Kane, Colloquium: Topological insulators, *Reviews of Modern Physics* **82**, 3045 (2010).

4. T. Mizushima, Y. Tsutsumi, M. Sato and K. Machida, Symmetry protected topological superfluid ^{3}He-B, *Journal of Physics-Condensed Matter* **27**, 56 (2015).

5. T. Mizushima, Y. Tsutsumi, T. Kawakami, M. Sato, M. Ichioka and K. Machida, Symmetry-protected topological superfluids and superconductors – From the basics to ^{3}He, *Journal of the Physical Society of Japan* **85**, 74 (2016).

6. A. J. Leggett, A theoretical description of the new phases of liquid ^{3}He, *Reviews of Modern Physics* **47**, 331 (1975).

7. A. J. Leggett, *Quantum Liquids* (Oxford University Press, Oxford, 2006).

8. D. Vollhardt and P. Wölfle, *The Superfluid Phases of Helium-3* (Dover, 2013).

9. G. E. Volovik, *The Universe in a Helium Droplet* (Clarendon, Oxford, 2003).

10. P. W. Anderson and W. F. Brinkman, Anisotropic superfluidity in ^{3}He: A possible interpretation of its stability as a spin-fluctuation effect, *Physical Review Letters* **30**, 1108 (1973).

11. G. E. Volovik, An analog of the quantum Hall effect in a superfluid ^{3}He film, *JETP (USSR)* **67**, 1084 (1988).

12. G. E. Volovik, Quantum Hall state and chiral edge state in thin ^{3}He-A film, *JETP Letters* **55**, 368 (1992).

13. G. E. Volovik, *Exotic Properties of Superfluid ^{3}He* (World Scientific, Singapore, 1992).

14. G. E. Volovik, Fermion zero modes at the boundary of superfluid ^{3}He-B, *JETP Letters* **90**, 440 (2009).

15. G. E. Volovik, Topological invariant for superfluid ^{3}He-B and quantum phase transitions, *JETP Letters* **90**, 587 (2009).

16. G. E. Volovik, Topological superfluid ^{3}He-B in magnetic field and Ising variable, *JETP Letters* **91**, 201 (2010).

17. H. Godfrin and H.-J. Lauter, Experimental properties of ^{3}He adsorbed on graphite, *Progress in Low Temperature Physics* **XIV**, 213 (1995).

18. A. Casey, H. Patel, J. Nyéki, B. P. Cowan and J. Saunders, Evidence for a Mott-Hubbard transition in a two-dimensional ^{3}He fluid monolayer, *Physical Review Letters* **90**, 115301 (2003).

19. D. Vollhardt, Normal ^{3}He: an almost localized Fermi liquid, *Reviews of Modern Physics* **56**, 99 (1984).

20. A. V. Chubukov, D. L. Maslov, S. Gangadharaiah and L. I. Glazman, Thermodynamics of a Fermi liquid beyond the low-energy limit, *Physical Review Letters* **95**, 026402 (2005).

21. P. W. Anderson, Singular forward scattering in the 2D Hubbard model and a renormalized Bethe ansatz ground state, *Physical Review Letters* **65**, 2306 (1990).

22. C. P. Lusher, B. P. Cowan and J. Saunders, Quasiparticle interactions in two-dimensional fluid ^{3}He films adsorbed on graphite, *Physical Review Letters* **67**, 2497 (1991).

23. F. Arnold, J. Nyéki, B. Cowan and J. Saunders, The density-driven Wigner-Mott transition in two dimensional ^{3}He into a gapless spin-liquid, *In preparation*.

24. J. W. Clark, V. A. Khodel and M. V. Zverev, Anomalous low-temperature behavior of strongly correlated Fermi systems, *Physical Review B* **71**, 012401 (2005).

25. M. Neumann, J. Nyéki, B. Cowan and J. Saunders, Bilayer ^{3}He: A simple two-dimensional heavy-fermion system with quantum criticality, *Science* **317**, 1356 (2007).

26. M. Vojta, Orbital-selective Mott transitions: Heavy fermions and beyond, *Journal of Low Temperature Physics* **161**, 203 (2010).
27. A. Benlagra and C. Pépin, Model of quantum criticality in ^{3}He bilayers adsorbed on graphite, *Physical Review Letters* **100**, 176401 (2008).
28. C. Pépin, Selective Mott transition and heavy fermions, *Physical Review B* **77**, 245129 (2008).
29. A. Rançon-Schweiger, A. Benlagra and C. Pépin, Activation gap in the specific heat measurements for ^{3}He bilayers, *Physical Review B* **83**, 073102 (2011).
30. K. S. D. Beach and F. F. Assaad, Orbital-selective Mott transition and heavy-fermion behavior in a bilayer Hubbard model for ^{3}He, *Physical Review B* **83**, 045103 (2011).
31. S. Sen and N. S. Vidhyadhiraja, Quantum critical Mott transitions in a bilayer Kondo insulator-metal model system, *Physical Review B* **93**, 155136 (2016).
32. N. Neumann, J. Nyéki, L. Levitin, A. Casey, B. Cowan and J. Saunders, Kondo breakdown and frustrated magnetism in heavy fermion ^{3}He bilayer films on graphite, *In preparation*.
33. J. Werner and F. F. Assaad, Ring-exchange periodic Anderson model for ^{3}He bilayers, *Physical Review B* **90**, 205122 (2014).
34. P. A. Crowell and J. D. Reppy, Superfluidity and film structure in ^{4}He adsorbed on graphite, *Physical Review B* **53**, 2701 (1996).
35. J. Nyéki, R. Ray, B. Cowan and J. Saunders, Superfluidity of atomically layered ^{4}He films, *Physical Review Letters* **81**, 152 (1998).
36. R. M. Bowley, A. D. Armour, J. Nyéki, B. P. Cowan and J. Saunders, Universal features of the superfluid transition of a thin ^{4}He film on an atomically flat substrate, *Journal of Low Temperature Physics* **113**, 399 (1998).
37. M. C. Gordillo and D. M. Ceperley, Path-integral calculation of the two-dimensional ^{4}He phase diagram, *Physical Review B* **58**, 6447 (1998).
38. M. Pierce and E. Manousakis, Path-integral Monte Carlo simulation of the second layer of ^{4}He adsorbed on graphite, *Physical Review B* **59**, 3802 (1999).
39. R. Hallock, The proprties of multilayer ^{3}He $-^4$He mixture films, *Progress in Low Temperature Physics* **XIV**, 321 (1995).
40. R. B. Hallock, The two-dimensional world of ^{3}He in ^{3}He $-^4$He mixture films, *Journal of Low Temperature Physics* **121**, 441 (2000).
41. N. Yokoshi and S. Kurihara, Effects of fermion-boson interaction in neutral atomic systems, *Physical Review B* **68**, 064501 (2003).
42. A. Waterworth, J. Nyéki, B. Cowan and J. Saunders, Instabilities of two dimensional ^{3}He on atomically layered superfluid ^{4}He film, *In preparation*.
43. M. Dann, J. Nyéki, B. P. Cowan and J. Saunders, Quasiparticle interactions in two-dimensional ^{3}He on ^{4}He films, *Physical Review Letters* **82**, 4030 (1999).
44. J. R. Engelbrecht, M. Randeria and L. Zhang, Landau f function for the dilute Fermi gas in two dimensions, *Physical Review B* **45**, 10135 (1992).
45. J. Engelbrecht and M. Randeria, Low-density repulsive Fermi gas in two dimensions: Bound-pair excitations and Fermi-liquid behavior, *Physical Review B* **45**, 12419 (1992).
46. A. V. Chubukov, Kohn-Luttinger effect and the instability of a two-dimensional repulsive Fermi liquid at $T = 0$, *Physical Review B* **48**, 1097 (1993).
47. A. Chubukov and A. Sokol, Theory of p-wave pairing in a two-dimensional Fermi gas, *Physical Review B* **49**, 678 (1994).
48. K. Miyake, Fermi liquid theory of dilute submonolayer ^{3}He on thin ^{4}He II film, *Prog. Theor. Phys.* **69**, 1794 (1983).

49. M. Y. Kagan, Fermi-gas approach to the problem of superfluidity in three- and two-dimensional solutions of ^{3}He in ^{4}He, *Physics-Uspekhi* **37**, 69 (1994).

50. S. Qin, J. Kim, Q. Niu and C. K. Shih, Superconductivity at the two-dimensional limit, *Science* **324**, 1314 (2009).

51. J. F. Ge, Z. L. Liu, C. Liu, C. L. Gao, D. Qian, Q. K. Xue, Y. Liu and J. F. Jia, Superconductivity above 100 K in single-layer FeSe films on doped SrTiO$_3$, *Nat. Mater.* **14**, 285 (2015).

52. G. Profeta, M. Calandra and F. Mauri, Phonon-mediated superconductivity in graphene by lithium deposition, *Nature Physics* **8**, 131 (2012).

53. B. M. Ludbrook, G. Levy, P. Nigge, M. Zonno, M. Schneider, D. J. Dvorak, C. N. Veenstra, S. Zhdanovich, D. Wong, P. Dosanjh, C. Strasser, A. Stohr, S. Forti, C. R. Ast, U. Starke and A. Damascelli, Evidence for superconductivity in Li-decorated monolayer graphene, *Proc. Natl. Acad. Sci. USA* **112**, 11795 (2015).

54. A. T. Sommer, L. W. Cheuk, M. J. Ku, W. S. Bakr and M. W. Zwierlein, Evolution of fermion pairing from three to two dimensions, *Physical Review Letters* **108**, 045302 (2012).

55. M. Feld, B. Frohlich, E. Vogt, M. Koschorreck and M. Kohl, Observation of a pairing pseudogap in a two-dimensional Fermi gas, *Nature* **480**, 75 (2011).

56. A. A. Abrikosov and I. M. Khalatnikov, Concerning a model for a non-ideal Fermi gas, *Soviet Physics JETP* **6**, 888 (1958).

57. A. A. Abrikosov and I. M. Khalatnikov, The theory of a Fermi liquid (the properties of liquid ^{3}He at low temperatures), *Reports on Progress in Physics* **22**, 329 (1959).

58. S. Kurihara, Singlet superfluidity in ^{3}He film on ^{4}He film, *Journal of the Physical Society of Japan* **52**, 1311 (1983).

59. F. Arnold, B. Yager, J. Nyéki, A. J. Casey, A. Shibahara, B. P. Cowan and J. Saunders, Application of low frequency SQUID NMR to the ultra-low temperature study of atomically layered ^{3}He films adsorbed on graphite, *Journal of Physics: Conference Series* **568**, 032020 (2014).

60. A. Waterworth, J. Nyéki, B. Cowan and J. Saunders, Interactions in two dimensional ^{3}He on atomically layered superfluid ^{4}He film, *In preparation.*

61. J. Leonard, A. Morales, P. Zupancic, T. Esslinger and T. Donner, Supersolid formation in a quantum gas breaking a continuous translational symmetry, *Nature* **543**, 87 (2017).

62. J. R. Li, J. Lee, W. Huang, S. Burchesky, B. Shteynas, F. C. Top, A. O. Jamison and W. Ketterle, A stripe phase with supersolid properties in spin-orbit-coupled Bose-Einstein condensates, *Nature* **543**, 91 (2017).

63. L. Balents, Spin liquids in frustrated magnets, *Nature* **464**, 199 (2010).

64. L. Savary and L. Balents, Quantum spin liquids: A review, *Rep. Prog. Phys.* **80**, 016502 (2017).

65. J. Nyéki, A. Phillis, A. Ho, D. Lee, P. Coleman, J. Parpia, B. Cowan and J. Saunders, Intertwined superfluid and density wave order in two-dimensional ^{4}He, *Nature Physics* **13**, 455 (2017).

66. J. Nyéki, A. Phillis, B. Cowan and J. Saunders, On the "supersolid" response of the second layer of ^{4}He on graphite, *Journal of Low Temperature Physics* **187**, 475 (2017).

67. E. Fradkin, S. A. Kivelson and J. M. Tranquada, Colloquium: Theory of intertwined orders in high temperature superconductors, *Reviews of Modern Physics* **87**, 457 (2015).

68. A. J. Leggett, Can a solid be "superfluid"?, *Physical Review Letters* **25**, 1543 (1970).

69. S. Balibar, The enigma of supersolidity, *Nature* **464**, 176 (2010).

70. M. Boninsegni and N. V. Prokof'ev, Supersolids: What and where are they?, *Reviews of Modern Physics* **84**, 759 (2012).

71. R. Hallock, Is solid helium a supersolid?, *Physics Today* **68**, 30 (2015).

72. A. F. Andreev, Quantum crystals, *Progress in Low Temperature Physics* **VIII**, 67 (1982).

73. J. Saunders, A glassy state of supersolid helium, *Science (New York, N.Y.)* **324**, 601 (2009).

74. Y. Vekhov and R. B. Hallock, Mass flux characteristics in solid ^{4}He for $T > 100$ mK: Evidence for bosonic Luttinger-liquid behavior, *Phys Rev Lett* **109**, 045303 (2012).

75. J. Shin, D. Y. Kim, A. Haziot and M. H. W. Chan, Superfluidlike mass flow through $8\,\mu$m thick solid ^{4}He samples, *Phys Rev Lett* **118**, 235301 (2017).

76. M. Boninsegni, A. B. Kuklov, L. Pollet, N. V. Prokof'ev, B. V. Svistunov and M. Troyer, Luttinger liquid in the core of a screw dislocation in helium-4, *Phys Rev Lett* **99**, 035301 (2007).

77. E. J. Mueller, T.-L. Ho, M. Ueda and G. Baym, Fragmentation of Bose-Einstein condensates, *Physical Review A* **74**, 033612 (2006).

78. R. A. Guyer, R. C. Richardson and L. I. Zane, Excitations in quantum crystals (A survey of NMR experiments in solid helium), *Reviews of Modern Physics* **43**, 532 (1971).

79. F. Ziouzia, H. Patel, J. Nyéki, B. Cowan and J. Saunders, Thermodynamic evidence for two-dimensional tunnelling excitations, *Physica B: Condensed Matter* **329-333**, 254 (2003).

80. J. Nyéki, H. Patel, F. Ziouzia, A. Waterworth, B. Cowan and J. Saunders, Thermodynamic evidence for two-dimensional ^{3}He tunnelling excitations in 2D supersolid ^{4}He, *In preparation*.

81. H. Godfrin, R. R. Ruel and D. D. Osheroff, Experimental observation of a two-dimensional Heisenberg nuclear ferromagnet, *Physical Review Letters* **60**, 305 (1988).

82. D. S. Greywall, Heat capacity of multilayers of ^{3}He adsorbed on graphite at low millikelvin temperatures, *Physical Review B* **41**, 1842 (1990).

83. M. Siqueira, J. Nyéki, B. Cowan and J. Saunders, Frustration by multiple spin exchange in 2D solid ^{3}He films, *Physical Review Letters* **78**, 2600 (1997).

84. H. Fukuyama, Nuclear magnetism in two-dimensional solid helium-three on graphite, *Journal of the Physical Society of Japan* **77**, 111013 (2008).

85. E. Collin, S. Triqueneaux, R. Harakaly, M. Roger, C. Bäuerle, Y. M. Bunkov and H. Godfrin, Quantum frustration in the "spin liquid" phase of two-dimensional ^{3}He, *Physical Review Letters* **86**, 2447 (2001).

86. M. Roger, Frustration due to competing cyclic ring exchanges in two-dimensional solid ^{3}He, *Physical Review Letters* **64**, 297 (1990).

87. D. J. Thouless, Exchange in solid ^{3}He and the Heisenberg hamiltonian, *Proceedings of the Physical Society* **86**, 893 (1965).

88. M. Roger, C. Bäuerle, Y. M. Bunkov, A. S. Chen and H. Godfrin, Multiple-spin exchange on a triangular lattice: A quantitative interpretation of thermodynamic properties of two-dimensional solid ^{3}He, *Physical Review Letters* **80**, 1308 (1998).

89. M. Siqueira, J. Nyéki, B. Cowan and J. Saunders, Heat capacity study of the quantum antiferromagnetism of a ^{3}He monolayer, *Physical Review Letters* **76**, 1884 (1996).

90. N. D. Mermin and H. Wagner, Absence of ferromagnetism or antiferromagnetism in one- or two-dimensional isotropic Heisenberg models, *Physical Review Letters* **17**, 1133 (1966).

91. A. Waterworth, J. Nyéki, B. Cowan and J. Saunders, Direct evidence for RKKY exchange interactions in ^{3}He films, *In preparation*.

92. D. Sato, K. Naruse, T. Matsui and H. Fukuyama, Observation of self-binding in monolayer ^{3}He, *Physical Review Letters* **109**, 235306 (2012).

93. A. Casey, M. Neumann, B. Cowan, J. Saunders and N. Shannon, Two-dimensional ferromagnetism of a ^{3}He film: Influence of weak frustration, *Physical Review Letters* **111**, 125302 (2013).

94. M. Siqueira, C. P. Lusher, B. P. Cowan and J. Saunders, Two-dimensional antiferromagnetism of solid ^{3}He adsorbed on plated graphite, *Physical Review Letters* **71**, 1407 (1993).

95. A. Casey, H. Patel, J. Nyéki, B. P. Cowan and J. Saunders, Nuclear magnetism of two dimensional solid ^{3}He adsorbed on plated graphite, *Journal of Low Temperature Physics* **113**, 265 (1998).

96. R. Masutomi, Y. Karaki and H. Ishimoto, Gapless spin liquid behavior in two-dimensional solid ^{3}He, *Physical Review Letters* **92**, 025301 (2004).

97. S. Murakawa, Y. Tamura, Y. Wada, M. Wasai, M. Saitoh, Y. Aoki, R. Nomura, Y. Okuda, Y. Nagato, M. Yamamoto, S. Higashitani and K. Nagai, New anomaly in the transverse acoustic impedance of superfluid ^{3}He-B with a wall coated by several layers of ^{4}He, *Physical Review Letters* **103**, 155301 (2009).

98. Y. Okuda and R. Nomura, Surface Andreev bound states of superfluid ^{3}He and Majorana fermions, *Journal of Physics-Condensed Matter* **24**, 343201 (2012).

99. P. Heikkinen, A. Casey, L. Levitin, X. Rojas, A. Vorontsov, T. Abhilash, N. Zhelev, J. Parpia and J. Saunders, Tuned pair-breaking at the surface of topological superfluid ^{3}He, *In preparation.*

100. L. V. Levitin, R. G. Bennett, A. Casey, B. Cowan, J. Saunders, D. Drung, T. Schurig and J. M. Parpia, Phase diagram of the topological superfluid ^{3}He confined in a nanoscale slab geometry, *Science* **340**, 841 (2013).

101. L. V. Levitin, R. G. Bennett, A. Casey, B. P. Cowan, C. P. Lusher, J. Saunders, D. Drung and T. Schurig, A nuclear magnetic resonance spectrometer for operation around 1 MHz with a sub-10 mK noise temperature, based on a two-stage dc superconducting quantum interference device sensor, *Applied Physics Letters* **91**, 262507 (2007).

102. S. Dimov, R. G. Bennett, A. Corcoles, L. V. Levitin, B. Ilic, S. S. Verbridge, J. Saunders, A. Casey and J. M. Parpia, Anodically bonded submicron microfluidic chambers, *Rev Sci Instrum* **81**, 013907 (2010).

103. A. J. Leggett, The spin dynamics of an anisotropic Fermi superfluid (^{3}He?), *Annals of Physics* **85**, 11 (1974).

104. A. J. Leggett, NMR lineshifts and spontaneously broken spin-orbit symmetry, *Journal of Physics C-Solid State Physics* **6**, 3187 (1973).

105. M. R. Freeman, R. S. Germain, E. V. Thuneberg and R. C. Richardson, Size effects in thin-films of superfluid ^{3}He, *Physical Review Letters* **60**, 596 (1988).

106. M. R. Freeman and R. C. Richardson, Size effects in superfluid ^{3}He films, *Physical Review B* **41**, 11011 (1990).

107. L. V. Levitin, B. Yager, L. Sumner, B. Cowan, A. Casey, N. Zhelev, R. Bennett, J. Parpia and J. Saunders, Evidence for a spatially-modulated superfluid phase of ^{3}He under confinement, *In preparation.*

108. N. Zhelev, T. S. Abhilash, E. N. Smith, R. G. Bennett, X. Rojas, L. Levitin, J. Saunders and J. M. Parpia, The A-B transition in superfluid helium-3 under confinement in a thin slab geometry, *Nat Commun* **8**, 15963 (2017).

109. J. A. Sauls and J. W. Serene, Potential-scattering models for the quasiparticle interactions in liquid ^{3}He, *Physical Review B* **24**, 183 (1981).

110. A. B. Vorontsov and J. A. Sauls, Thermodynamic properties of thin films of superfluid ^{3}He-A, *Physical Review B* **68**, 064508 (2003).
111. L. V. Levitin, R. G. Bennett, E. V. Surovtsev, J. M. Parpia, B. Cowan, A. J. Casey and J. Saunders, Surface-induced order parameter distortion in superfluid ^{3}He-B measured by nonlinear NMR, *Physical Review Letters* **111**, 235304 (2013).
112. L. V. Levitin, R. G. Bennett, A. Casey, B. Cowan, J. Saunders, D. Drung, T. Schurig, J. M. Parpia, B. Ilic and N. Zhelev, Study of superfluid ^{3}He under nanoscale confinement, *Journal of Low Temperature Physics* **175**, 667 (2014).
113. Y. Nagato, A-B transition of superfluid ^{3}He in a slab with rough surfaces, *Physica B: Condensed Matter* **284-288**, 269 (2000).
114. K. Nagai, Y. Nagato, M. Yamamoto and S. Higashitani, Surface bound states in superfluid ^{3}He, *Journal of the Physical Society of Japan* **77**, 111003 (2008).
115. P. Fulde and R. A. Ferrell, Superconductivity in a strong spin-exchange field, *Physical Review* **135**, A550 (1964).
116. A. I. Larkin and Y. N. Ovchinnikov, Nonuniform state of superconductors, *JETP (USSR)* **20**, 762 (1965).
117. Y. Matsuda and H. Shimahara, Fulde-Ferrell-Larkin-Ovchinnikov state in heavy fermion superconductors, *Journal of the Physical Society of Japan* **76**, 051005 (2007).
118. D. Y. Kim, S.-Z. Lin, F. Weickert, M. Kenzelmann, E. D. Bauer, F. Ronning, J. D. Thompson and R. Movshovich, Intertwined orders in heavy-fermion superconductor CeCoIn$_5$, *Physical Review X* **6**, 041059 (2016).
119. H. Mayaffre, S. Krämer, M. Horvatić, C. Berthier, K. Miyagawa, K. Kanoda and V. F. Mitrović, Evidence of Andreev bound states as a hallmark of the FFLO phase in κ-(BEDT-TTF)$_2$Cu(NCS)$_2$, *Nature Physics* **10**, 928 (2014).
120. M. M. Salomaa and G. E. Volovik, Cosmiclike domain walls in superfluid ^{3}He-B, instantons and diabolical points in $(\mathbf{k}, \mathbf{r})$ space, *Physical Review B* **37**, 9298 (1988).
121. A. B. Vorontsov and J. A. Sauls, Crystalline order in superfluid ^{3}He films, *Physical Review Letters* **98**, 045301 (2007).
122. P. Schiffer and D. D. Osheroff, Nucleation of the AB transition in superfluid ^{3}He: Surface effects and baked Alaska, *Reviews of Modern Physics* **67**, 491 (1995).
123. Z. Tesanovic, M. V. Jaric and S. Maekawa, Quantum transport and surface scattering, *Physical Review Letters* **57**, 2760 (1986).
124. N. Trivedi and N. W. Ashcroft, Quantum size effects in transport properties of metallic films, *Physical Review B* **38**, 12298 (1988).
125. A. Casey, J. Parpia, R. Schanen, B. Cowan and J. Saunders, Interfacial friction of thin ^{3}He slabs in the Knudsen limit, *Physical Review Letters* **92**, 255301 (2004).
126. P. Sharma, A. Corcoles, R. G. Bennett, J. M. Parpia, B. Cowan, A. Casey and J. Saunders, Quantum transport in mesoscopic ^{3}He films: Experimental study of the interference of bulk and boundary scattering, *Physical Review Letters* **107**, 196805 (2011).
127. H. Kawamura, Successive transitions and intermediate chiral phase in a superfluid ^{3}He film, *Physical Review Letters* **82**, 964 (1999).
128. S. E. Korshunov, Phase transitions in two-dimensional systems with continuous degeneracy, *Physics-Uspekhi* **49**, 225 (2006).
129. J. J. Wiman and J. A. Sauls, Superfluid phases of ^{3}He in a periodic confined geometry, *Journal of Low Temperature Physics* **175**, 17 (2013).
130. S. Autti, V. V. Dmitriev, J. T. Makinen, A. A. Soldatov, G. E. Volovik, A. N. Yudin, V. V. Zavjalov and V. B. Eltsov, Observation of half-quantum vortices in topological superfluid ^{3}He, *Physical Review Letters* **117**, 255301 (2016).

131. J. J. Wiman and J. A. Sauls, Superfluid phases of ^{3}He in nanoscale channels, *Physical Review B* **92**, 144515 (2015).

132. J. A. Sauls, Surface states, edge currents, and the angular momentum of chiral p-wave superfluids, *Physical Review B* **84**, 214509 (2011).

133. S. Murakawa, Y. Wada, Y. Tamura, M. Wasai, M. Saitoh, Y. Aoki, R. Nomura, Y. Okuda, Y. Nagato, M. Yamamoto, S. Higashitani and K. Nagai, Surface Majorana cone of the superfluid ^{3}He B phase, *Journal of the Physical Society of Japan* **80**, 013602 (2011).

134. H. Wu and J. A. Sauls, Majorana excitations, spin and mass currents on the surface of topological superfluid ^{3}He-B, *Physical Review B* **88**, 184506 (2013).

135. M. A. Silaev, Majorana states and longitudinal NMR absorption in a ^{3}He-B film, *Physical Review B* **84**, 144508 (2011).

136. M. A. Silaev, Quasiclassical theory of spin dynamics in superfluid ^{3}He: Kinetic equations in the bulk and spin response of surface Majorana states, *arXiv:1710.04468* (2018).

137. T. Mizushima and J. A. Sauls, Bosonic surface states and acoustic spectroscopy of confined superfluid ^{3}He-B, *arXiv:cond-mat/1801.02277* (2018).

138. H. Ikegami, Y. Tsutsumi and K. Kono, Chiral symmetry breaking in superfluid ^{3}He-A, *Science* **341**, 59 (2013).

Topological gauge theory of the superconductor-insulator transition

M. C. Diamantini

NiPS Laboratory, INFN and Dipartimento di Fisica e Geologia, University of Perugia,
via Pascoli, I-06100 Perugia, Italy
E-mail: cristina.diamantini@pg.infn.it
www.nipslab.org

C. A. Trugenberger

SwissScientific Technologies SA,
rue du Rhone 59, 1204 Geneva, Switzerland
E-mail: c.trugenberger@swissscientific.com
www.swissscientific.com

V. M. Vinokur

Materials Science Division, Argonne National Laboratory,
9700 S. Cass Avenue, Argonne, Illinois 60637, USA
E-mail: vinokour@anl.gov
www.anl.gov

We develop a topological gauge theory of the superconductor-insulator transition (SIT) in the context of Josephson junction arrays (JJA) and disordered superconducting films. We show that the universal long-distance physics of these systems is governed by the BF topological gauge theory, which, in two dimensions, reduces to the doubled Chern-Simons gauge theory. This approach reveals that the quantum phase structure of the SIT is governed by the competition between three quantum orders: (i) Bose condensate of Cooper pairs, which is the superconductor (SC) with no Higgs field, (ii) its dual Bose condensate of magnetic monopoles, corresponding to the superinsulator (SI), a state having infinite resistance at finite temperatures, and (iii) No-condensate state, usually called a Bose (or quantum) metal (BM). We demonstrate that this BM state, described by the purely topological BF term, is a topological insulator (TI), in which the large topological mass suppresses bulk currents and the electronic transport in the self-dual approximation is carried only by the edge excitations. We derive the quantitative criterion for the direct vs the transition via the intermediate metallic state SIT (SMIT) and find the critical strength of quantum fluctuations necessary for the SMIT scenario. The developed approach establishes a deep connection between the exotic phases emerging in the vicinity of the SIT and QED.

Keywords: Superconductor-insulator transition, Superinsulator, Bose metal, Topological insulator, Topological gauge theory, Confinement.

1. Introduction

The dual quantum Aharonov–Bohm and Aharonov–Casher effects[1,2] and topological phase transitions introduced by Berezinskii, Kosterlitz, and Thouless (BKT)[3-6], are two paradigms that brought topology to prominence in condensed matter physics. Gripping both in a whole, the superconductor-insulator transition (SIT) in strongly disordered superconducting films and Josephson junction arrays (JJA)[7-18], has emerged as a trove of fascinating quantum phases that harness intertwined charges and vortices and their topological interactions. The nature of the SIT critical region and the phases it harbors still remain mysterious and are the subject of intense research and debate. Here we develop a long-distance topological gauge theory of the SIT and find, in its critical vicinity, three competing quantum orders: the superconductor, the superinsulator[14–17], and the quantum Bose metal (BM)[19,20], which, as we will show, is a Bose topological insulator (TI)[21–23]. We find that, as long as quantum fluctuations are weak, the SIT occurs as a first-order direct superconductor-superinsulator transition with the granular superconducting-insulating texture that forms around the phase boundary. Strong quantum fluctuations, drive the SIT via the intermediate TI phase. The three quantum phases near the SIT epitomize the possible mechanisms for a gauge field mass without Higgs fields, with the TI and the superconductor realizing topologically massive gauge models[24] and the superinsulator implementing[25] Polyakov's instanton-driven linear confinement with neutral mesons as excitations[26].

In the framework of the gauge theory, the superconductor-insulator transition (SIT)[7–13] appears as a material realization of the field-theoretical S-duality, which, in its simplest form, expresses the invariance of Maxwell's equations under the interchange of electric and magnetic fields in the presence of magnetic monopoles[27]. To gain an insight into the role of S-duality in the physics of the SIT, let us consider its simplest quantum mechanical example, a system endowed with the first-order Lagrangian[28] $L = (\varphi \dot{N} - N\dot{\varphi})/2 - H(N, \varphi)$, having as canonically conjugate pair of variables an amplitude N and the 2π-periodic phase φ. Depending on the parameters of the Hamiltonian $H(N, \varphi)$, one finds three possible outcomes of the uncertainty principle following from the commutation relation, $[N, \varphi] = i\hbar$: (i) sharp $\varphi = x$ and plane waves $\sim e^{-ixN}$ for $H \equiv H(\varphi)$; (ii) sharp $N = y$ and plane waves $\sim e^{-iy\varphi}$ for $H \equiv H(N)$; and (iii) eigenstates with fixed quantized uncertainties ΔN and $\Delta \varphi$ in the general case. Going over to an infinite number of degrees of freedom, encoded in a field $\Psi = N \exp(i\varphi)$, one finds that all three realizations of the uncertainty principle appear manifestly as distinct quantum states around the quantum tri-critical point of the SIT: a superconductor, comprising N-plane waves and fixed φ (Cooper pair condensate); a superinsulator, φ-plane waves with sharp N (vortex condensate)[14–17]; and a phase harboring both frozen charge and vortex fluctuations (neither Cooper pairs nor vortices condense), which is the Bose topological insulator[21–23]. Accordingly, the SIT is a quantum phase transition between the generic dual superconducting and superinsulating states, which can be either a direct transition or and indirect one through the intermediate bosonic topological insulator state.

2. Gauge theory of JJA

We consider a lateral JJA, which sets the perfect stage for a gauge theory and adequately models superconducting disordered films[12,29]. The properties of the JJA are controlled by the competition between the Josephson coupling energy E_J and the charging energy $E_c = (2e^2)/2C$ of a single Josephson junction, C being the junction capacitance[7]. At $E_J > E_c$, superconducting correlations win and the system is a superconductor. At $E_J < E_c$, the Coulomb blockade turns the system insulating.

The Hamiltonian for a lateral JJA, modeled as a 2D grid of superconducting granules coupled by weak links is

$$\mathcal{H} = \frac{1}{2}\sum_x V(C_0 - C\ell^2\Delta)V + \sum_{x,l} E_J(1 - \cos(2\ell\nabla_l\phi)) , \tag{1}$$

where the sum is taken over all the points x of the grid, ℓ is the lattice spacing, l is the vector connecting adjacent superconducting granules, V is the electric potential in the granule x, $\nabla_l\phi$ is the superconducting phase difference between adjacent granules, C_0 is the granule capacitance to the ground, and we assume $C \gg C_0$.

The JJA partition function $\mathcal{Z} = \int \mathcal{D}V\mathcal{D}\varphi \exp(-\mathcal{H}/T)$ can be expressed via an Euclidean action of a coupled Coulomb gas of charges and vortices[12]. This allows for the construction of a topological gauge theory of JJA[14] in the self-dual approximation. To do so, we express the Cooper pair, j_μ, and vortex, ϕ_μ, currents in terms of gauge fields a_μ and b_μ as $j_\mu = (\sqrt{2}/2\pi)\epsilon_{\mu\nu\alpha}\partial_\nu b_\alpha$ and $\phi_\mu = (\sqrt{2}/2\pi)\epsilon_{\mu\nu\alpha}\partial_\nu a_\alpha$, with $\epsilon_{\mu\nu\alpha}$ being unity antisymmetric tensor and current conservation constraints (Bianchi identities), $\partial_\nu j_\nu = 0$, $\partial_\nu\phi_\nu = 0$. The large-scale physics of the SIT is governed by the infrared-dominant gauge-invariant terms in a_μ and b_μ, leading, up to non-relativistic effects, to the doubled topologically massive gauge theory with the action[24] (we use natural units $c = 1$, $\hbar = 1$ but we will be restoring physical units when necessary)

$$S = \int d^3x \left[\frac{1}{8\pi^2 E_J} f^b_{\mu\nu}f^b_{\mu\nu} + \frac{1}{16 E_C} f^a_{\mu\nu}f^a_{\mu\nu} + \frac{i}{\pi}a_\mu\epsilon^{\mu\nu\alpha}\partial_\nu b_\alpha \right] , \tag{2}$$

where $f^a_{\mu\nu} = \partial_\mu a_\nu - \partial_\nu a_\mu$ (and similarly for $f^b_{\mu\nu}$). The coupling constants are uniquely determined by the charge of Cooper pairs, by the requirement that the topological mass[24] is the plasma frequency, and by the duality $\pi^2 E_J \leftrightarrow 2E_c$ of JJA.

This simple consideration, however, does not take into account the topological defects arising from the compactness of the two U(1) gauge fields. The compactness implies that the kinetic terms in (2) become non-perturbatively relevant and can drive the system away from the naive topological phase towards other fixed points. The full topological gauge theory of JJA, which can be derived directly from the JJA Hamiltonian[14] takes these into account, and the full action assumes the form

$$S = \sum_x \left[\frac{\ell^3}{8\pi^2 E_J}f_\mu f_\mu + i\frac{\ell^3}{\pi}a_\mu k_{\mu\nu}b_\nu + \frac{\ell^3}{16 E_C}g_\mu g_\mu + i\ell\sqrt{2}\left(a_\mu Q_\mu + b_\mu M_\mu\right) \right] , \tag{3}$$

where the sum runs over the Euclidean 3D discrete lattice. The first and the third terms in this action describe Josephson coupling and Coulomb energies in the JJA expressed through the gauge fields, $f_\mu = k_{\mu\nu}b_\nu$, $g_\mu = k_{\mu\nu}a_\nu$, and $k_{\mu\nu} = S_\mu\epsilon_{\mu\alpha\nu}d_\alpha$ is the lattice Chern-Simons operator, formulated in terms of the lattice derivative d_α and shift operator $S_\mu f(x) = f(x+\ell\hat\mu)$ in a form guaranteeing the gauge invariance[14]. The second term, the so called mixed Chern-Simons term, describes the Aharonov-Bohm coupling between charges and vortices. Note that, contrary to the pure Chern-Simons term used to model topological states in strong magnetic fields[30], this mixed, or doubled Chern-Simons term does not violate parity and time-reversal (in Minkowski space) since it involves a vector and a pseudovector gauge field. The topological mass $\sqrt{8E_cE_J}$ is the JJA plasma frequency. The last two terms, comprising integer-valued fields Q_μ and M_μ, describe the topological excitations – electric and magnetic strings [26] – arising from the compactness of the U(1) gauge group and represent (Cooper pair) charge and vortex excitations, respectively. Strings can be closed, in which case they are the "world-lines" of charge-anticharge and vortex-antivortex quantum fluctuations over the ground state, or else, infinitely long, representing then the "world-lines" of point charges and vortices. Infinitely long strings can also end on electric or magnetic (monopole) instantons, describing tunneling events [26].

In 3D an analogous construction leads to the BF model [31]

$$
S = \sum_x \frac{\ell^3}{16E_c} f_{\mu\nu} f_{\mu\nu} - i\frac{\ell^4}{2\pi} a_\mu k_{\mu\alpha\beta} b_{\alpha\beta} + \frac{\ell^5}{24\pi^2 E_J} h_{\mu\nu\alpha} h_{\mu\nu\alpha}
$$

$$
+ i\ell\sqrt{2} a_\mu Q_\mu + i\ell^2 \frac{\sqrt{2}}{2} b_{\mu\nu} M_{\mu\nu} ,
\tag{4}
$$

where $k_{\mu\alpha\beta}$ is the lattice version of the topological BF coupling [14]. Here a_μ describes an ordinary gauge field with field strength $f_{\mu\nu} = d_\mu a_\nu - d_\nu a_\mu$. The action is invariant under gauge transformations $a_\mu \to a_\mu + d_\mu\lambda$ and thus a_μ embodies the usual two degrees of freedom. The third term in (4) represents the corresponding kinetic term. The second gauge field, $b_{\mu\nu}$, is an antisymmetric Kalb-Ramond [32] gauge field (gauge field of the second kind) whose field strength is given by the three-form $h_{\mu\nu\rho} = d_\mu b_{\nu\rho} + d_\nu b_{\rho\mu} + d_\rho b_{\mu\nu}$. The first term in (4) is the kinetic term for this Kalb-Ramond gauge field. In addition to the usual gauge invariance the action of the BF model is also invariant under gauge transformations of the second kind, $b_{\mu\nu} \to b_{\mu\nu} + d_\mu\chi_\nu - d_\nu\chi_\mu$. As a consequence, the Kalb-Ramond gauge field describes a single scalar degree of freedom. The two gauge fields are coupled by the topological BF term [31], that gives its name to the model. This coupling induces a topological mass $m = \sqrt{8E_cE_J}$ for all degrees of freedom [33,34]. Note that, while a_μ is a vector gauge boson as usual, $b_{\mu\nu}$ is a pseudotensor, so that the BF coupling respects the discrete symmetries of parity (P) and time-reversal invariance (T).

Introducing the conserved charge and vortex currents with the dual field strengths according to

$$
j_\mu = \frac{\sqrt{2}}{2\pi} h_\mu = \frac{\sqrt{2}}{4\pi} k_{\mu\nu\rho} b_{\nu\rho} ,
$$

$$
\phi_{\mu\nu} = \frac{\sqrt{2}}{2\pi} \tilde{f}_{\mu\nu} = \frac{\sqrt{2}}{2\pi} \hat{k}_{\mu\nu\rho} a_\rho .
\tag{5}
$$

one arrives at the description of the universal long-distance physics of 3D Josephson junction arrays. In 3D, the integer topological excitations Q_μ and $M_{\mu\nu}$ represent electric strings and magnetic surfaces. When the gauge symmetry is $U(1)$ these topological excitations are closed or infinitely extended, representing the world-lines of charge-anticharge fluctuations and the world-surfaces of vortex fluctuations or the world-lines and the world-surfaces of charges and vortices, respectively. When the symmetry is broken down to $\mathbb{Z}$, with the gauge parameters restricted to the values $\lambda = (2\pi/\sqrt{2})n$ and $\lambda_\mu = (2\pi\sqrt{2})n_\mu$, with $n, n_\mu \in \mathbb{Z}$, instantons $\hat{d}_\mu Q_\mu$ at the end of electric strings [26] and magnetic monopole world-lines $\hat{d}_\mu M_{\mu\nu}$

at the boundary of magnetic surfaces [26] are allowed, since $\exp(-S)$ is invariant under these restricted gauge transformations.

3. The structure of quantum phases

The structure of quantum phases at $T = 0$ is governed by the competition between the topological excitations. Integrating out the gauge fields in Eq. (4), one arrives at the effective action for the transverse (closed or infinitely extended) topological excitations as

$$S_{\text{top}} = \sum_x \frac{4E_C}{\ell} Q_\mu \frac{\delta_{\mu\nu}}{m^2 - \nabla^2} Q_\nu + \frac{\pi^2 E_J}{2\ell} M_{\mu\nu} \frac{\delta_{\mu\alpha}\delta_{\nu\beta} - \delta_{\mu\beta}\delta_{\nu\alpha}}{m^2 - \nabla^2} M_{\alpha\beta}$$

$$+ i\frac{\pi m^2}{l} Q_\mu \frac{k_{\mu\alpha\beta}}{\nabla^2 (m^2 - \nabla^2)} M_{\alpha\beta} . \tag{6}$$

In 2D, magnetic topological excitations M_μ are the strings as well, and the BF operator has one index less. Note that only closed loops and surfaces contribute to the last term. On scales much larger than $1/m$, where the denominator reduces to $m^2\nabla^2$, this term represents the topological "linking number" of loops and surfaces of typical "width" $1/m$ in Euclidean 4D. By expressing $Q_\mu = \ell k_{\mu\alpha\beta}Y_{\alpha\beta}$ one can recognize that it is given by $i2\pi$ − integer and drops thus out off the partition function. This reflects the absence of Aharonov-Bohm and Aharonov-Casher phases between charges ne and magnetic fluxes $2\pi/ne$. This explains also why the Dirac strings and surfaces "attached" to electric monopole instantons and magnetic monopoles are unobservable by linking interactions with physical magnetic fluxes and charges, respectively.

In order to proceed we follow [35] and retain only self-interaction terms. Loops and surfaces are thus assigned with the "energies" (formally Euclidean actions in the present statistical field theory setting and thus dimensionless in our natural units) proportional to their length N and area A measured as numbers of links and plaquettes involved,

$$S_N = \pi(m\ell)G(m\ell) \sqrt{\frac{2E_c}{\pi^2 E_J}} Q^2 N ,$$

$$S_A = \pi(m\ell)G(m\ell) \sqrt{\frac{\pi^2 E_J}{2E_c}} M^2 A , \tag{7}$$

where $G(ml)$ is the diagonal element of the lattice kernel $G(x - y)$ representing the inverse of the operator $\ell^2 \left(m^2 - \nabla^2\right)$ and Q and M are the integer quantum numbers carried by the two kinds of topological defects. However, also the entropy of link strings and plaquette surfaces is proportional to their length and area [36], $\mu_N N$ and $\mu_A A$. Both coefficients μ are non-universal: $\mu_N \simeq \ln(7)$ since at each step the non-backtracking string can choose among 7 possible directions on how to continue, while μ_A does not have such a simple interpretation but can be estimated numerically [36]. This gives for both types of topological defects a "free energy" proportional to their dimension and with coefficient that can be positive or negative depending on the parameters of the theory,

$$\mathcal{F} = \eta \left[\left(\frac{1}{g} Q^2 - \frac{1}{\eta} \right) \mu_N N + \left(g M^2 - \frac{1}{\eta} \right) \mu_A A \right] , \tag{8}$$

where we have defined

$$g = \sqrt{\frac{\pi^2 E_J \mu_N}{2 E_c \mu_A}} \, ,$$

$$\eta = \frac{\pi(m\ell)G(m\ell)}{\sqrt{\mu_N \mu_A}} \, . \tag{9}$$

If the coefficients in Eq. (8) are positive, the self-energy dominates and large string/surface configurations are suppressed in the partition function. If the coefficient, instead is negative the entropy dominates and topological excitations "condense", which means that large configurations will dominate the partition function. In 2D, $\mu_A = \mu_N = \mu \approx \ln(5)$ and analysis of the phase diagram can be conveniently exercised via the "ellipsoid technique"[25]. Namely, the condition that $\mathcal{F}$ is negative, i.e. can be written as

$$(1/g)Q^2 + gM^2 < 1/\eta \, , \tag{10}$$

and the proliferation of loops of an arbitrary size, infinitely long strings and instantons, i.e. the Bose condensation of charges and/or vortices, becomes energetically advantageous. The condensation condition (10) implies that a particular condensate forms if the pair $\{Q, M\}$ on a square lattice of integer electric and magnetic charges falls within the interior of an ellipse with semi-axes $r_Q = (g/\eta)^{1/2}$ and $r_M = 1/(g\eta)^{1/2}$. Graphic representation of various phases that emerge in the vicinity of the SIT are presented in Fig. 1. The panel (a) shows the conditions for the formation of the Cooper pair condensate comprising quantum fluctuations carrying 'unit' charges $\pm 2e$. In the superconducting state a charge current does not induce any voltage since the condensate of electric strings prevents the formation of vortex loops that would cause a finite resistance. In the dual, superinsulating state, see Fig. 1b, $M = \pm 1$ and vortex-antivortex quantum fluctuations form a Bose condensate which blocks the propagation of Cooper pairs. The conditions shown in Fig. 1c, refer to coexisting Cooper pair and vortex condensates, $Q, M = \pm 1$. Finally, Fig. 1d illustrates the situation where none of the condensates can form, $Q = M = 0$. This is the intermediate state often referred to as a quantum (Bose) metal[19,20]. Under the condition of exact duality and charge-hole symmetry, the Bose metal is characterized by the universal quantum resistance $R_Q = h/(4e^2)$ at $T = 0$[11]. In real materials the resistance can significantly deviate from the quantum value.

At variance to the 2D case, in 3D "dyonic excitations" with $Q \neq 0$ and $M \neq 0$ must be excluded since the two types of excitations are different, only pairs $\{0, M\}$ or $\{Q, 0\}$ have to be considered. The phase diagram is found simply by analyzing which integer charges lie within the ellipse when the semi-axes are varied,

$$\eta < 1 \rightarrow \begin{cases} g > 1, \text{ superconductor}, \\ g < 1, \text{ superinsulator}, \end{cases}$$

$$\eta > 1 \rightarrow \begin{cases} g < \eta, \text{ superconductor}, \\ \frac{1}{\eta} < g < \eta, \text{ topological insulator}, \\ g < \frac{1}{\eta}, \text{ superinsulator}, \end{cases} \tag{11}$$

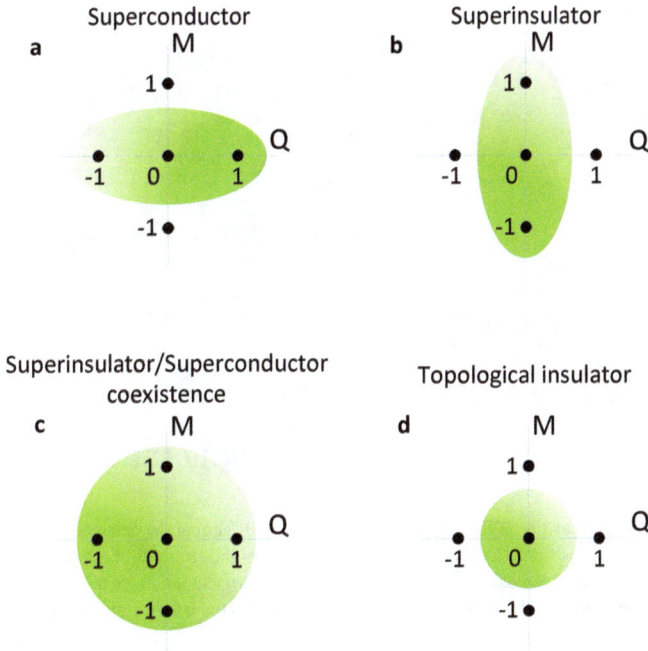

Fig. 1. The configurations that minimize the string gas free energy are determined by the integer quantum numbers that fall in the interior of an ellipse with semi-axes given by the model parameters. (a) Superconductor: strings with electric quantum numbers condense. (b) Superinsulator: Strings with magnetic quantum numbers condense. (c) Coexistence of long electric and magnetic strings: this is an unstable configuration near the first-order direct transition from a superinsulator to a superconductor, the strings with the quantum numbers that minimize their self-energy are the stable configurations. (d) Bose topological insulator/quantum metal: all strings are suppressed by their high self-energy.

where we have anticipated that a phase with a condensation of electric strings corresponds to a superconductor, a phase with a condensate of magnetic surfaces is a superinsulator and a phase without any condensation describes a bosonic topological insulator.

The zero-temperature phase diagram of the close vicinity of the SIT shown in Fig. 2, emerges thus from the competition of three orders: Bose condensates of either (i) Cooper pairs, or (ii) magnetic vortices, and (iii) 'quantumness' quantified by the magnitude of η, and this geometric consideration holds in both 2D and 3D systems. The phase structure described by Eq. (11) resolves two outstanding fundamental puzzles of the SIT: first, the condition for the direct SIT vs. the transition via an intermediate Bose- or quantum metal state[20] is $\eta < 1$; secondly this intermediate Bose metal is identified as a bosonic topological insulator[21-23], as we now show. The lines $\eta = 1/g$ and $\eta = g$, separating the intermediate topological insulator from the superinsulator and the superconductor, mark continuous phase transitions[37]. The point $g = \eta = 1$ where they merge with the first-order direct SIT is the tri-critical point.

To reveal the nature of the phases that emerge near the tri-critical point, we will derive electromagnetic response in every respective phase. To that end, we compute the induced

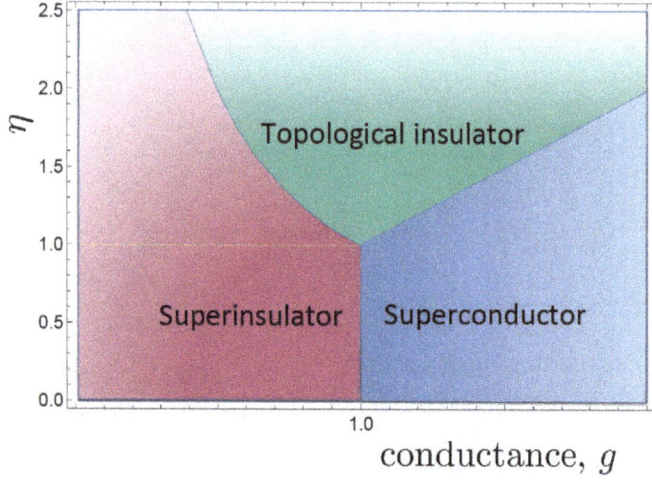

Fig. 2. **Phase diagram near the SIT in two dimensions at** $T = 0$. Tuning the parameter $g = \sqrt{\pi^2 E_J \mu_N / 2 E_C \mu_A}$, one drives the system across the SIT. The quantity η characterizes the strength of quantum fluctuations. When these are strong, $\eta > 1$, an intermediate quantum state, the b topological insulator phase, opens up between the superconductor and the superinsulator. This phase diagram describes the bosonic phases in the critical vicinity of the SIT.

effective action for the minimally coupled electromagnetic gauge potential A_μ. Then, the induced current in the various phases is obtained, in 2D, as

$$j^\mu = \frac{1}{\ell^3} \frac{\delta}{\delta A_\mu} S_M^{\mathrm{eff}}(A_\mu) \,, \tag{12}$$

where S_M^{eff} is the effective action, defined by

$$e^{-S_{\mathrm{eff}}(A_\mu)} = \sum_{Q_\mu, M_\mu} \int_{a_\mu, b_\mu} \mathcal{D}a_\mu \mathcal{D}b_\mu \, e^{-S\left(a_\mu, b_\mu, Q_\mu, M_\mu + \frac{\ell^2 e}{\pi} F_\mu\right)} . \tag{13}$$

with S the action (3) and $F_\mu = \hat{k}_{\mu\nu} A_\nu$ the dual electromagnetic field strength. Since the different phases are driven only by the condensation (or lack thereof) of topological excitations we integrate out right away the two gauge fields a_μ and b_μ to obtain

$$e^{-S_{\mathrm{eff}}(A_\mu)} = \sum_{Q_\mu, M_\mu} e^{-S\left(Q_\mu, M_\mu + \frac{\ell^2 e}{\pi} F_\mu\right)} , \tag{14}$$

with

$$S\left(Q_\mu, M_\mu + \frac{\ell^2 e}{\pi} F_\mu\right) = \sum_x \frac{4 E_C}{\ell} Q_\mu \frac{\delta_{\mu\nu}}{m^2 - \nabla^2} Q_\nu$$

$$+ \frac{\pi^2 E_J}{2\ell} \left(M_\mu + \frac{\ell^2 e}{\pi} F_\mu\right) \frac{\delta_{\mu\nu}}{m^2 - \nabla^2} \left(M_\nu + \frac{\ell^2 e}{\pi} F_\nu\right)$$

$$+ i \frac{2\pi m^2}{\ell} Q_\mu \frac{k_{\mu\nu}}{\nabla^2 (m^2 - \nabla^2)} \left(M_\nu + \frac{\ell^2 e}{\pi} F_\nu\right) . \tag{15}$$

In 3D, the corresponding result would be

$$
S\left(Q_\mu, M_{\mu\nu} + \frac{\ell^2 e}{\pi}\tilde{F}_{\mu\nu}\right) = \sum_x \frac{4E_C}{\ell} Q_\mu \frac{\delta_{\mu\nu}}{m^2 - \nabla^2} Q_\nu
$$

$$
+ \frac{\pi^2 E_J}{2\ell}\left(M_{\mu\nu} + \frac{\ell^2 e}{\pi}\tilde{F}_{\mu\nu}\right)\frac{\delta_{\mu\alpha}\delta_{\nu\beta} - \delta_{\mu\beta}\delta_{\nu\alpha}}{m^2 - \nabla^2}\left(M_{\alpha\beta} + \frac{\ell^2 e}{\pi}\tilde{F}_{\alpha\beta}\right)
$$

$$
+ i\frac{\pi m^2}{\ell} Q_\mu \frac{k_{\mu\alpha\beta}}{\nabla^2(m^2 - \nabla^2)}\left(M_{\alpha\beta} + \frac{\ell^2 e}{\pi}\tilde{F}_{\alpha\beta}\right), \tag{16}
$$

where $\tilde{F}_{\mu\nu} = \hat{k}_{\mu\nu\alpha}A_\alpha$ is again the dual electromagnetic tensor.

4. The superconductor

Let us start by showing how the best understood state, the superconducting phase, arises in our gauge formalism. We will do this explicitly in 2D, although the 3D computation follows exactly the same line and gives an analogous result. Since electric topological defects condense, we have first to explicitly perform the sum over Q_μ in (15)

$$
e^{-S_{\text{eff}}(A_\mu)} = \sum_{\{Q_\mu, M_\mu\}} e^{-\frac{e_\ell^2}{\ell}\sum_x M'_\mu \frac{\delta_{\mu\nu}}{m^2 - \nabla^2} M'_\nu - \frac{e_g^2}{\ell}\sum_x Q_\mu \frac{\delta_{\mu\nu}}{m^2 - \nabla^2} Q_\nu + i\frac{2\pi m^2}{\ell}\sum_x Q_\mu \frac{k_{\mu\nu}}{\nabla^2(m^2 - \nabla^2)} M'_\mu}, \tag{17}
$$

where we have introduced, for simplicity, the shorthand notation $M'_\mu = M_\mu + (e\ell^2/\pi)F_\mu$. Note that, in this case, only closed loops satisfying $\hat{d}_\mu Q_\mu = 0$ couple to the electromagnetic field, since $d_\mu M'_\mu = 0$. We can thus perform explicitly the sum over Q_μ by solving first the constraint $\hat{d}_\mu Q_\mu = 0$ with the representation $Q_\mu = \ell k_{\mu\nu} n_\nu$ and summing over $\{n_\mu\}$

$$
e^{-S_{\text{eff}}(A_\mu)} = \sum_{\{n_\mu, M_\mu\}} e^{-\frac{e_\ell^2}{\ell}\sum_x M'_\mu \frac{\delta_{\mu\nu}}{m^2 - \nabla^2} M'_\nu - e_g^2\ell\sum_x n_\mu \frac{-\delta_{\mu\nu}\nabla^2 + (1+\lambda)d_\mu d_\nu}{m^2 - \nabla^2} n_\nu + i2\pi m^2\sum_x n_\mu \frac{\delta_{\mu\nu}}{m^2 - \nabla^2} M'_\mu}. \tag{18}
$$

The representation $Q_\mu = \ell k_{\mu\nu} n_\nu$ is, however redundant since gauge transformed $n_\mu \to n_\mu + \ell d_\mu \chi$ correspond to the same Q_μ. In the above expression we have already gauge fixed this symmetry by introducing the standard parameter λ in the Maxwell operator. We now turn the sum over $\{n_\mu\}$ into an integral by the usual Poisson formula,

$$
\sum_{n_\mu} f\left(n_\mu\right) = \sum_{k_\mu}\int dn_\mu f\left(n_\mu\right) e^{i2\pi n_\mu k_\mu}, \tag{19}
$$

where the new integer link variables $\{k_\mu\}$ must satisfy $\hat{d}_\mu k_\mu = 0$ in order to guarantee gauge invariance under transformations $n_\mu \to n_\mu + \ell d_\mu t$. At this point one can explicitly perform the Gaussian integration over $\{n_\mu\}$. In doing so, the Maxwell kernel can be inverted due to the gauge fixing term. As usual, the term containing λ in the inverse operator drops out since it is applied on the gauge invariant field M'_μ,

$$
e^{-S_{\text{eff}}(A_\mu)} = \sum_{\{k_\mu, M_\mu\}} e^{-\frac{e_\ell^2}{\ell}\sum_x M'_\mu\left(\frac{\delta_{\mu\nu}}{m^2 - \nabla^2} + \frac{m^2\delta_{\mu\nu}}{-\nabla^2(m^2 - \nabla^2)}\right)M'_\nu - \frac{e_\ell^2}{\ell}\sum_x\left[2M'_\mu \frac{\delta_{\mu\nu}}{-\nabla^2}k_\nu + k_\mu \frac{1 - \frac{\nabla^2}{m^2}}{\nabla^2}\delta_{\mu\nu}k_\nu\right]}. \tag{20}
$$

Note first that the kernels in the first term of the exponential above combine to produce a massless Coulomb interaction. The result becomes particularly transparent for $m\ell = O(1)$ so that the kernel $1/(m^2 - \nabla^2)$ becomes essentially a contact term on the lattice scale,

$$e^{-S_{\text{eff}}(A_\mu)} = \sum_{\{k_\mu, M_\mu\}} e^{-\frac{e_f^2}{\ell} \Sigma_x (M'_\mu + k_\mu) \frac{\delta_{\mu\nu}}{-\nabla^2} (M'_\nu + k_\nu)}. \tag{21}$$

We can now reabsorb the integers $\{k_\mu\}$ by a shift of the magnetic topological excitations M_μ and drop the sum over k_μ,

$$e^{-S_{\text{eff}}(A_\mu)} = \sum_{\{M_\mu\}} e^{-\frac{e_f^2}{\ell} \Sigma_x \left(M_\mu + \frac{e\ell^2}{\pi} F_\mu\right) \frac{\delta_{\mu\nu}}{-\nabla^2} \left(M_\nu + \frac{e\ell^2}{\pi} F_\nu\right)}. \tag{22}$$

Finally, remembering that this is the phase in which magnetic topological excitations are dilute, we set $M_\mu = 0$, thereby obtaining the final result

$$S_{\text{eff}}(A_\mu) = \sum_x \ell^3 \frac{e^2 e_f^2}{\pi^2} A_\mu \left(\delta_{\mu\nu} - \frac{d_\mu \hat{d}_\nu}{\nabla^2}\right) A_\nu, \tag{23}$$

from where we obtain the induced current

$$j_\mu = \frac{2e^2 e_f^2}{\pi^2} \left(\delta_{\mu\nu} - \frac{d_\mu \hat{d}_\nu}{\nabla^2}\right) A_\nu. \tag{24}$$

This is the non-local form of the London equations. The usual representation is recovered, e.g. by choosing the Coulomb gauge $\hat{d}_\mu A_\mu = 0$, in which case one obtains readily

$$\partial_t \mathbf{j} = \frac{2e^2 e_f^2}{\pi^2} \mathbf{E},$$

$$\text{rot } \mathbf{j} = \frac{2e^2 e_f^2}{\pi^2} B, \tag{25}$$

which shows that this phase is a superconductor with London penetration depth $\lambda \propto 1/(2e\sqrt{E_J/d})$ where d is the thickness of the film, which must be taken into account in the present 2D setting. This amounts to a Pearl screening length $\lambda_\perp \equiv \lambda^2/d \propto 1/(4e^2 E_J)$.

5. The Bose metal/Bose topological insulator

In this section we will tackle the nature of the phase with no condensation that can appear between the superconducting and the superinsulating states, a phase usually called the Bose metal[20]. We will do so again explicitly in 2D, although the conclusions hold in 3D as well. Since no topological excitations condense, $Q_\mu = 0$ and $M_\mu = 0$, we obtain

$$S^{\text{eff}}(A_\mu, 0) = \sum_x \frac{\ell^4 e^2 g\eta\mu}{\pi^2} A_\mu \left(-\delta_{\mu\nu} \nabla^2 + d_\mu \hat{d}_\nu\right) A_\nu. \tag{26}$$

The functional derivative with respect to the field A_μ yields the current induced by external electromagnetic fields $F_{\mu\nu} = \partial_\mu A_\nu - \partial_\nu A_\mu$ as

$$j^\mu = \ell \frac{2e^2 g\eta\mu}{\pi^2} \partial_\nu F^{\mu\nu}, \tag{27}$$

showing that only *variations* of external electric and magnetic fields, but not homogenous field themselves, can cause an induced current, which is the typical electromagnetic response of an insulator. Therefore, the bulk conductances, both the longitudinal and the Hall ones, vanish.

Is this the whole story, however? To answer this, let us consider the infrared-dominant terms in the action. In the phase with no topological defects these are

$$S = \frac{1}{\pi} \int d^3x \, a_\mu \epsilon^{\mu\nu\alpha} \partial_\nu b_\alpha + \frac{2e\sqrt{2}}{2\pi} \int d^3x \, A_\mu \epsilon^{\mu\nu\alpha} \partial_\nu b_\alpha , \tag{28}$$

where we have restored the continuum notation, less cumbersome to work with (note that only the presence of topological defects requires a lattice formulation).

The catch is that these terms are not gauge-invariant under transformations that do not vanish on the boundary: these produce a boundary term. Exactly as in the analogous treatment of the quantum Hall effect [38], one must add edge degrees of freedom so that gauge invariance of the complete theory is restored. In order to do this one considers the $a_0 = 0$, $b_0 = 0$ gauge (the 0 components of gauge fields are not dynamic degrees of freedom since they never come with a time derivative) and one sets $a_i = \partial_i \lambda$ and $b_i = \partial_i \chi$ to discover what is the boundary term arising from gauge transformations. This is

$$\begin{aligned}
S_{\text{edge}} &= -\frac{1}{\pi} \int d^3x \, a_i \epsilon^{ij} \partial_0 b_j + \frac{2e\sqrt{2}}{2\pi} \int d^3x \left(A_0 \epsilon^{ij} \partial_i b_j - A_i \epsilon^{ij} \partial_0 b_j \right) \\
&= \frac{1}{2\pi} \int d^2x \, (\partial_0 \lambda \partial_s \chi + \partial_s \lambda \partial_0 \chi) + \frac{2e\sqrt{2}}{2\pi} \int d^2x \, (A_0 \partial_s \chi - A_s \partial_0 \chi) , \tag{29}
\end{aligned}$$

where s denotes the space coordinate along the one-dimensional boundary.

We now introduce the two new fields $\lambda = \xi + \eta$ and $\chi = \xi - \eta$. In terms of these, the edge action decouples,

$$S_{\text{edge}} = \frac{1}{\pi} \int d^2x \, (\partial_0 \xi \partial_s \xi - \partial_0 \eta \partial_s \eta) + 2e \int d^2x \, A_0 \left(\frac{\sqrt{2}}{2\pi} \partial_s \chi \right) , \tag{30}$$

where we have taken, for simplicity's sake $A_s = 0$ and from which we can identify

$$\rho = \frac{\sqrt{2}}{2\pi} \partial_s \chi \tag{31}$$

as the one-dimensional charge density of edge excitations.

If we add this edge action to the model, complete gauge invariance is restored. However, there is still no dynamics embodied in this action since it is the edge action corresponding to a topological bulk model. As in the case of the quantum Hall effect, a non-universal dynamics for the edge modes is assumed to be generated by boundary effects [38]. These contribute a Hamiltonian

$$H = \frac{1}{\pi} \int ds \left[-v \, (\partial_s \xi)^2 - v \, (\partial_s \eta)^2 \right] , \tag{32}$$

where v is the velocity of propagation of the edge modes along the boundary. Adding this term, the total edge action becomes

$$S_{edge} = \frac{1}{\pi} \int d^2x \; [(\partial_0 - v\partial_s) \xi \partial_s \xi - (\partial_0 + v\partial_s) \eta \partial_s \eta] + 2e \int d^2x \, A_0 \left(\frac{\sqrt{2}}{2\pi} \partial_s \chi \right) . \quad (33)$$

This is a well known field theory model[39] describing two chiral bosons circulating along the edge in two opposite directions (the different sign in the velocity term).

Let us now compute the equations of motion of this field theory model:

$$(\partial_0 - v\partial_s) \, \partial_s \xi = -\frac{2e}{2\sqrt{2}} E ,$$

$$(\partial_0 + v\partial_s) \, \partial_s \eta = -\frac{2e}{2\sqrt{2}} E , \quad (34)$$

where $E = \partial_s A_0^{el}$ is the electric field. These equations can be re-expressed in terms of λ and χ as

$$\partial_0 \partial_s \lambda - v\partial_s^2 \chi = -\frac{2e}{\sqrt{2}} E ,$$

$$\partial_0 \partial_s \chi - v\partial_s^2 \lambda = 0 . \quad (35)$$

In a static situation (all time derivatives vanishing) the first of these equations becomes

$$v\partial_s^2 \chi = \frac{2e}{\sqrt{2}} E . \quad (36)$$

Finally, using the identification (31) we rewrite this

$$v\partial_s \rho = \frac{2e}{2\pi} E = \frac{2e}{2\pi} \partial_s V , \quad (37)$$

where we have used the voltage symbol V for A_0^{el}. The one-dimensional current being defined as $I = 2ev\rho$ we obtain the final equation

$$\frac{I}{V} = \frac{(2e)^2}{2\pi} , \quad (38)$$

i.e. the charge conduction with the quantum resistance $R_Q = h/(4e^2)$. Thus, while all bulk conductances are suppressed by the large Chern-Simons mass, there remains the ballistic edge conductance with the quantum resistance, which is characteristic of a topological insulator[40,41]. With the flux characterizing, this topological insulator is π, the charge is $2e$ rather than e, because it is a Cooper pair state. This has the consequence that this bosonic topological insulator is actually a level 1 topological insulator with no ground state degeneracy on the torus. The edge modes in this topological insulator carry the charge $2e$ since the elemental carriers mediating the conductivity are Cooper pairs. Then, a straightforward calculation yields the TI sheet resistance $R_\square = h/(4e^2) \equiv R_Q$, which corresponds to perfect duality, where exactly one fluxon for each Cooper pair traverses the system. This coincides with the result by M. P. A. Fisher[10] obtained by an elegant qualitative consideration. In experiments, the resistance can significantly differ from R_Q signaling deviations from strict duality[42]. Remarkably, our prediction that in the Bose metal the Hall resistance should disappear is in a full accord with the recent observation by Ref. 44.

The nature of topological insulators in 3D remains the same: they are states in which all bulk currents are suppressed by a large topological mass gap and conduction is restricted to edge modes. They are described by a purely topological field theory[40], the doubled Chern-Simons theory in 2D and the BF theory in 3D. In 3D, however, topological insulators may display a magnetoelectric polarizability $\partial M/\partial E = \partial P/\partial B = \theta(e^2/2\pi h)$, with M the magnetization, P the polarization and E and B the electric and magnetic fields[41]. This can be accounted for by an additional axion electrodynamics term[45], also called θ-term, $S_{\text{axion}} = i \sum_x \ell^4(\theta/16\pi^2)F_{\mu\nu}\tilde{F}_{\mu\nu}$ in their electromagnetic response. This is the surface term, since the partition function $\exp(-S_{\text{axion}})$ is invariant under shifts $\theta \to \theta + 2\pi$. Time reversal, $\mathcal{T}$, maps $\theta \to -\theta$. So the only values of θ compatible with $\mathcal{T}$-invariance are $\theta = 0$ and $\theta = \pi$, modulo 2π. The former case defines normal topological insulators, the latter case is realized for so-called strong topological insulators.

6. Disordered films and material parameters

We now extend the gauge theory of the SIT and the analysis of the corresponding emergent phases onto disordered superconducting continuous media. To do so, we transcribe the above results obtained in terms of the JJA parameters E_C and E_J into the language of material characteristics of superconducting films. The role of the tuning parameter driving the film across the SIT at zero magnetic field is taken by the resistance per square, $R_\square$ (or, equivalently, by the dimensionless conductance $g = 4R_Q/R_\square$). The disorder-driven SIT in films is expected to occur at $g = g_c = 1$ which corresponds to the condition $\sqrt{2E_C/\pi^2 E_J} = 1$. Further connection between the material characteristics of the superconducting films and those of JJA is established by the relation[29] $\lambda_\perp = c\Phi_0/(8\pi^2 I_c)$, where $\lambda_\perp = \lambda_L^2/d$ is the Pearl screening length, λ_L is the bulk London penetration depth, d is the film thickness, and $I_c \equiv (2eE_J/\hbar)$ is the critical current of a single Josephson junction. Then, our dimensionless parameter $\natural \equiv m\ell$ acquires a remarkable form

$$\natural = \frac{1}{8\alpha} \frac{\ell d}{\lambda_L^2}, \tag{39}$$

where $\alpha = e^2/(\hbar c)$ is the fine structure constant, and ℓ now plays the role of the characteristic microscopic cutoff length of order of the superconducting coherence length ξ. Therefore, the parameter η describes the strength of quantum fluctuations: an intermediate quantum metal/topological insulator phase occurs when these are strong enough, i.e. $\eta > 1$. In disordered superconductors, where the coherence length ξ exceeds the mean free path, $\lambda_L^2 = [4\pi n_s e^2/(mc^2)](\tau\Delta/\hbar)$, where n_s is the density of superconducting electrons, which at $T = 0$ is equal to the total electron density n_e and τ is the transport scattering time. Accordingly, $\natural \simeq (\alpha/4)(\ell d n_e^{2/3})(\tau\Delta/\hbar)$. If we associate the ultraviolet cutoff ℓ with the minimal superconducting scale ξ, the dependence $\natural \sim n_e^{2/3}$ reflects correctly the trend of crossing over from the direct SIT to SMIT upon an increase of the electron density in films.

For a typical value $\mu = 0.25\ln(5)$, the magnitude of the parameter $\natural_c$ where all three phases meet, i.e. its magnitude corresponding to the tri-critical point, is $\natural = \natural_c \approx 0.65$.

However, when adopting the JJA paradigm to model thin films, one must ultimately take into account also the (relative) electric permittivity ε_P and the (relative) magnetic permeability μ_P of these materials, that determine together the light velocity in these media. These parameters will typical deviate from their relativistic values 1. Both in the original JJA Hamiltonian and in its Coulomb gas version, however it is not a priori clear how to include these parameters. The gauge theory formulation, instead offers the advantage to make this evident, the electric permittivity being a multiplicative constant for the square of electric fields in the action (energy) while the inverse of the magnetic permeability being a multiplicative constant for the square of magnetic fields in the action (energy)[48]. Note that different coupling constants for electric and magnetic fields constitute the most generic non-relativistic gauge theory action since they distinguish between electric fields, which are tensor components with one time-like and one space-like index from magnetic fields, which are tensor components with purely space-like indices.

We thus consider the generic non-relativistic action in 2D

$$
S = \sum_x \frac{\ell^3}{2e_f^2\mu_P}f_0f_0 + \frac{\ell^3\varepsilon_P}{2e_f^2}f_if_i + i\frac{\ell^3\kappa}{2\pi}a_\mu k_{\mu\nu}b_\nu + \frac{\ell^3}{2e_g^2\mu_P}g_0g_0
$$

$$
+ \frac{\ell^3\varepsilon_P}{2e_g^2}g_ig_i + i\ell\sqrt{\kappa}a_\mu Q_\mu + i\ell\sqrt{\kappa}b_\mu M_\mu , \tag{40}
$$

with a magnetic permeability μ_P and an electric permittivity ε_P. Note that the Chern-Simons term, being topological, is unaltered. We have introduced a generic charge unit κ ($\kappa = 2$ for Cooper pairs) for reasons to become clear in a moment. This action can be rewritten as

$$
S = \sum_x \frac{\ell^3}{2e_f^2}\frac{1}{\mu_P}\left[b_i\left(-\frac{1}{v^2}d_0\hat{d}_0 - \nabla_2^2\right)\delta_{ij}b_j + b_id_i\hat{d}_jb_j\right]
$$

$$
+ \frac{\ell^3}{2e_f^2}\varepsilon_P\left[b_0\left(-\frac{1}{v^2}d_0\hat{d}_0 - \nabla_2^2 + \frac{1}{v^2}d_0\hat{d}_0\right)b_0 + b_0d_0\hat{d}_ib_i + b_id_i\hat{d}_0b_0\right]
$$

$$
+ i\frac{\ell^3\kappa}{2\pi}a_\mu k_{\mu\nu}b_\nu
$$

$$
+ \frac{\ell^3}{2e_g^2}\frac{1}{\mu_P}\left[a_i\left(-\frac{1}{v^2}d_0\hat{d}_0 - \nabla_2^2\right)\delta_{ij}a_j + a_id_i\hat{d}_ja_j\right]
$$

$$
+ \varepsilon_P\left[a_0\left(-\frac{1}{v^2}d_0\hat{d}_0 - \nabla_2^2 + \frac{1}{v^2}d_0\hat{d}_0\right)a_0 + a_0d_0\hat{d}_ia_i + a_id_i\hat{d}_0b_0\right]
$$

$$
+ i\ell\sqrt{\kappa}a_\mu Q_\mu + i\ell\sqrt{\kappa}b_\mu M_\mu , \tag{41}
$$

with $v = 1/\sqrt{\mu_P\varepsilon_P}$ the light velocity in the material and ∇_2 being the Laplacian over "spatial" indices, denoted by Latin letters (the repeated latin letters stand, correspondingly, for the summation over "spatial indices" only). By introducing auxiliary rescaled fields

$$\tilde{a}_0 = \sqrt{\varepsilon_P}\, a_0\,, \qquad \tilde{a}_i = \frac{1}{\sqrt{\mu_P}}\, a_i\,,$$

$$\tilde{b}_0 = \sqrt{\varepsilon_P}\, b_0\,, \qquad \tilde{b}_i = \frac{1}{\sqrt{\mu_P}}\, b_i\,,$$

$$\tilde{Q}_0 = \sqrt{v}\, Q_0\,, \qquad \tilde{Q}_i = \frac{1}{\sqrt{v}}\, Q_i\,, \tag{42}$$

a rescaled time derivative $\tilde{d}_0 = (1/v)d_0$, a correspondingly modified Chern-Simons operator $\tilde{k}_{\mu\nu}$ containing this time derivative and a rescaled Chern-Simons coupling $\tilde{\kappa} = \sqrt{\mu_P/\varepsilon_P}\, \kappa$, the action (41) can be reformulated exactly as the original relativistic action (3), but now expressed entirely in terms of the rescaled quantities. Integrating out the rescaled gauge fields leads thus to the same action (6) expressed in terms of the rescaled topological excitations $\tilde{Q}_\mu$ and $\tilde{M}_\mu$ and with the modified kernel

$$m^2 - \nabla^2 \to \tilde{m}^2 - \left(\frac{1}{v^2} d_0 \tilde{d}_0 + \nabla_2^2 \right)\,, \tag{43}$$

with $\tilde{m} = \sqrt{\mu_P/\varepsilon_P}(\kappa e_f e_g / 2\pi) = \sqrt{\mu_P/\varepsilon_P}(e_f e_g/\pi)$ for the relevant case of Cooper pairs. In terms of the original topological defects the real part of this action reads

$$S_{\text{top}}^{\text{real}} = \sum_x \frac{\sqrt{\mu_P} e_g^2}{\sqrt{\varepsilon_P}\ell} \left[v Q_0 \frac{1}{\tilde{m}^2 - \frac{1}{v^2} d_0 \tilde{d}_0 - \nabla_2^2} Q_0 + \frac{1}{v} Q_i \frac{\delta_{ij}}{\tilde{m}^2 - \frac{1}{v^2} d_0 \tilde{d}_0 - \nabla_2^2} Q_j \right]$$

$$+ \frac{\sqrt{\mu_P} e_f^2}{\sqrt{\varepsilon_P}\ell} \left[v M_0 \frac{1}{\tilde{m}^2 - \frac{1}{v^2} d_0 \tilde{d}_0 - \nabla_2^2} M_0 + \frac{1}{v} M_i \frac{\delta_{ij}}{\tilde{m}^2 - \frac{1}{v^2} d_0 \tilde{d}_0 - \nabla_2^2} M_j \right]\,, \tag{44}$$

where we have reintroduced the value $\kappa = 2$ for Cooper pairs.

The main difference with respect to the relativistic case, apart from a light velocity $v < 1$, is that the contributions of the components Q_0 and M_0 are now different from those of the components Q_i and M_i, as could have been expected. Consider a closed string made of N bonds, with integer quantum numbers $Q_\mu = Q$ and $M_\mu = M$ on all the lattice bonds forming the string and zero elsewhere. This corresponds to a fluctuation in which a charge-anticharge or vortex-antivortex pair is created from the vacuum, lives for a "time" proportional to its length in the 0 direction and is then annihilated in the vacuum again. We are interested in very long-lived such fluctuations, in which the dominant contribution to the action comes from the "time" terms, first and third terms in the above action.

We can repeat verbatim the steps leading to (8), with the only difference that the parameter η is modified to

$$\eta = \frac{\mu_P \pi m \ell \tilde{G}(m\ell)}{\mu}\,, \tag{45}$$

with

$$\tilde{G}(m\ell) = \frac{1}{(2\pi)^3} \int_{-\pi}^{\pi} d^3k \frac{1}{\frac{\mu_P}{\varepsilon_P}(m\ell)^2 + \frac{4}{v^2}\sin\left(\frac{k^0}{2}\right)^2 + \sum_{i=1}^{2} 4\sin\left(\frac{k^i}{2}\right)^2} . \tag{46}$$

The entropy parameter μ in this non-relativistic case must be presumably lowered to the range $\ln(5) > \mu > \ln(3)$ because of the predominance of the "spatial" string components. The uncertainty about its exact value reflects an approximation: it can lead to a shift of the entire curve $\eta(m\ell)$ but is expected to be predominantly material-independent. The magnetic permeability μ_P and the electric permittivity ε_P, instead embody the non-universal, material-dependent elements of the $\eta(m\ell)$ relation (from now on we shall consider only the case $\mu_P = 1$ and drop the subscript "P", introduced to distinguish the magnetic permeability form the string entropy, on ε). Accordingly, the relation $\eta(\flat)$ ceases to be universal since ε depends on the proximity of the system to the SIT and diverges as $g \to 1$[17]. Since η decreases with increasing ε, the scenario of the SIT, direct vs. via the intermediate quantum metal phase, depends on the magnitude of $\eta(\varepsilon_{max})$ where ε_{max} is the maximal dielectric constant achieved at the value of $g = 1$, where the correlation length associated with the SIT compares to the lateral dimension of the film. Taking an estimate of $\varepsilon_{max} \simeq 10^4$ as a characteristic value for the NbTiN film, where the divergent ε was observed on approach to the magnetic field-driven SIT[49], and, using a corresponding estimate of $\flat \approx 0.26$ for this material, one obtains $\eta < 1$ and hence one expects this NbTiN film to exhibit a direct SIT, as indeed is observed experimentally. Analogously, one can observe that near the SIT $\eta < 1$ for TiN, for which $\flat \approx 2.63$, which thus also follows the direct SIT scenario, as observed. Instead, for NbSi, one does not expect a ε divergence and, correspondingly, the estimate $\flat \approx 21.6$ gives $\eta > 1$ confirming a transition via an intermediate quantum (Bose) metal phase.

7. Transitions induced by the magnetic field

In concrete experimental settings the phase transitions in thin films are driven either by varying the thickness of the films (disorder-driven transitions) or by varying an applied magnetic field. Varying the film thickness amounts effectively to varying the parameters η and E_J/E_C of the corresponding array. In the latter situation, instead there is one additional parity (P) and time-reversal (T) breaking external parameter that must be accounted for in the gauge theory formulation. This is the subject of this section.

A uniform external magnetic field can be simply incorporated by the minimal coupling $i\ell^3(2e\sqrt{2}/2\pi)A_\mu f_\mu$ to the charge current ($\sqrt{2}/2\pi)f_\mu$. By a summation by parts this amounts to the following additional term in the gauge theory action,

$$S \to S + \sum_x i\ell\sqrt{2}\, b_0 f , \tag{47}$$

where f denotes the number of elementary fluxes π/e per plaquette piercing the array. This modification amounts simply to shifting the integers

$$M_\mu \to M_\mu + \Phi_\mu , \tag{48}$$

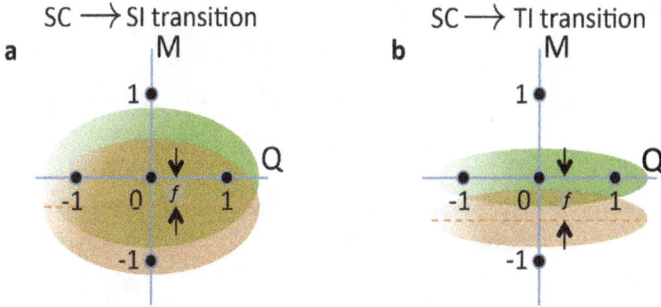

Fig. 3.   Phase transitions induced by an external magnetic field. The frustration parameter $0 \leq f < 1$ displaces the original ellipse along the magnetic axis: (a) Direct transition from a superconductor to a superinsulator for $\eta < 1$. (b) Transition from a superconductor to a topological insulator for $\eta > 1$.

in the original gauge theory model, where Φ_μ represents infinitely long strings in the Euclidean time direction at each lattice point, such that $\Phi_i = 0$ and $\Phi_0 = f$. This shows that f is a periodic parameter defined modulo an integer: it is normally called the magnetic frustration, $0 \leq f < 1$, which explains why it is usually denoted by f.

An external magnetic field corresponds thus to a special case of condensed magnetic strings with non-integer quantum number. To incorporate the additional frustration parameter f we modify thus the string free energy to

$$F = \pi m \ell G(m\ell) \left[\frac{1}{g} Q^2 + g\,(M + f)^2 - \frac{1}{\eta} \right] N, \qquad (49)$$

and to compare with the experimentally relevant situation let us assume we start with $f = 0$ in the superconducting phase with condensation of electric strings,

$$\eta < 1 \rightarrow g > 1,$$
$$\eta > 1 \rightarrow g > \eta. \qquad (50)$$

In this case the original $f = 0$ ellipse is elongated along the electric quantum number axis. Turning on an external magnetic field amounts to increasing the frustration parameter f and, as consequence the ellipse moves down along the magnetic quantum number axis. There are two important thresholds in this downward movement. The points on the ellipse corresponding to $Q = \pm 1$ have M coordinate $M = r_M \sqrt{1 - (1/r_Q^2)}$, where r_Q and r_M are the semi-axes. The M-negative ellipse point with $Q = 0$, instead lies at a distance r_M from the origin by definition. A transition can be caused either by the fact that the two points $Q = \pm 1$ will "exit" the interior of the ellipse before the point $M = -1$ has "entered" it, in which case the initial superconductor turns into a topological insulator/quantum metal, or by the fact that the point $M = -1$ has "entered" the ellipse interior while the two points $Q = \pm 1$ are still inside. In this case we have a coexistence regime of electric and magnetic strings. There will thus be a first-order direct transition to a superinsulator when the frustration reaches a point such that the energy of the magnetic strings becomes smaller than that of the electric ones.

Clearly the two thresholds are given by the frustrations

$$f_1 = r_M \sqrt{1 - \frac{1}{r_Q^2}} \, ,$$

$$f_2 = 1 - r_M \, , \tag{51}$$

and the condition for an intermediate topological insulator/quantum metal phase is $f_1 < f_2$. Using the explicit expressions for the semiaxes in terms of array parameters we can translate this conditions into

$$F(g) \equiv \frac{1}{g^2} - \frac{2}{\sqrt{\eta}} \sqrt{\frac{1}{g}} + 1 > 0 \, . \tag{52}$$

Let us first consider the case $\eta < 1$ and let us express

$$\sqrt{\frac{1}{g}} = 1 - \epsilon \, , \tag{53}$$

to satisfy (50). In this case the function F reduces to

$$F(g) = 2 \left(1 - \frac{1}{\sqrt{\eta}} \right) - O(\epsilon) < 0 \, , \tag{54}$$

which shows that, in this case there is a direct transition from a superconductor to a superinsulator. For small f, the transition takes place when the new effective magnetic semiaxis $r_M = r_M(1 + f)$ of the ellipse becomes larger than the originally larger electric semiaxis r_Q. This gives the critical frustration

$$f_{\text{crit}} = \frac{1}{2} \left(g^2 - 1 \right) \, . \tag{55}$$

Let us now consider the second case $\eta > 1$. In this case (50) requires

$$\sqrt{\frac{1}{g}} = \frac{1}{\eta} - \epsilon \, , \tag{56}$$

and the function F becomes

$$F(g) = \left(\frac{1}{\eta} - 1 \right)^2 - O(\epsilon) > 0 \, , \tag{57}$$

confirming that in this case, instead the transition from the superconductor is to a topological insulator/quantum metal phase. In this case the critical frustration is determined by the value at which the original electric topological excitations exit the shifted ellipse,

$$f_{\text{crit}} = f_1 = \sqrt{\frac{1}{g}} \sqrt{\frac{1}{\eta} - \frac{1}{g}} \, . \tag{58}$$

Note that, knowing the ratio of the array parameters E_C/E_J, a measurement of this critical frustration amounts to a measurement of the remaining array parameter η.

We have derived that an external magnetic field can induce the transition from a super-conductor to a topological insulator. Let us now analyze what happens to the topological insulator if the magnetic field is further increased. In the topological insulator phase the topological excitations are absent: as a consequence we can diagonalize the doubled Chern-Simons model (2) by the transformation

$$\tilde{a} = \frac{1}{\sqrt{2}} \left(\sqrt{\frac{e_g}{e_f}} b - \sqrt{\frac{e_f}{e_g}} a \right) ,$$

$$\tilde{b} = \frac{1}{\sqrt{2}} \left(\sqrt{\frac{e_g}{e_f}} b + \sqrt{\frac{e_f}{e_g}} a \right) , \tag{59}$$

where we have adopted, for simplicity, the continuum notation and we have introduced $e_g = \sqrt{4E_c}$ and $e_f = \sqrt{2\pi^2 E_J}$. In these new variables the action decouples into two free modes of mass $m = \kappa e_f e_g / 2\pi$,

$$S = \int d^3x \, \frac{1}{4e_f e_g} \tilde{f}_{\mu\nu} \tilde{f}_{\mu\nu} + i \frac{2\kappa}{2\pi} \tilde{b}_\mu \epsilon^{\mu\nu\alpha} \partial_\nu \tilde{b}_\alpha$$

$$+ \int d^3x \, \frac{1}{4e_f e_g} \tilde{g}_{\mu\nu} \tilde{g}_{\mu\nu} - i \frac{2\kappa}{2\pi} \tilde{a}_\mu \epsilon^{\mu\nu\alpha} \partial_\nu \tilde{a}_\alpha , \tag{60}$$

where we have introduced an explicit charge unit κ ($\kappa = 2$ for Cooper pairs). These modes describe fluctuations of composites of a charge κ and fluxes $\pm 2\pi/\kappa$, the pure Chern-Simons term describing the Aharonov-Bohm phase 2π picked up when one of these composites encircles another, as described after Eq. (6). When an external magnetic field is introduced, however, a charge looping around an elementary plaquette picks up an additional phase $\pm 2\pi f$ depending on the sign of the frustration. In other words, the "effective flux" of one of the composites is increased while the other is decreased, an effect captured by two effective $\kappa_\pm = \kappa(1 \pm f)$. An external magnetic field leads thus to a mass splitting

$$m \to m_\pm = \sqrt{8E_c E_J} (1 \pm f) . \tag{61}$$

In particular, the topological mass gap characterizing the topological insulator is lowered by an external magnetic field. When this is sufficiently strong we expect thus a "defrosting transition" leading to normal metallic behavior.

Using these results it is possible to obtain a qualitative description of the transformation of the superconducting/superinsulating phases into the TI and further into a bulk insulator/vortex liquid. To this end, let us first introduce the dimensionless temperature $\tau = T\ell/(\hbar v_c)$, with v_c the light velocity in the material. As derived in a previous section, the critical field for the superconductor-to-TI transition is $f_c = \sqrt{\epsilon}/\eta^{3/2} = (g - \eta)^{1/2}/\eta^{3/2}$. The important point is then that, at finite temperatures, the parameter η scales as $\eta(T) = \eta_0 S(T)$, with a scale factor $S(T) > 1$ and $\eta_0 = \eta(T = 0)$, as derived above. Using this scaling we obtain the equation that defines the SC$\leftrightarrow$TI transition temperature τ_{TI} as a function of the applied magnetic field,

$$f(\tau_{\text{TI}}) = \frac{\sqrt{\eta_0(1 - S(\tau_{\text{TI}})) + \epsilon_0}}{\eta_0^{3/2} S^{3/2}(\tau_{\text{TI}})} , \tag{62}$$

where $\epsilon_0 = g - \eta_0$. The critical temperature for the SC↔TI transition at $f = 0$ is then defined by the equation

$$S(\tau_{TI}) = 1 + \epsilon_0/\eta_0 .\tag{63}$$

The dimensionless thawing temperature, instead, is determined by the topological mass splitting caused by the external magnetic field (see SM),

$$\tau_{th} = \frac{m\ell}{\hbar v_c}(1 - f),\tag{64}$$

and the intermediate phase appears at zero magnetic field if $m\ell/(\hbar v_c) > \tau_{TI}$.

8. Emergent granularity

Being an exemplary laboratory for theoretical and experimental study of the SIT and emerging phases in its critical vicinity, Josephson junction arrays offer a perfect model for strongly disordered superconducting films [12,29]. Moreover, a striking quantitative agreement between the experimentally measured characteristics of the superinsulating state in disordered TiN films [50] and its description in terms of a regular JJA [18] suggests that the close connection between these two systems goes well beyond the similarity of coarse-grained JJA and disordered film. The idea of emergent electronic granularity, i.e. that even in structurally homogeneous films the electronic texture in the critical vicinity of the SIT is an array of superconducting puddles connected by weak links and immersed in an insulating matrix was pioneered in [51]. Over two past decades, this hypothesis evolved into a paradigmatic attribute of the SIT, see [52] and references therein. It was conjectured [16,50] that it is this emergent granularity that serves as a material platform for the superinsulating state. Yet, while being generally accepted, the concept of emergent granularity is not thoroughly justified. Our gauge approach to the SIT enables us to put this concept on a firm field-theoretical foundation.

To see this, note that there is a finite width strip embracing the line of the direct SIT at $g = g_c = 1$ and $\eta < 1$, where Cooper pairs and vortex condensates coexist. Calculating at $g > 1$ and $\eta < 1$ the energy E of a superconducting droplet of perimeter L immersed into a superinsulating matrix of area A, one can show

$$E \propto A - \frac{1}{4\pi}\left(1 - \frac{1}{g^2}\right)L^2 + \sigma L,\tag{65}$$

where the boundary contribution arises due to fluctuation-induced charge and vortex excitations within the opposite condensates, respectively, and σ is a numerical coefficient. The energy has a global minimum $E_{global} \propto A/g^2$ when the superconductor droplet fills the whole area A, but also a local minimum $E_{local} \propto A$ at $L = 0$. These two minima are separated by a maximum at $L \propto g^2/(g^2 - 1)$. If close enough to the SIT, this maximum spreads over a scale exceeding the radius of the droplet. Hence, it becomes energetically advantageous to fragment a superconducting droplet into smaller ones. The fragmentation stops at the minimal dimension of order ξ setting the scale of the self-induced granularity. The structure of the emergent near the SIT electronic granularity is illustrated in Fig. 4.

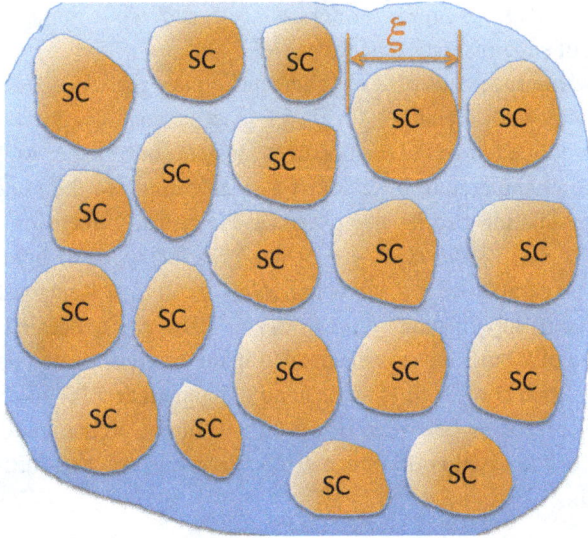

Fig. 4. Self-induced electronic 'granular' structure in the critical vicinity of the SIT. At the first order direct superconductor-superinsulator transition ($\eta < 1$) the phase separation occurs and the system breaks into the superconducting droplets immersed into an insulating matrix and connected by weak Josephson links. The characteristic size of a droplet is $\gtrsim \xi$.

9. Asymptotic freedom and experimental evidence for linear confinement in superinsulators

The experimental implications of our results are far reaching. The correspondence between JJA and disordered 2D and 3D superconductors is given by the relations $E_c = 4e^2/a$ and $E_J \simeq 1/(16\pi e^2 \lambda)$, where $a \simeq r\xi$ in 3D and $a \simeq d$ in 2D, ξ is the superconducting coherence length, d is the film thickness, and r is a numerical factor of order unity, $\lambda = \lambda_L^2/d$ in 2D and $\lambda = \lambda_L$ in 3D, λ_L being the London penetration depth. Furthermore, it was shown that at low temperatures the granular JJA array is effectively described by the model of a continuous medium with the conductance g[53] while E_c near the SIT renormalizes to Δ/g (Δ is the superconducting gap in a single granule). This justifies our definition of g.

It is experimentally established that in TiN and NbTiN disordered films the superinsulating critical temperature coincides with the temperature T_{CBKT} of the charge BKT transition[17,49]. Since in 2D the deconfinement temperature coincides with T_{CBKT}, this strongly supports our finding that the mechanism behind the superinsulation is the Mandelstam-'t Hooft-Polyakov confinement. At the same time, there has been a striking observation in InO films[54] where the precursor of the emergence of the finite temperature insulator, the divergent resistance, demonstrated the so-called Vogel-Fulcher-Tamman (VFT) criticality, $R_\square \propto \exp[-\mathrm{const}/(T_0 - T)]$, rather than the expected Kosterlitz behavior $R_\square \propto \exp[-\mathrm{const}/(\sqrt{T_{BKT} - T})]$. According to the recently developed gauge theory of the BKT transition in the disordered XY-model[55], the standard BKT criticality transforms into the VFT behavior in 3D. One thus can speculate that InO films, which are much more

3D than TiN and NbTiN films (in the former usually $\xi \ll d$, while in the latter $\xi \gtrsim d$), offer a first observation of superinsulation in 3D.

In QCD, the string mechanism of confinement implies the so-called asymptotic freedom at scales smaller than the string size[56], i.e. the unconstrained dynamics of quarks. A corresponding implication of the confinement mechanism of superinsulation, would be the metallic-like low temperature behavior of a small sample of the material that should have turned superinsulating had its size exceeded the typical dimension of the confining string. The latter is $d_{string} \lesssim \hbar v_c / k_B T_{CBKT}$, where v_c is the speed of light in the material. Using the TiN films parameters[16,17] we obtain $d_{string} \lesssim 60\ \mu m$. Remarkably, the study of the size dependence of superinsulating properties in TiN films[57] revealed that in the films with the lateral size of 20 μm and less the insulating thermally activated behavior saturates to the metallic one upon cooling to 'superinsulating temperatures.' This complies with the expected asymptotic freedom behavior. However, it would be premature to take it as a conclusive evidence for the asymptotic freedom in superinsulators, and further experimental research is needed.

Acknowledgments

We are delighted to thank N. Nekrasov, M. Vasin, and Ya. Kopelevich for illuminating discussions. M. C. D. thanks CERN, where she completed this work, for kind hospitality. The work of V. M. V. was supported by the U.S. Department of Energy, Office of Science, Materials Sciences and Engineering Division.

Appendix A. Lattice Chern-Simons and BF terms

The formulation of the Chern-Simons term on a lattice of spacing ℓ[14] requires particular care since it is not entirely trivial how to maintain discrete gauge invariance. We introduce the forward and backward derivatives and shift operators

$$d_\mu f(x) = \frac{f(x + \ell\hat\mu) - f(x)}{\ell}, \qquad S_\mu f(x) = f(x + \ell\hat\mu),$$

$$\hat d_\mu f(x) = \frac{f(x) - f(x + \ell\hat\mu)}{\ell}, \qquad \hat S_\mu f(x) = f(x - \ell\hat\mu). \tag{A.1}$$

Summation by parts on the lattice interchanges both the two derivatives (with a minus sign) and the two shift operators. Gauge transformations are defined by using the forward lattice derivative. In terms of these operators one can then define two lattice Chern-Simons terms

$$k_{\mu\nu} = S_\mu \epsilon_{\mu\alpha\nu} d_\alpha, \qquad \hat k_{\mu\nu} = \epsilon_{\mu\alpha\nu} \hat d_\alpha \hat S_\nu, \tag{A.2}$$

where no summation is implied over equal indices. Summation by parts on the lattice interchanges also these two operators (without any minus sign). Gauge invariance is then guaranteed by the relations

$$k_{\mu\alpha} d_\nu = \hat d_\mu k_{\alpha\nu} = 0, \qquad \hat k_{\mu\nu} d_\nu = \hat d_\mu \hat k_{\mu\nu} = 0. \tag{A.3}$$

Note that the product of the two Chern-Simons terms gives the lattice Maxwell operator

$$k_{\mu\alpha}\hat{k}_{\alpha\nu} = \hat{k}_{\mu\alpha}k_{\alpha\nu} = -\delta_{\mu\nu}\nabla^2 + d_\mu\hat{d}_\nu \,, \tag{A.4}$$

where $\nabla^2 = \hat{d}_\mu d_\mu$ is the 3D Laplace operator.

The BF lattice operator in 3D can be formulated in an analogous way, by introducing the operators

$$k_{\mu\nu\rho} \equiv S_\mu \epsilon_{\mu\alpha\nu\rho} d_\alpha \qquad \hat{k}_{\mu\nu\rho} \equiv \epsilon_{\mu\nu\alpha\rho} \hat{d}_\alpha \hat{S}_\rho \,. \tag{A.5}$$

These operators are interchanged (no minus sign) upon summation. Moreover they are gauge invariant, in the sense that they obey the following equations:

$$k_{\mu\nu\rho}d_\nu = k_{\mu\nu\rho}d_\rho = \hat{d}_\mu k_{\mu\nu\rho} = 0 \,,$$
$$\hat{k}_{\mu\nu\rho}d_\rho = \hat{d}_\mu \hat{k}_{\mu\nu\rho} = \hat{d}_\nu \hat{k}_{\mu\nu\rho} = 0 \,. \tag{A.6}$$

Finally, they satisfy also the equations

$$\hat{k}_{\mu\nu\rho}k_{\rho\lambda\omega} = -\left(\delta_{\mu\lambda}\delta_{\nu\omega} - \delta_{\mu\omega}\delta_{\nu\lambda}\right)\nabla^2 + \left(\delta_{\mu\lambda}d_\nu\hat{d}_\omega - \delta_{\nu\lambda}d_\mu\hat{d}_\omega\right) + \left(\delta_{\nu\omega}d_\mu\hat{d}_\lambda - \delta_{\mu\omega}d_\nu\hat{d}_\lambda\right) \,,$$
$$\hat{k}_{\mu\nu\rho}k_{\rho\nu\omega} = k_{\mu\nu\rho}\hat{k}_{\rho\nu\omega} = 2\left(\delta_{\mu\omega}\nabla^2 - d_\mu\hat{d}_\omega\right) \,, \tag{A.7}$$

where $\nabla^2 = \hat{d}_\mu d_\mu$ is the 4D lattice Laplacian.

References

1. Y. Aharonov and D. Bohm, Significance of Electromagnetic Potentials in the Quantum Theory. *Phys. Rev.* **115**, 485–491 (1959).
2. Y. Aharonov and A. Casher, Topological Quantum Effects for Neutral Particles. *Phys. Rev. Lett.* **53**, 319–321 (1984).
3. V. L. Berezinskii, Destruction of long-range order in one-dimensional and two-dimensional systems having a continuous symmetry group I. Classical systems. *Sov. Phys.–JETP* **32**, 493–500 (1970).
4. V. L. Berezinskii, Destruction of Long-range Order in One-dimensional and Two-dimensional Systems Possessing a Continuous Symmetry Group. II. Quantum Systems. *Zh. Eksp. Theor. Fiz.* **61**, 1144 (1971). (*Sov. Phys.– JETP*, **34**, 610–616 (1971)).
5. J. M. Kosterlitz and D. J. Thouless, Long range order and metastability in two dimensional solids and superfluids. (Application of dislocation theory). *Journal of Physics C: Solid State Physics* **5**, L124 (1972).
6. J. M. Kosterlitz and D. J. Thouless, Ordering, metastability and phase transitions in two-dimensioal systems. *J. Phys. C: Solid State Phys.* **6**, 1181–1203 (1973).
7. K. B. Efetov, Phase transition in granulated superconductors. *Sov. Phys. JETP* **51**, 1015–1022 (1980).
8. D. Haviland, Y. Liu and A. Goldman, Onset of superconductivity in the two-dimensional limit. *Phys. Rev. Lett.* **62**, 2180–2183 (1989).
9. A. Hebard and M. A. Paalanen, Magnetic-field-tuned superconductor-insulator transition in two-dimensional films. *Phys. Rev. Lett.* **65**, 927–930 (1990).
10. M. P. A. Fisher, G. Grinstein and S. M. Girvin, Presence of quantum diffusion in two dimensions: Universal resistance at the superconductor-insulator transition. *Phys. Rev. Lett.* **64**, 587–590 (1990).

11. M. P. A. Fisher, Quantum Phase Transitions in Disordered Two-Dimensional Superconductors. *Phys. Rev. Lett.* **65**, 923–926 (1990).

12. R. Fazio and G. Schön, Charge and Vortex Dynamics in Arrays of Tunnel Junctions. *Physical Review* B **43**, 5307–5320 (1991).

13. A. M. Goldman, Superconductor-Insulator Transitions. *Int. J. Mod. Phys.* B**24**, 4081–4101 (2010).

14. M. C. Diamantini, P. Sodano and C. A. Trugenberger, Gauge theories of Josephson junction arrays. *Nuclear Physics* B**474**, 641–677 (1996).

15. A. Krämer and S. Doniach, Superinsulator phase of two-dimensional superconductors. *Phys. Rev. Lett.* **81**, 3523–3527 (1998).

16. V. M. Vinokur *et al.*, Superinsulator and quantum synchronization. *Nature* **452**, 613–615 (2008).

17. T. I. Baturina and V. M. Vinokur, Superinsulator-superconductor duality in two dimensions. *Ann. Phys.* **331**, 236–257 (2013).

18. M. V. Fistul, V. M. Vinokur and T. I. Baturina, Collective Cooper-Pair Transport in the Insulating State of Josephson-Junction Arrays. *Phys. Rev. Lett.* **100**, 086805 (2008).

19. D. Das and S. Doniach, Existence of a Bose metal at $T = 0$. *Phys. Rev.* B**60**, 1261–1275 (1999).

20. D. Das and S. Doniach, Bose metal: gauge field fluctuations and scaling for field-tuned quantum phase transitions. *Phys. Rev.* B**64**, 134511 (2001).

21. C. L. Kane and E. J. Mele, Z_2 Topological Order and the Quantum Spin Hall Effect. *Phys. Rev.* B**95**, 146802 (2005).

22. L. Fu, C. L. Kane and E. J. Mele, Topological insulators in three dimensions. *Phys. Rev. Lett.* **98**, 106803 (2007).

23. J. Moore and L. Balents, Topological invariants of time-reversal-invariant band structures. *Phys. Rev.* B**75**, 121306(R) (2007).

24. S. Deser, R. Jackiw and S. Templeton, Three-Dimensional Massive Gauge Theories. *Phys. Rev. Lett.* **48**, 975 (1982).

25. M. C. Diamantini, C. A. Trugenberger and V. M. Vinokur, Confinement and Asymptotic Freedom with Cooper pairs, under consideration for publication.

26. A. M. Polyakov, *Gauge Fields and Strings*, Harwood Academic Publisher, Chur (Switzerland) (1987).

27. P. Goddard and D. I. Olive, Magnetic monopoles in gauge field theories. *Rep. Prog. Phys.* **41**, 1357 (1978).

28. G. Dunne, R. Jackiw and C. A. Trugenberger, "Topological" (Chern-Simons) quantum mechanics. *Phys. Rev.* D**41**, 661–666 (1989).

29. M. Tinkham, Introduction to Superconductivity. *McGraw-Hill, Inc.* (1996).

30. X.-G. Wen and A. Zee, Classification of Abelian Quantum Hall States and Matrix Formulation of Topological Fluids, *Phys. Rev.* B**46**, 2290–2301 (1992).

31. D. Birmingham, M. Blau, M. Rakowski and G. Thompson, Topological Field Theory. *Phys. Rep.* **209**, 129–340 (1991).

32. M. Kalb and P. Ramond, Classical Direct Interstring Action, *Phys. Rev.* D**9**, 2273–2284 (1974).

33. T. Allen, M. Bowick and A. Lahiri, Topological Mass Generation in 3+1 Dimensions. *Mod. Phys. Lett.* A**6**, 559–571 (1991).

34. A. P. Balachandran and P. Teotonio-Sobrinho, *Int. J. Mod. Phys.* A**8**, 723 (1993).

35. T. Banks, R. Myerson and J. Kogut, Phase Transitions in Abelian Lattice Gauge Theories. *Nuclear Physics* B**129**, 493–510 (1977).

36. D. Nelson, T. Piran and S. Weinberg, *Statistical Mechanics of Membranes and Surfaces*, World Scientific, Singapore (2004).

37. K. Su-Peng, Yu Jing and X.-G. Wen, Mutual Chern-Simons Landau-Ginzburg Theory for Continuous Quantum Phase Transitions of Z_2 Topological Order, *Phys. Rev.* B**80**, 125101 (2009).

38. X.-G. Wen, Theory of the Edge States in Fractional Quantum Hall Effects. *Int. J. Mod. Phys.* **B6**, 1711–1762 (1992).

39. R. Floreanini and R. Jackiw, Self-dual fields as charge-density solitons, *Phys. Rev. Lett.* **59**, 1873–1876 (1987).

40. G. Y. Cho and J. E. Moore, Topological BF Field Theory Description of Topological Insulators. *Ann. Phys.* **326**, 1515–1535 (2011).

41. M. Z. Hasan and C. L. Kane, Topological Insulators. *Rev. Mod. Phys.* **82**, 3045 (2010).

42. A. Kapitulnik, S. A. Kivelson and B. Spivak, Anomalous metals – failed superconductors. arXiv:1712.07215v1 [cond-mat.supr-con] (2017).

43. B. Svetitsky and L. G. Yaffe, Critical behavior at finite temperature confinement transitions. *Nucl. Phys.* **B210**, 423–447 (1982).

44. N. P. Breznay and A. Kapitulnik, Particle-hole symmetry reveals failed superconductivity in the metallic phase of two-dimensional superconducting films. *Sci. Adv.* **3**, e1700612 (2017).

45. F. Wilczek, Two Applications of Axion Electrodynamics. *Phys. Rev. Lett.* **58**, 1799–1802 (1987).

46. M. C. Diamantini, F. Quevedo and C. A. Trugenberger, Confining Strings with Topological Term. *Phys. Lett.* **B396**, 115–121 (1997).

47. A. M. Polyakov, Fine structure of strings. *Nucl. Phys.* **B268**, 406–412 (1986).

48. J. D. Jackson, *Classical Electrodynamics*. John Wiley & Sons, New York (USA) (1962).

49. A. Yu. Mironov, D. M. Silevitch, T. Proslier, S. V. Postolova, M. V. Burdastyh, A. K. Gutakovskii, T. F. Rosenbaum, V. M. Vinokur and T. I. Baturina, Charge Berezinskii-Kosterlitz-Thouless transition in superconducting NbTiN films. *Scientific Reports*, **8**, 4082 (2018). DOI:10.1038/s41598-018-22451-1.

50. T. I. Baturina, A. Yu. Mironov, V. M. Vinokur, M. R. Baklanov and C. Strunk, Localized superconductivity in the quantum-critical region of the disorder-driven superconductor-insulator transition in TiN thin films. *Phys. Rev. Lett.* **99**, 257003 (2007).

51. D. Kowal and Z. Ovadyahu, Disorder induced granularity in an amorphous superconductor. *Solid State Commun.* **90**, 783–786 (1994).

52. R. Ganguly, I. Roy, A. Banerjee, H. Singh, A. Ghosa and P. Raychaudhuri, Magnetic field induced emergent inhomogeneity in a superconducting film with weak and homogeneous disorder. *Phys. Rev.* **B96**, 054509 (2017).

53. I. S. Beloborodov, A. V. Lopatin, V. M. Vinokur and K. B. Efetov, Granular electronic system, *Rev. Mod. Phys.* **79**, 469–518 (2007).

54. M. Ovadia *et al.*, Evidence for a finite-temperature insulator, *Scientific Reports* **5**, 13503 (2015).

55. M. G. Vasin, V. N. Ryzhov, V. M. Vinokur and Berezinskii-Kosterlitz-Thouless and Vogel-Fulcher-Tammann criticality in XY model, arXiv:1712.00757 (2017).

56. D. Gross, Twenty Five Years of Asymptotic Freedom, *Nucl. Phys. B: Proceedings Supplements* **74**, 426–446 (1998).

57. D. Kalok, A. Bilušić, T. I. Baturina, V. M. Vinokur and C. Strunk, arXiv:1004.5153v2 (2010).

BKT stability against disorder, external magnetic fields, classical and quantum fluctuations and quasi-particle tunneling dissipation

Jorge V. José

*Department of Physics, Indiana University, Bloomington Indiana, USA**
CAS Key Laboratory of Theoretical Physics, Institute of Theoretical Physics,
Chinese Academy of Sciences, China

I present a brief overview of work done with my students and collaborators during a period of several years after the seminal publications by Berezinskii-Kosterlitz-Thouless (BKT) [1–3], and the José-Kadanoff-Kirkpatrick-Nelson (JKKN) paper [4]. We proceeded to study the stability of the BKT scenario and the impact in the KT topological phase structure against the presence of different types of disorder [5–7], classical and quantum fluctuations, with and without constant external magnetic fields [8–23], including also quasi-particle tunneling dissipation [13]. The models considered have different types classical and quantum duality symmetries plus gauge invariances. I will also discuss the interesting case of having two capacity coupled JJA layers, one dominated by quantum charge fluctuations and the other by thermal vortex fluctuations. In these studies, we used different analytical and numerical approaches: e.g. the Replica Trick [5, 35], Classical and Quantum Renormalization Group calculations [5–7], WKB and variational calculations plus Classical and Quantum Monte Carlo simulations [8–12]. Adding these extra perturbations produced small but also significant changes to the BKT phase transition structure in particular at low temperatures, including reentrant putative Quantum Induced Transitions (QUIT) [8–23]. Good experimental representation of these model variations are the Josephson Junction arrays (JJA) and granular superconducting films [24–31].

1. Introduction

Soon after people realized the importance of the seminal BKT papers, they asked a series of questions about the stability and possible changes to the BKT phase structure against experimental and theoretical perturbations. To start, often experimental systems are not ideal since they include a number of imperfections not present in the original theoretical models. These could have important consequences when testing and extending the validity of the theory. For example, JJKN considered the presence of p-symmetric magnetic fields that may be produced by the underlying geometrical properties of the substrate. These perturbations led to significant changes in the phase transition structure of the BKT scenario. Often, these perturbations are random values of the coupling constants, or by applied external magnetic fields that can generate random frustration. Generally, within the context of Universality [32], quantum fluctuations are not supposed to be important close to T_{BKT}. But there are experimental systems where quantum fluctuations can play important roles when the temperature is significantly lowered below T_{BKT}, that can

---

*Permanent address.

lead to low temperatures quantum critical fluctuations. Good examples of this occur in submicron Josephson Junction (JJ) Arrays (JJA) and in granular superconducting films with small micrometer grains [24–31]. The review presented here relates mostly to work and papers that I published with my students and collaborators some time ago, but that still remain relevant due to the recent interest in quantum topological phase transitions, general duality transformations and different types of gauge invariance with possible applications to quantum computers.

The quantized Josephson Ginsburg-Landau order parameter phase, $\hat{\varphi}$, and the Copper pair number operator, $\hat{n}$, are conjugate quantum variables satisfying the commutation relation, $[\hat{n}(\vec{r}), \hat{\varphi}(\vec{r}')] = -i\delta_{\vec{r},\vec{r}'}$, with $i = \sqrt{-1}$. The general model Hamiltonian considered in this paper, having a JJA and superconducting films as an experimental model systems, is defined by the quantum Hamiltonian:

$$\hat{H} = \hat{H}_C + \hat{H}_J = \sum_{\langle \vec{r}, \vec{r}' \rangle} \frac{(2e)^2}{2} \hat{n}_{\vec{r}} C^{-1}_{\vec{r},\vec{r}'} \hat{n}_{\vec{r}'} + E_J(\vec{r}, \vec{r}')[1 - \cos(\hat{\varphi}_{\vec{r}} - \hat{\varphi}_{\vec{r}'} - f_{\vec{r},\vec{r}'})]. \quad (1)$$

Here $C^{-1}_{\vec{r},\vec{r}'}$ is the inverse capacitance matrix between the junctions, $E_J(\vec{r}, \vec{r}')$ is the Josephson coupling between the junction faces, e is the electron charge and, $f_{\vec{r},\vec{r}'}$, is the frustration link parameter defined by the line integral,

$$f_{\vec{r},\vec{r}'} = \frac{2\pi}{\Phi_0} \int_{\vec{r}}^{\vec{r}'} \vec{A} \cdot d\vec{l}. \quad (2)$$

$\vec{A}$ is the vector potential with Φ_0 the quantum of flux. The frustration parameter is given by the sum over a plaquette P:

$$f = \sum_{P} f_{\vec{r}.\vec{r}'} = \Phi/\Phi_0, \quad (3)$$

with Φ the magnetic flux through P.

The many body quantum Hamiltonian given in Eq. (1) has a very rich set of possible solutions as a function of its parameters. It is, however, an insoluble problem both analytically and via numerical simulations. It can only be studied within different types of approximations, while preserving the Hamiltonian's different symmetries present in the models.

The fluctuations in the number of Cooper pairs and in the phase of the superconducting wave function obey the Heisenberg's uncertainty principle, $\Delta\hat{n}\Delta\hat{\varphi} \geq \frac{1}{2}$. This uncertainty relation was demonstrated experimentally [30]. We will consider different limits of this Hamiltonian, most of the time having in mind its JJA representation, but also superconducting granular films. In so doing, each array can be represented by a competition between the Josephson coupling energy E_J, and the charging energy $E_C = \frac{e^2}{2C_m}$ of the junctions. Here C_m the mutual capacitance of a junction. In the semi-classical limit, $E_J \gg E_C$, having the phases of the JJ's superconducting order parameters well defined, while the average Cooper pair number is undefined. This being a consequence of the Heisenberg commutation relations. In the semi-classical regime, the vortex excitations are pinned by the intrinsic lattice potential with the array being in a superconducting (SC) state. In the extreme quantum limit, $E_J \ll E_C$, the electrostatic charging energy needed to add one Cooper pair between the two islands forming the junctions is much larger than the thermal Josephson tunneling energy. The electric field localizes the Cooper pairs in the islands while the phases are delocalized by the vortex excitations. The zero-point charge quantum fluctuations in the junction islands drive the array into an insulating state.

The superconducting-insulating (S-I) phase transition induced by the charging energy in the arrays of this type has $\alpha \gg 1$, with $\alpha \equiv E_C/E_J$. Experimentally it was measured as a function of α by different groups [24, 31]. Their junction sizes had constant values, while they varied the normal state junction resistance to change the Josephson coupling energy. This allowed them to fabricate arrays with α in the range [0.13-4.55], or as high as 33 [27].

JJA were also studied in connection with quantum phase transitions [7–23]. JJA have the advantage, over films, that their internal structure can be carefully controlled experimentally. The drawback is that, as in numerical simulations, the array sizes are limited.

In this review I will discuss results from different limits of his Hamiltonian, both in the purely semi-classical limit as well as in the full quantum regime. In Section 2. I will start by considering possible stability changes against disorder in the KT transition by adding small amounts of disorder to the coupling constant or by having an external constant magnetic field in a random lattice producing random frustration. The latter is a representation of a *superconducting gauge glass model*. Solving the statistical mechanics of models with *quenched* disorder is not easy analytically, or numerically. I'll first describe the case of having random coupling constants using the *replica-trick* calculation scheme [35]. This allowed us to treat analytically a quenched disordered system in equilibrium. For small disorder, I derived the corresponding corrections to the T_{KT} Renormalization group equations [5].

In Section 3, I consider the effect of adding constant magnetic fields to periodic and random lattice JJA geometries. Adding a constant external magnetic field to a periodic array lattice introduces complicated frustration parameter phase effects. This depends on the number theoretic nature of the total frustration as well as on the corresponding magnetic gauge invariance. This latter magnetic gauge invariance is different from the duality transformation gauge invariance shown by JKKN to be the source of the vortex pair excitations in the 2D XY-model. This type of periodic model with arbitrary frustration values is still not yet fully understood [35].

Next, we considered the case of the superconducting glass model represented by having random lattice sites in the presence of a constant external magnetic field introducing random frustration. Using the classical Monte Carlo method, we simulated the metastable properties of the model when it is zero-field cooled or via finite field cooling the system [6]. The motivation for this work was to try to understand the metastable properties found early on High Temperature superconductors [33].

When the junctions in the JJA are made of submicron size, *zero-point quantum fluctuations* become important due to non-zero charging energy effects in the junctions. I started considering this problem by doing Renormalization Group (RG) calculation within a semi-classical WKB perturbative expansion in $\beta\hbar$, with $\beta = 1/k_B T$, to ascertain the BKT stability against quantum fluctuations. I used the Feynman path integral representation of the model. It becomes a quasi-3-dimensional system with the extra finite imaginary-time dimension τ, with $\tau \varepsilon [0, \beta\hbar]$. The imaginary time dimension increases as the temperature is lowered. This leads to a quantum nucleation of vortices along the imaginary time axis. Close to BKT, the problem remains stable. However, a new low temperature instability arose in the analytic RG calculations indicating the possible presence of lower re-entrant phase transitions induced by the proliferation of quantum zero-point induced vortices. To further ascertain this conjecture, we developed a non-perturbative Quantum Monte Carlo (QMC) approach to treat this problem. There we found that in fact there is indeed a re-entrant ***first order*** *Quantum Induced Transition* (QUIT) between two superfluid phases [8–15]. The T_{QUIT} temperature was relatively low making the simulations CPU time consuming. We then added an external constant magnetic field to the model having half-frustration. In this case we also found a QUIT(f) instability, but at higher temperatures. In the experimental JJA and superconducting granular films, they may also involve dissipation. There are two possible ways to include dissipation. By adding Ohmic dissipation or by quasi-particle tunneling dissipation. I'll discuss both possibilities briefly, but include only results for quasi-particle dissipation [13]. To further complete the analyzes with dissipation we also added a constant full frustration external magnetic field to the Hamiltonian.

Further, motivated by experimental results [43] we also considered the case of having two capacitive coupled layers of JJA. Each array may be either in the semi-classical limit or in the quantum regime. The semi-classical limit is dominated by vortex excitations while the quantum regime by Cooper pair excitations. We analyzed analytically, within the WKB approximation, the case when both arrays are in the semi-classical regime. The more interesting case consists of having one of the arrays in the semi-classical limit and interacting with the other being in the full quantum regime. We analyzed this problem approximately analytically leading to some interesting results about a bound state made of a vortex in the semi-classical array and charge in the other array in the quantum regime with gauge invariance coupling that looks like the one found in the fractional Quantum Hall Effect [44, 45]. We then proceeded to do Quantum Monte Carlo simulations to derive the corresponding charge vortex phase diagram. I review our results in Section 4. Section 5 gives a recap of the results reviewed in this paper and the questions that still remain to be answered.

2. Quenched Disorder

How stable are the RG KT equations against adding disorder, in particular where the disorder is *quenched*? I first considered the problem of adding **disorder in the exchange coupling constant** in the 2-D XY model defined by the classical limit of Eq. (1) [4]:

$$\beta H = \sum_{\vec{r}, \vec{r}'} E_{\vec{r}, \vec{r}'} \cos(\varphi_{\vec{r}} - \varphi_{\vec{r}'}), \tag{4}$$

here the random exchange coupling constant is defined as:

$$E_{\vec{r}, \vec{r}'} = E_0 \exp(-z_{\vec{r}, \vec{r}'}). \tag{5}$$

With the randomness given by a bounded Gaussian distribution function:

$$P(z_{\vec{r}, \vec{r}'}) = \left(\frac{2}{x\pi}\right)^{1/2} \theta(z_{\vec{r}, \vec{r}'}) \exp(-z_{\vec{r}, \vec{r}'}^2 / 2x). \tag{6}$$

Where $\theta(z_{\vec{r}, \vec{r}'})$ is the Heaviside step function. As in the JKKN paper, I expressed the Hamiltonian in the Villain approximation [34]:

$$\beta H_V = -\frac{1}{2} \sum_{\vec{r}, \vec{r}'} E_{\vec{r}, \vec{r}'} (\varphi_{\vec{r}} - \varphi_{\vec{r}'} + 2\pi m_{\vec{r}, \vec{r}+1} + i z_{\vec{r}, \vec{r}+1})^2, \tag{7}$$

when there is no disorder we return to the Villain Hamiltonian with the $m_{\vec{r}, \vec{r}+1}$ being the vortex integer link gauge variables in the dual representation. Here $m = 0, \pm 1, \pm 2, \pm 3, \ldots$ and $\varphi \varepsilon [0, 2\pi]$ [4]. The random contribution is given by an imaginary link contribution in the Hamiltonian H_V, $i z_{\vec{r}, \vec{r}+1}$. In this representation the Gaussian nature of the probability distribution function combines with the quadratic nature of the Villain approximation in the Hamiltonian allowing for analytical calculations. Next I needed to evaluate the quenched averaged Free energy. To calculate its quench average, I used the replica trick [35]. It consists in taking the $n \to 0$ limit:

$$\langle \beta F_V \rangle_z = -\lim_{n \to 0} \frac{1}{n} \int dz \theta(z) \ln \int d\{\vec{U}\} e^{\beta \tilde{H}_V (\vec{U}, z)}. \tag{8}$$

The two traces are evaluated with respect to $P(z)$ and $\{\varphi, m\}$. The n-dimensional $\vec{U}$ vector is defined as $\vec{U} = (U^1, U^2, U^3, \ldots, U^n)$, with components: $U_{\vec{r}}^n = (\varphi_{\vec{r}}^n - \varphi_{\vec{r}+1}^n + 2\pi m_{\vec{r}, \vec{r}+1}^n)$. After carrying out the integrals and taking the $n \to 0$ limit, for small x, I got a Hamiltonian that has the same form as the non-random classical Hamiltonian but with a renormalized coupling constants,

$$E_0(x) = 1 \Big/ \left[\frac{1}{2}\pi + x - \frac{1}{2}x^2 + O(x^3)\right] \tag{9}$$

and vortex density

$$y(x) \sim \exp\left[-\frac{\pi^2}{2} E_0(1 + J_0 x)\right]. \tag{10}$$

This counterintuitive result shows that $T_{KT}(x)$ increases with disorder, for small disorder. The average coupling constant decreases in value with x. $E_0(x)$ measures the strength of the interaction between renormalized vortex pairs, thus the critical temperature increases. However, the critical exponents remain the same, as expected from Universality [32]. Of course, this result for the critical temperature depends on the way the disorder was introduced, but often the changes are quantitative not qualitative as per the universality hypothesis. Of importance here is that the BKT phase structure remains invariant.

Next we considered a **superconducting gauge spin glass** model introducing randomness in the lattice sites locations. In this case the classical limit of Hamiltonian (1) reads:

$$H = E_J \sum_{\langle ij \rangle} [1 - \cos(\varphi_{\vec{r}}(\tau) - \varphi_{\vec{r}'}(\tau) + 2\pi f_{\vec{r},\vec{r}'})]. \tag{11}$$

Here $f_{\vec{r},\vec{r}'}$ is the frustration link parameter and the total plaquette frustration, defined in Eq. (2) and Eq. (3). In the classical limit there is the special case of full frustration, as mentioned above. This case has been studied extensively but even now its critical properties are not fully understood. For the classical case having rational frustration, $f = \frac{p}{q}$, with (p, q) prime numbers or irrational f the phase structure is rather complex [36]. An analytical calculation of the corresponding two-point correlation functions in the fully frustrated case was rather complex [37].

Disorder was introduced via the random location of the JJA junctions, or grains in a superconducting film. In the case we studied the randomness was chosen along the x-axis site locations given by, $x_i = ia + \delta a$, with a the lattice sites with δ a random variable determined by a uniform probability distribution function, $P(\delta) = 1/2\delta$, in the interval $[a - \delta, a + \delta]$. We tested that adding the same type of randomness in the y-direction did not change the qualitative nature of the results nor when using a Gaussian distribution function. This quenched problem is difficult to attack analytically, even within the replica trick method. We used the classical Monte Carlo simulation approach to calculate the magnetization M and the Helicity modulus Υ. Choosing the Landau gauge:

$$f_{i,j} = \frac{2\pi}{\Phi_0} H \frac{x_i + x_j}{2} (y_i - y_j), \tag{12}$$

with H the constant external magnetic field applied. The quenched magnetization is then given by:

$$M(T) = -\frac{1}{N} \left\langle \sum_{\langle ij \rangle} \frac{\pi E_J}{\Phi_0} \sin(\varphi_i(\tau) - \varphi_j(\tau) + 2\pi f_{i,j})(x_i + x_j) \right\rangle_c \tag{13}$$

With N, the total number of lattice sites. The *helicity modulus*, Υ, measures the response of the system to a twist of all angles along a special direction, say x.

Adding to the cosine argument in the Hamiltonian a kx term Υ is defined as:

$$\Upsilon \equiv \frac{\partial^2 \mathcal{F}_c}{\partial k^2}\bigg|_{k=0}. \tag{14}$$

Here $\mathcal{F}_c$ is the quenched free energy given by $\beta \mathcal{F}_c = -\langle \ln Z[f,\delta] \rangle_c$. Υ is proportional to the superfluid mass density, $\rho_S(T)$. The explicit expression for Υ is:

$$
\begin{aligned}
\Upsilon^x(T) = \frac{1}{N} \Bigg[&\left\langle \sum_{i,j} x_{i,j}^2 \cos[\varphi_i - \varphi_j + 2\pi f_{i,j}] \right\rangle_c \\
& - \beta \left\langle \left(\sum_{i,j} x_{i,j} \sin[\varphi_i - \varphi_j + 2\pi f_{i,j}] \right)^2 \right\rangle_c \\
& - \beta \left\langle \sum_{i,j} x_{i,j} \sin[\varphi_i - \varphi_j + 2\pi f_{i,j}] \right\rangle_c^2 \Bigg]
\end{aligned}
\tag{15}
$$

Following a similar approach as in the experiments, we calculated the **zero-field-cooled** (ZFC) and **field-cooled** (FC) magnetization and the helicity modulus. We found that, as in experiment, there is a hysteresis loop that is due to the random flux trapping in the arrays quenched by the presence of random frustration [6]. It was remarkable that our results were qualitatively similar in many respects to those found experimentally by Muller *et al.*, on high temperature superconductors [33].

3. Quantum Fluctuations

The Josephson Junctions act like capacitors. When the junctions are of submicron size the charging energy of the electric field between the faces of the junctions can induce zero-point quantum fluctuations. The corresponding quantum Hamiltonian for a JJA is given in Eq. (1). There $C_{\vec{r},\vec{r}'}^{-1}$ represents the inverse capacitance matrix and $E_J(\vec{r},\vec{r}')$ the Josephson coupling, with $f_{\vec{r},\vec{r}'}$ the frustration link parameter defined in Eq. (2) and Eq. (3).

The quantum partition function is given by the trace, $Z = Tr\{e^{-\beta \hat{H}}\}$. Using the complete set of states

$$\langle n(\vec{r})|\varphi(\vec{r}')\rangle = \delta_{\vec{r},\vec{r}'} \frac{\exp\{in(\vec{r})\varphi(\vec{r}')\}}{2\pi}, \tag{16}$$

the operator partition function reads:

$$Z = \prod_{\vec{r}} \int_0^{2\pi} d\varphi(0,\vec{r}) \langle \varphi(0,\vec{r})|e^{-\beta \hat{H}}|\varphi(0,\vec{r}')\rangle \tag{17}$$

The corresponding partition function in the imaginary time representation becomes,

$$Z = \int D[\varphi(\tau)] e^{-S[\varphi(\tau)]/\hbar}, \tag{18}$$

where the action is given by:

$$S = \int_0^{\beta\hbar} d\tau \left(\frac{1}{2}\left(\frac{\hbar}{2e}\right)^2 \sum_{\vec{r},\vec{r}'} \frac{\partial\varphi_{\vec{r}}(\tau)}{\partial\tau} C_{\vec{r},\vec{r}'}^{-1} \frac{\partial\varphi_{\vec{r}'}(\tau)}{\partial\tau} \right.$$

$$\left. + E_J \sum_{\langle ij \rangle} [1 - \cos(\varphi_{\vec{r}}(\tau) - \varphi_{\vec{r}'}(\tau) + f_{\vec{r},\vec{r}'})] \right). \tag{19}$$

The angular integrals here are taken with $\varphi_{\vec{r}}(\tau)\varepsilon[-\pi,\pi]$. The angles satisfy the quantization periodicity boundary condition, $\varphi_{\vec{r}}(\tau) = \varphi_{\vec{r}}(\tau + \beta\hbar)$. We will next discuss different limits of this imaginary time Action.

Let's start discussing results for the **self-capacitance matrix case** $C_{\vec{r},\vec{r}'}^{-1} = \delta_{\vec{r},\vec{r}'}/C$, defined by the operator Hamiltonian:

$$\hat{H} = \sum_{\vec{r}} \frac{E_C}{2}\hat{n}_{\vec{r}}^2 + \sum_{\langle \vec{r},\vec{r}' \rangle} E_J[1 - \cos(\hat{\varphi}_{\vec{r}} - \hat{\varphi}_{\vec{r}'} - f_{\vec{r},\vec{r}'})]. \tag{20}$$

The parameter $\alpha \equiv E_C/E_J$, measures the competition between quantum and thermal fluctuation effects. Using the Feynman path integral representation, we get a quasi-3-D problem.

I began to look at the leading quantum corrections to the BKT scenario, within a WKB semi-classical expansion in $\beta\hbar$ [8]. Expanding the action about the semi-classical solution, obtained by averaging the quantum fluctuations of the phase along the imaginary time axis,

$$\bar{\varphi}_{\vec{r}} = \frac{1}{\beta\hbar} \int_0^{\beta\hbar} d\tau\varphi_{\vec{r}}(\tau). \tag{21}$$

Expanding the action about this to second order in $\beta\hbar$, I obtained:

$$Z_{sc} \sim \int d[\bar{\varphi}_{\vec{r}}]e^{-K(1-Kx)[[1-\cos(\bar{\varphi}_{\vec{r}}-\bar{\varphi}_{\vec{r}'})]}, \tag{22}$$

where $x = \alpha/24$, and $K = \beta E_J$. This partition function has the same form as the 2D-XY classical one but with a different coupling constant. Thus we can immediately write down the corresponding KT RG equations:

$$\frac{dK}{d\ell} = 4\pi^3 K^2 \tilde{y}^2 \frac{(1-xK)^2}{(2Kx-1)}$$

$$\frac{d\tilde{y}}{d\ell} = [2 - \pi K(1-xK)]\tilde{y}, \tag{23}$$

with the renormalized vortex pair density given by,

$$\tilde{y} \sim \exp\left[-\left(\frac{\pi^2}{2}K(K-xK)\right)\right]. \tag{24}$$

These RG equations reduce to the KT-RG equations when there are no quantum fluctuations, i.e. $x = 0$. However, the nature of the fixed points structure of the model changes significantly at lower temperatures. We have that the vortex pair excitations are irrelevant in the interval $K_+^{-1} \leq K^{-1} \leq K_-^{-1}$, where K_+^{-1}, K_-^{-1} are the following two solutions to the equation $[2 - \pi K(1 - xK)] = 0$,

$$K_\pm^{-1} = \frac{2x}{(1 \pm \sqrt{1 - 8x/\pi})}. \tag{25}$$

This has a stable line of fixed points below $T_{BKT}(x)$. This line of fixed points shrinks as x increases, becoming unstable as K_+^{-1} and K_-^{-1} approach each other. However, the instability at K_-^{-1} appears to indicate a re-entrant phase transition. Although the WKB approximation is supposed to be valid close to $T_{BKT}(x)$, it is often in fact an asymptotic expansion. A higher order calculation in $\beta\hbar$, would not resolve the low temperature instability problem.

We thus decided instead to develop a non-perturbative Quantum Monte Carlo (QMC) approach to do numerical calculations of the model. To do the QMC simulations we needed to rewrite the partition function given in Eq. (20) in a discretize form along the imaginary time axis. The explicit functional integral trace $\int D[\varphi(\tau)]$ was obtained by expanding the quantum evolution operator using Trotter's formula in powers of $1/L_\tau$, where L_τ is the imaginary time dimension. Using the complete set of states given in Eq. (18), plus the Poisson summation formula we got:

$$Z = \prod_{\tau=0}^{L_\tau-1} \sum_{n(\tau,\vec{r})=-\infty}^{n(\tau,\vec{r})=+\infty} \sum_{m(\tau,\vec{r},\vec{r}')=-\infty}^{m(\tau,\vec{r},\vec{r}')=+\infty} \prod_{\vec{r}} \int_0^{2\pi} \frac{d\varphi(\tau,\vec{r})}{2\pi}$$

$$\cdot \exp\left\{ i \sum_{\tau=0}^{L_\tau-1} \sum_{\vec{r}} n(\tau,\vec{r})[\varphi(\tau+1,\vec{r}) - \varphi(\tau,\vec{r})] \right\} x$$

$$\cdot \exp\left[-\frac{\beta}{L_\tau} \left\{ \sum_{\tau=0}^{L_\tau-1} \sum_{\langle \vec{r},\vec{r}'\rangle} E_J[1 - (\varphi(\tau,\vec{r}) - \varphi(\tau,\vec{r}') + 2\pi m(\tau,\vec{r},\vec{r}'))^2] \right.\right.$$

$$\left.\left. + \sum_{\vec{r},\vec{r}'} \frac{q^2}{2} n(\tau,\vec{r}) \mathbf{C}^{-1}(\vec{r},\vec{r}') n(\tau,\vec{r}') \right\} \right] + O\left(\frac{1}{L_\tau^2}\right) \tag{26}$$

Here we have defined the integer site variables $n(\tau,\vec{r})$ representing the Copper pair charges in the junction at $\vec{r}$. We did the Villain approximation in the Josephson cosine terms adding the sum over the link integer vortex variables $m(\tau,\vec{r})$ that preserve the rotational gauge invariance of the model defined at each point along the imaginary time axis. The first set of site integers are induced by quantum fluctuations and the link integer variables represent the thermal fluctuations. These two integer degrees of freedom are **dual** to each other. One dominates at high temperatures and the other at low temperatures. Following the conjecture from the analytic WKB calculation that there may be an instability at low temperatures, we did QMC at relatively low temperatures, while always making sure that gauge invariance and duality symmetries were preserved at all stages of the calculations.

We carried out extensive numerical calculations for different lattice sizes, making sure that the results were size stable and reliable. The results are described below.

Quantum Induced Transition (QUIT)

In the case of zero frustration we calculated several thermodynamic quantities in QMC simulations. In particular the quantum helicity modulus, Υ, which measures the response of the system to a twist of all angles along a special direction, say x. Within the discretized path integral representation, Υ^x is given by:

$$\frac{1}{E_J L_x L_y}\Upsilon^x(T) = \frac{1}{L_x L_y L_\tau}\left[\left\langle\sum_{\tau=0}^{L_\tau-1}\sum_{\vec{r}}\cos[\phi(\tau,\vec{r})-\phi(\tau,\vec{r}+\hat{x})]\right\rangle\right.$$

$$-\frac{E_J\beta}{L_\tau}\left\{\left\langle\left(\sum_{\tau=0}^{L_\tau-1}\sum_{\vec{r}}\sin[\phi(\tau,\vec{r})-\phi(\tau,\vec{r}+\hat{x})]\right)^2\right\rangle\right.$$

$$\left.\left.-\left\langle\sum_{\tau=0}^{L_\tau-1}\sum_{\vec{r}}\sin[\phi(\tau,\vec{r})-\phi(\tau,\vec{r}+\hat{x})]\right\rangle^2\right\}\right] \qquad (27)$$

Here $\hat{x}$ is the unitary vector along the x-direction and (L_x, L_y, L_τ) are the spatial and imaginary time lattice size dimensions. We also calculated the dielectric constant for the Cooper pair charge degrees of freedom:

$$\epsilon^{-1} = \lim_{\vec{k}\to 0}\left[1 - \beta^2 q^2 \frac{1}{\mathbb{C}(\vec{k})}\langle n(\vec{k})n(-\vec{k})\rangle\right], \qquad (28)$$

here $\mathbb{C}(\vec{k})$ is the Fourier transform of the capacitance matrix. We carried out several tests to ascertain the reliability and robustness of the numerical results, both in the high temperature limit where vortex fluctuations dominate and in the low temperature limit where quantum charge fluctuations do.

We first confirmed that the BKT critical temperature results were quantitatively the same as those obtained from the WKB quantum renormalization group analytic calculations. The high temperature phase being a superconducting phase.

More remarkable we found that at low temperatures an instability developed. For example the helicity modulus, as well as the specific heat, for $\alpha = 0.3$, showed clear *discontinuities* typical of *first order* phase transitions. The discontinuity temperature for $\alpha = 0.3$ was quit low, i.e. $T_{QUIT}(\alpha = 0.3) \sim 0.03$. We named this phase transition the *Quantum-Induced-Transition = QUIT*. For low values of α the transition is between *a normal to a superconducting state* (SC), S_1, dominated by thermally vortex pairs fluctuations to a lower temperature SC phase, S_2, dominated by quantum charge fluctuations from induced quantum vortices along the imaginary time dimension. We found that as α increases there appears to be a boundary between S_1 and S_2 going into a normal state. We found some evidence for a critical value of $\alpha \sim 2.5$, below which there is a T_{QUIT}, but above which $T_{QUIT} \to 0$ and the system only shows a normal phase state. The latter results are just tentative since we had limitation on the imaginary time axis sizes, even using extensively Cray computers to do the calculations.

Given that the $T_{QUIT}(f = 0)$ is rather low, making the QMC calculations more CPU demanding, we decided to do some calculations for the fully frustrated case, $f = 1/2$. Since we now have a vector potential we calculated the helicity modules given by the expression:

$$\frac{1}{E_J L_x L_y}\Upsilon^x(T) = \frac{1}{L_x L_y L_\tau}\left[\left\langle\sum_{\tau=0}^{L_\tau-1}\sum_{\vec{r}}\cos[\phi(\tau,\vec{r}) - \phi(\tau,\vec{r}+\hat{x}) - f_{\vec{r},\vec{r}+\hat{x}}]\right\rangle\right.$$

$$-\frac{E_J\beta}{L_\tau}\left\{\left\langle\left(\sum_{\tau=0}^{L_\tau-1}\sum_{\vec{r}}\sin[\phi(\tau,\vec{r}) - \phi(\tau,\vec{r}+\hat{x}) - f_{\vec{r},\vec{r}+\hat{x}}]\right)^2\right\rangle\right.$$

$$\left.\left.-\left\langle\sum_{\tau=0}^{L_\tau-1}\sum_{\vec{r}}\sin[\phi(\tau,\vec{r}) - \phi(\tau,\vec{r}+\hat{x}) - f_{\vec{r},\vec{r}+\hat{x}}]\right\rangle^2\right\}\right] \qquad (29)$$

We started by testing our QMC algorithms by calculating the critical temperatures for different small values of α in the zero and fully frustrated cases. Using a jack knife calculational approach to estimate the critical temperatures, plus using the experimentally determined critical temperatures, we found that the leading corrections to the critical temperatures as a function of α and f were [14, 15]:

$$\left(\frac{k_B T(\alpha)}{E_J}\right)_{f=0} \approx (0.9787 \pm 0.0070) - (0.256 \pm 0.017)\alpha + O(\alpha^2) \qquad (30)$$

$$\left(\frac{k_B T(\alpha)}{E_J}\right)_{f=1/2} \approx (0.3188 \pm 0.0070) - (0.2929 \pm 0.015)\alpha + O(\alpha^2). \qquad (31)$$

The $f = 0$ agrees experimentally with renormalized measured values. From our QMC simulations, in the zero field case, we found that at lower temperatures there is a discontinuity in the helicity modulus as well as in the specific heat at $T_{QUIT}(\alpha = 0.3, f = 0) \approx 0.018$, with a helicity modulus discontinuity $\Delta\Upsilon(\alpha = 0.3, f = 0) \approx 0.03$. In the fully frustrated case there is also an upsilon discontinuity at $T_{QUIT}(\alpha = 0.3, f = 1/2) \approx 0.15$, about 6 times larger than the critical temperature in the zero field case. $\Delta\Upsilon(\alpha = 0.3, f = 1/2) \sim 0.17$, an order of magnitude larger. These values allowed us to do more extensive calculations in the full frustration case than we could not do in the zero frustration case. However, it is of importance to highlight the existence of $T_{QUIT}(\alpha = 0.3, f = 0) \neq 0$ and $T_{QUIT}(f = 1/2, \alpha) \neq 0$, supporting the hypothesis for the existence of putative low temperature reentrant transitions that can be checked in submicron JJA systems as well as in granular superconducting films.

Dissipation Effects on the QUIT

In trying to compare with JJA experiments or with metallic granular films like in Sn, Pb or Ga, there are two types of dissipation mechanisms that could be at play due to their explicit dependence on the normal sheet resistance $R_\square^N$ [38, 39]. Dissipation could mask the possible observation of the T_{QUIT}, for example, within the context of the idealized models considered here. It was found experimentally [38] that superconductivity can only occur for systems with $R_\square^N \leq 6k\Omega$. This value of sheet resistance is close to the Quantum of resistance, $R_Q = \frac{h}{4e^2} \cong 6.5k\Omega$. Similar results were also found in JJA with junction with sizes down to $0.01 \ \mu m^2$ [38]. Two types of mechanisms can produce dissipation: Either due to quasiparticle tunneling [40] or due to Ohmic dissipation [41]. The latter case was introduced by Caldeira and Leggett. It consists on having a set of harmonic oscillators in contact with a thermal bath. No microscopic derivation of the Ohmic model was given but it does reduce to the quasiparticle model one in some limits. We considered the first type of dissipation that involves quasi-particle tunneling between the two faces of a junction. This involves adding the following dissipation term to the Lagrangian,

$$L_D = \frac{1}{2} \frac{\delta C}{C} \sum_{\langle \vec{r}, \vec{r}' \rangle} \left[\frac{\partial \varphi_{\vec{r}}(\tau)}{\partial \tau} - \frac{\partial \varphi_{\vec{r}'}(\tau)}{\partial \tau} \right]^2 \tag{32}$$

Where $\frac{\delta C}{C}$ is the capacitance change due to the quasiparticle tunneling, given by [40]

$$\frac{\delta C}{C} = \frac{3\pi\hbar}{32\Delta_0 R_N} C, \tag{33}$$

where Δ_0 is the superconducting energy gap. This leads to the discretized imaginary time dissipation contribution to the action given by [13],

$$S_D = \frac{AL_\tau}{\beta\alpha E_C} \sum_{\langle \vec{r}, \vec{r}' \rangle} \left[1 - \cos\frac{1}{2}(\varphi_{\vec{r}}(\tau) - \varphi_{\vec{r}'}(\tau) - \varphi_{\vec{r}}(\tau+1) + \varphi_{\vec{r}'}(\tau+1)) \right] \tag{34}$$

Here the important parameter, A, measuring dissipation is defined as:

$$A = \frac{3}{16} \frac{E_C}{\Delta_0} \frac{R_Q}{R_N}. \tag{35}$$

The total action is now the sum:

$$S_{tot} = S + S_D. \tag{36}$$

Where S is given by the discretized version of Eq. (21), within the self-capacitive approximation by:

$$S = \sum_{\vec{r}, \tau} \left\{ \frac{\beta E_J}{L_\tau} \right\} [1 - \cos(\varphi_{\vec{r}}(\tau) - \varphi_{\vec{r}'}(\tau) + f_{\vec{r}, \vec{r}'}(\tau))]$$

$$+ \sum_{\vec{r}, \tau} \left\{ \frac{L_\tau}{\beta\alpha E_J} \right\} [1 - \cos(\varphi_{\vec{r}}(\tau+1) - \varphi_{\vec{r}}(\tau))]. \tag{37}$$

The discretized dissipation contribution to the action then reads:

$$S_D = \frac{AL_\tau}{\beta E_C} \sum_{\langle \vec{r}, \vec{r}' \rangle} \left[1 - \cos\frac{1}{2}(\varphi_{\vec{r}}(\tau) - \varphi_{\vec{r}'}(\tau) - \varphi_{\vec{r}}(\tau+1) + \varphi_{\vec{r}'}(\tau+1)) \right]. \tag{38}$$

Note the important $\frac{1}{2}$ multiplicative factor in the cosine argument in S_D. It implies a 4π periodicity in the dissipation term emphasizing the single quasiparticle tunneling compared to the 2π periodicity present in the Josephson and capacitive terms describing Cooper pair tunneling. Note that S_D has a four point interaction reminiscent of the similar Baxter model. The total partition function including a Josephson plus capacitive and dissipation terms is rather complex to solve analytically or numerically as well.

We decided then to extend our previous Quantum Monte Carlo approach to the action including the dissipation parameter. We calculated the corresponding Helicity Modulus, Υ, plus other thermodynamic properties. In the simulations we used the fact that, for fixed L_τ, the charging and dissipation terms are multiplied by the factor, $1/\beta$, while the Josephson term by β. This *"duality"* was used to optimize the convergence of the simulations. We tested the correctness of the calculational QMC scheme by considering first T_{BKT} for small values of α, say $\alpha = 0.3$, in the fully frustrated case with $f = 1/2$, while varying A in the interval $0 \leq A \leq 1$. For small values of A we found that T_{BKT} increases as A increases, i.e. the dissipation quenches the charging energy effects that lower T_{BKT}. In contrast, T_{QUIT} decreases as A increases. Studying Υ for larger values of A within the interval $1 \leq A \leq 1$, required larger imaginary time axes. Fixing the temperature at $T = 0.02$, we found hysteretic effects when calculating Υ while increasing A and then decreasing it. The hysteretic loop has a significant jump as a function of A, with a critical value of $A \sim 0.78$. These two results mean that dissipation diminishes the charging energy effects at high temperatures while they quench the lower T_{QUIT} temperature at lower temperatures [13].

4. Coupled Quantum Josephson Junctions Arrays

A novel experimental system composed of *two capacitively coupled* JJA made with ultra-small junctions was initially discussed in a paper by the Delft group [43]. Each array was produced with different α_i parameter values, with $i = 1, 2$. We decided to study this very interesting system. As we shall see at the end of this section this configuration has very interesting gauge duality properties that are of contemporary research interest.

The Hamiltonian describing the set of two coupled arrays can be generally written as:

$$\hat{H} = \hat{H}_{E_J}(1) + \hat{H}_{E_J}(2) + \hat{H}_C(1,2). \tag{39}$$

Here $\hat{H}_{E_J}(i)$, gives the Josephson Hamiltonian for each array. $H_C(1,2)$ describes the full capacitive interaction within and between the two arrays. $H_C(1,2)$ includes the total charging energy matrix, including the self- and mutual capacitive terms in each plane, plus the arrays interaction via an ultra-small nearest neighbor self-capacitive coupling. The derived effective Hamiltonian is:

$$\hat{H} = \frac{q^2}{2} \sum_{\langle \vec{r}_1, \vec{r}_2 \rangle} \sum_{\mu=1, \nu=1}^{2} \hat{\mathbf{n}}_\mu(\vec{r}_1) \mathbf{C}_{\mu,\nu}(\vec{r}_1, \vec{r}_2) \mathbf{n}_\nu(\vec{r}_2)$$

$$+ E_{J1} \sum_{\langle \vec{r}_1, \vec{r}_2 \rangle} (1 - \cos(\hat{\varphi}_1(\vec{r}_1) - \hat{\varphi}_1(\vec{r}_2) + 2\pi f_{\vec{r}_1, \vec{r}_2}))$$

$$+ E_{J2} \sum_{\langle \vec{r}_1, \vec{r}_2 \rangle} (1 - \cos(\hat{\varphi}_2(\vec{r}_1) - \hat{\varphi}_2(\vec{r}_2) + 2\pi f_{\vec{r}_1, \vec{r}_2})) \tag{40}$$

Here $\tilde{\mathbf{C}}_{\mu,\nu}$ is the inverse of the full interacting capacitance matrix derived by solving the discrete Poisson equation with proper boundary conditions, resulting in the capacitance matrix:

$$\mathbf{C}_{\mu,\nu}(\vec{r}_1, \vec{r}_2) = \begin{cases} (C_s^\mu + C_m^\mu + C_{\text{int}}), & \text{if } \mu = \nu \text{ and } \vec{r}_1 = \vec{r}_2, \\ -C_m^\mu, & \text{if } \mu = \nu \text{ and } \vec{r}_1 = \vec{r}_2 + \vec{d}, \\ -C_{\text{int}}, & \text{if } \mu \neq \nu \text{ and } \vec{r}_1 = \vec{r}_2, \\ 0, & \text{otherwise.} \end{cases} \tag{41}$$

With $(\mu, \nu) = 1, 2$. The diagonal blocks of this matrix denote the interaction within the arrays. The off-diagonal components of the super-matrix are given by the identity matrices $-C_{\text{int}} I_{N,N}$, with C_{int} the capacitive coupling between the arrays defined at each lattice site with N the linear dimension of each array. We assumed square lattice geometry arrays.

This Hamiltonian is far more complicated to study than the single array Hamiltonian given in Eq. (1). Even the single array problem is non-trivial in itself. However, we had learned how to treat different limits in the one array problem so we proceeded to do the same for the coupled array system. Each array can be described by its ratio $\alpha_i = (E_{C_i}/E_{J_i})$. The inter-array capacitive coupling parameter is α_{int}. In the $\alpha_i \ll 1$ limit, the ith array is dominated by localized vortex excitations, V_i, while the Cooper pair excess charge excitations, Q_i, are in a superconducting state. In the $\alpha_i \gg 1$ regime, the system has the Q_i's localized in an insulating state while the V_i's are delocalized. There are different parameter regimes that could be considered. We started by studying the simpler case when both arrays are in the semi-classical regime. We did a WKB calculation expansion in $\beta\hbar$. In this case the calculation had to be done to higher order in $\beta\hbar$ than to first order as was done in the single array case. The next higher order correction in the effective action in $(\beta\hbar)^2$ is a non-local term, complicating the estimation of the critical temperature changes as a function of α_{int}. Thus we also used a MFT variational calculation in the effective Hamiltonian obtained from the WKB expansion in $\beta\hbar$. The important aspect of the results obtained were that, within the semi-classical limit, the interaction between the arrays as a function of α_{int} decreased the critical temperature of the coupled array system. Although not shown explicitly in detail here, the putative QUIT hypothesis was still possible in the coupled arrays within the WKB approximation.

We next proceeded to do extensive Quantum Monte Carlo simulations of the corresponding two array helicity modulus plus the dielectric constant. We were interested in evaluating the corresponding imaginary time partition function, Z,

$$
Z = \prod_{\tau=0}^{L_\tau-1} \prod_{\vec{r}} \sum_{n(\tau,\vec{r})} \int_0^{2\pi} \frac{d\varphi(\tau,\vec{r})}{2\pi}
$$

$$
\times \exp\left\{ -\int_0^{\beta\hbar} d\tau \left[\sum_{\vec{r}_1,\vec{r}_2} \frac{q^2}{2} n(\tau,\vec{r}_1) \mathbf{C}^{-1}(\vec{r}_1,\vec{r}_2) n(\tau,\vec{r}_2) + i\sum_{\vec{r}} n(\tau,\vec{r}) \frac{d\varphi}{d\tau}(\tau,\vec{r}) \right.\right.
$$

$$
\left.\left. + E_J \sum_{\langle\vec{r}_1,\vec{r}_2\rangle} [1 - \cos(\varphi(\tau,\vec{r}_1) - \varphi(\tau,\vec{r}_2))] \right] \right\}
\tag{42}
$$

to calculate the Helicity modulus, given in its discretized imaginary time expression:

$$
\frac{1}{E_{J_\nu} L_x L_y} \Upsilon_\nu^x(T) = \frac{1}{L_x L_y L_\tau} \left[\left\langle \sum_{\tau=0}^{L_\tau-1} \sum_{\vec{r}_\nu} \cos(\phi(\tau,\vec{r}_\nu) - \phi(\tau,\vec{r}_\nu+\hat{x}) + 2\pi f_{\vec{r}_\nu,\vec{r}_{\nu+\hat{x}}}) \right\rangle \right.
$$

$$
- \frac{E_{J_\nu}\beta}{L_\tau} \left\{ \left\langle \sum_{\tau=0}^{L_\tau-1} \sum_{\vec{r}_\nu} \sin(\phi(\tau,\vec{r}_\nu) - \phi(\tau,\vec{r}_\nu+\hat{x}) + 2\pi f_{\vec{r}_\nu,\vec{r}_{\nu+\hat{x}}})^2 \right\rangle \right.
$$

$$
\left.\left. - \left\langle \sum_{\tau=0}^{L_\tau-1} \sum_{\vec{r}_\nu} \sin(\phi(\tau,\vec{r}_\nu) - \phi(\tau,\vec{r}_\nu+\hat{x}) + 2\pi f_{\vec{r}_\nu,\vec{r}_{\nu+\hat{x}}}) \right\rangle^2 \right\} \right]
\tag{43}
$$

We also evaluated the corresponding dielectric constant associated with the charge degrees of freedom. To do so we calculated the inverse dielectric constant in terms of the imaginary time two-point charge correlation function expressed as:

$$
\langle n(\vec{r}_1)n(\vec{r}_2)\rangle = \frac{1}{\beta q^2}\mathbf{C}(\vec{r}_1,\vec{r}_2) + \left(\frac{2\pi}{\beta L_\tau}\right)^2 \sum_{\vec{r}_3,\vec{r}_4} \mathbf{C}(\vec{r}_1,\vec{r}_3)\mathbf{C}(\vec{r}_2,\vec{r}_4)\langle m(\vec{r}_3), m(\vec{r}_4)\rangle,
\tag{44}
$$

here, $\mathbf{C}(\vec{r}_1,\vec{r}_2)$ is the two array geometric capacitance, or Green function, between lattice sites $(\vec{r}_1,\vec{r}_2)$. It is calculated by solving Poisson's equation given the lattice geometry structure plus boundary conditions given in Eq. (41). We also defined the imaginary time averaged integer field, $m(\vec{r}) = \sum_{\tau=0}^{L_\tau-1} m(\vec{r},\tau)$, from which the corresponding Fourier transform of the inverse dielectric constant in the discretized imaginary time representation is finally given by [17, 18]:

$$
\epsilon^{-1} = \lim_{\vec{k}\to 0} \left[\frac{(2\pi)^2}{\beta q^2} \mathbf{C}(\vec{k}) < |m(\vec{k})|^2 \right].
\tag{45}
$$

We analyzed first the capacitive coupled arrays by using a combination of Villain approximations and different types of JKKN duality transformations. It was shown

[14–23] that the interaction between vortices and charges between the two arrays has a minimal gauge coupling form, with the interactions sharply localized in space. The effective action when we considered this limit is nonetheless complicated. The arrays have a vortex dominated phase while Cooper pairs interact only if they are localized at the same site in both arrays. Considering the case of one vortex in array 1 interacting locally with a charge on array 2, it was found to have a similar form to the vortex-charge bound state found in the fractional Quantum Hall effect [15, 42, 43].

We then carried out extensive path integral QMC simulations of the two capacitive coupled JJA [22]. We considered the interesting case when array 1 was in the semi-classical limit while array 2, was in the quantum dominated regime. We evaluated the helicity modulus and the inverse dielectric constant defined in Eqs. (44) and (45), as a function of temperature for different values of the interlayer interaction capacitances, as well as the corresponding intra array α_i's, parameter values. In the case when both arrays were in the semi-classical regime, regardless of the interlayer coupling considered, each array has its own superconducting-normal (SC-N) transition at finite temperature. There is also no insulator-conductor transition at finite temperatures. As array 2 enters the low temperature quantum regime, a SC-N reentrant phase transition is seen in Υ_2. At the same time the quantum array develops charge coherence with an insulator-conductor (I-C) transition at finite temperatures. We found this scenario for all the values considered for the interlayer capacitive coupling. As array 2 becomes more quantum ($3.0 \leq \alpha_2 < 4.0$), and for all values of the interlayer couplings considered, reentrant transitions occur not only in the phase degrees of freedom, but *also in the charge degrees of freedom*. When $\alpha_2 = 4.0$ in array 2 the SC-N transition is quenched for values of the interlayer coupling $C_{int}/C_m \leq 0.78261$, with C_m the intra-layer mutual capacitance. For these parameter values there is an insulating-conducting (I-C) transition present. Increasing the interlayer coupling further yields a SC-N reentrant phase transition in the low temperature interval $0.20 \lesssim T \lesssim 0.5$, together with a charge I-C phase transition. One way to understand these results is via the gauge interaction that develops between the phase in one array and a charge degrees of freedom in the other array induced by the interlayer coupling. We showed that there is a duality vortex-charge transformation symmetry in this model. This was pointed out in [42, 43].

The phase diagrams resulting from our QMC simulations had similar features as those seen in the Delft experiments at the time.

5. Conclusions

In this review paper I presented a brief description of work done with my students and collaborators were we asked a number of different interesting questions about the stability of the BKT scenario, with increasing degrees of complexity. The motivation for studying these models came from advances in photolithographic fabrication techniques of ultra-small Josephson Junctions in layered arrays as well as in superconducting granular films. They involved classical and quantum modelling descriptions. In the models considered we added different types of disorders with and without magnetic fields leading, for example, to superconducting metastable gauge glass phases. We unraveled different types of reentrant phase transition between superconducting, insulating and conduction phases from our studies. At all stages of analyses, it was important to preserve the different types of gauge invariances in the models as well as using duality symmetry transformations properties which helped illuminate the nature of the low temperature phase structures unraveled in our model calculations.

Acknowledgements

Many of the results and ideas I have described here come from Ph.D. theses work carried out by my students, Jungzae Choi, Guillermo Ramirez, and Cristian Rojas, as well as my colleagues, Allen Goldman, Laurence Jacobs, Tadeusz Kopec and Mark Novotny. I take this opportunity to thank the National Science Foundation that partially funded the work described here continuously for close to three decades. I also want to thank the President of the Chinese Academy of Sciences for supporting my visits to its Theoretical Physics Institute in Beijing, through its Presidential Fellowship Initiative under grant No. to CAS Fellowship 2016VMA004.

References

1. V. L. Berezinskii, Zhur. Eksp. Teor. Fiz. **59**, 907 (1970); Sov. Phys. JETP **32**, 493 (1970); V. L. Berezinskii, Zhur. Eksp. Teor. Fiz. **61**, 1144 (1971); Sov. Phys. JETP **34**, 610 (1971).
2. J. M. Kosterlitz and D. J. Thouless, J. Phys. C: Solid St. Phys. **5**, L124 (1972); J. M. Kosterlitz and D. J. Thouless, J. Phys. C: Solid St. Phys. **6**, 1181 (1973).
3. J. M. Kosterlitz, J. Phys. C: Solid St. Phys. **7**, 1046 (1974).
4. J. V. José, L. P. Kadanoff, S. Kirkpatrick and D. R. Nelson, Phys. Rev. B **16**, 1217 (1977); Errata in Phys. Rev. B **17**, 1477 (1978).
5. J. V. José, Physical Review Letters **46**, 1591, ibid. **47**, 1419 (1981).
6. J. Choi and J. V. José, Physical Review Letters **62**, 320 (1989).
7. G. Ramirez-Santiago and J. V. José, Journal of Physics **A** Letters 26: L535 (1993).
8. J. V. José, Physical Review **B29 (RC)**, 2836 (1984).
9. L. Jacobs, J. V. José, and M. Novotny, Phys. Rev. Lett. **53**, 2177 (1984).
10. A. Goldman, L. Jacobs, J. V. José, and M. Novotny, 1987: Europhysics Letters **3**, 1295 (1987).
11. A. Goldman, L. Jacobs, J. V. José, and M. Novotny, Physical Review **B38**, 4562 (1987).
12. L. Jacobs and J. V. José, Physica **B152**, 148 (1988).
13. J. Choi and J. V. José, Physical Review Letters **62**, 1904 (1989).
14. C. Rojas and J. V. José, Physica **B203**, 481 (1994).
15. C. Rojas, Ph. D. Thesis, Northeastern University (1995).
16. T. K. Kopec and J. V. José, Physical Review **B52**, 16140 (1996).
17. C. Rojas and J. V. José, Physical Review **B54(1)**, 12361 (1996).

18. J. V. José, J. Stat. Phys. **93**, 943 (1998).

19. T. Kopec and J. V. José, Physical Review **60**(10), 7473 (1999).

20. T. K. Kopec and J. V. José, Physical Review Letters **84**, 749 (2000).

21. J. V. José and T. K. Kopec, Physical Review B 6305:064504 (2001).

22. G. Ramirez-Santiago and J. V. José, Physical Review **B70**, 174516 (2004).

23. G. Ramirez-Santiago and J. V. José, Physical Review **B77**, 064513 (2008).

24. Proceedings of the NATO advanced Research Workshop on Coherence on Superconducting Networks, Delft Netherlands, 1987, Ed. J. E. Mooij and G. B. Schoen, Physica **B152**, 1 (1988). Proceedings of the second CTP Workshop on Statistical Physics: KT Transition and Superconducting Arrays, Ed. D. Kim et. al. (Min Eum Sa, Seoul Korea, 1993).

25. E. Mooij, B. J. et al., Phys. Rev. Lett. **65**, 645 (1990).

26. H. S. J. van der Zant, Ph. D. Thesis, Delft University (1991); H. S. J. van der Zant et al., Europhys. Lett. **18**, 343 (1992).

27. T. S. Tighe et al., Phys. Rev. **B47**, 1145 (1993); P. Delsing et al., Phys. Rev. **B50**, 3959 (1994).

28. W. J. Elion, Ph. D. Thesis, Delft University (1995).

29. H. S. J. van der Zant, W. J. Elion, L. J. Geerlings and J. E. Mooij, Phys. Rev. **B54**, 10081 (1996).

30. W. J. Ellion et al., Nature, **371**, 594 (1994).

31. For a review see: R. Fazio and H. S. J. van der Zant, Physics Reports, 235–334, 355 (2001).

32. L. P. Kadanoff, Critical Phenomena, Proc. Intl. School of Physics, "Enrico Fermi", Course LI, M. S. Green, ed. (Academic Press, New York, 1971. P. 101) R. B. Griffiths, Phys. Rev. Lett. 24, 1479 (1970).

33. K. A. Muller et al., Phys. Rev. Lett. **58**, 1143 (1987).

34. J. Villain, J. Phys. (Paris) **36**, 581 (1975).

35. S. Edwards and P. W. Anderson, J. Phys. **F** 5, 965 (1975).

36. S. Teitel and C. Jayaprakash, Phys. Rev. **B27**, 598 (1983).

37. J. V. José, Physical Review **B20**, 2167 (1980).

38. L. J. Geerlings and J. E. Moij, Physica (Amsterdam) **152B**, 212 (1988).

39. B. G. Orr et al., Physical Review Letters **56**, 378 (1986).

40. V. Amegaokar et al., Physical Review Letters **48**, 1745 (1982).

41. A. O. Caldeira and A. J. Leggett, Ann. Phys. (N.Y.) **149**, 373 (1984).

42. L. L. Sohn et al., Physica **B 194-196**, 125 (1994).

43. C. Rojas, J. V. José, and A. M. Tikofsky, Bull. Am. Phys. Soc. **40**, p. 68, B11-7 (1995).

44. Ya. Blanter and G. Schoen, Phys. Rev. **B53**, 14534 (1996).

Superfluidity, phase transitions, and topology

John Reppy

Cornell University
E-mail: jdr13@cornell.edu

The critical point and the superfluid transition of bulk liquid ^{4}He provide a unique opportunity for the study of influence of geometry on phase transitions. The variation of physical properties such as the superfluid density and heat capacity of the ^{4}He system near a phase transition can be characterized by exponential functions. The high purity and almost strain free condition (particularly under microgravity conditions) of bulk liquid helium samples allows an unparalleled determination of these exponents. Another approach has been the study of helium adsorbed as a thin film on various porous media where the topology of the substrate controls the character of the transition. Examples are the 3-D superfluid transition observed for a medium such a porous Vycor glass which exhibits a random 3-D structure and opposed the fractal structure of the aerogels. Another important case is that of the Kosterlitz-Thouless transition which is realized for ^{4}He adsorbed on a 2-D surface. Over the years, a range of experimental techniques have been applied to the problem of ^{4}He critical phenomena. These include the observation of persistent currents in ^{4}He, high resolution heat capacity, and superfluid density measurements with torsional oscillators.

Topological phenomena in the Moiré pattern of Van der Waals heterostructures

Wang Yao

The University of Hong Kong
E-mail: wangyao@hku.hk

In monolayer Transition Metal Dichalcogenides (TMDs), a newly emerged class of 2D semiconductors, the low energy carriers are described by massive Dirac cones located at K and -K corners of the hexagonal Brillouin zone. These massive Dirac Fermions at K and -K valleys, being time reversal of each other, have interesting phenomena associated with their valley index, including the valley optical transition selection rules, valley Hall effects, and valley magnetic moment, which enable the use of both valley and spin (via spin-valley coupling) as information carriers in electronics. Van der Waals stacking of 2D semiconductors into heterostructures further provides a powerful approach towards designer quantum materials that extend the exotic properties of the building blocks. As a generic aspect of these vdW heterostructures, the inevitable lattice mismatch always leads to the formation of Moiré pattern (i.e. periodic variation of local atomic registries). I will show that the vdW Moire can endow heterostructures unprecedented properties including: (i) electrically switchable lateral superstructures of topological insulators; (ii) nano-patterned spin optics, and spin-orbit coupled excitonic superlattices.

Heterogeneous interfaces for teasing out the physics of embedded surface states

Arthur Hebard

University of Florida
E-mail: Statesafh@phys.ufl.edu

The interface between two dissimilar materials in intimate contact can usually be probed by either current-voltage or capacitance-voltage measurements to reveal unique surface state manifestations of bulk physical phenomena occurring in either (or both) contacting materials. For example, we have shown that the Van der Waals interface between freshly prepared graphene and a variety of doped semiconductors manifests high quality Schottky barriers with pronounced barrier height (Fermi energy) modulations achieved by the application of electric fields or chemical coatings. Similar rectifying structures made with flakes of transition metal dichalcogenides (TaS_2, $TiSe_2$, $NbSe_2$) harbouring charge density waves (CDWs), Bi_2Se_3 with topologically protected surface states or high T_c superconducting cuprates (Bi-2212) pressed against appropriately doped semiconductors reveal surprising phenomenology including shifts in Fermi energies and the redistribution of interfacial charges. By replacing the semiconductor with freshly cleaved Bi-2212 flakes we have also studied proximity-induced high-T_c superconductivity in Bi-2212/1T-TaS_2 junctions driven by coupling to the metastable metallic phase coexisting within the Mott commensurate CDW phase of this material. Encouraged by these observations, we will discuss the insights gained by applying these techniques to a variety of layered materials with putative topological states residing at heterogeneous Van der Waals interfaces.

Emergent particle-hole symmetry in spinful bosonic quantum Hall systems

Nicolas Regnault

Ecole Normale Supérieure de Paris
E-mail: regnault@lpa.ens.fr

When a fermionic quantum Hall system is projected into the lowest Landau level, there is an exact particle-hole symmetry between filling fractions v and $1 - v$. We investigate whether a similar symmetry can emerge in bosonic quantum Hall states, where it would connect states at filling fractions v and $2 - v$. We will begin by showing that the particle-hole conjugate to a composite fermion 'Jain state' is another Jain state, obtained by reverse flux attachment. We will show how information such as the shift and the edge theory can be obtained for states which are particle-hole conjugates. Using the techniques of exact diagonalization and infinite density matrix renormalization group, we will study a system of two-component (i.e., spinful) bosons, interacting via a δ-function potential. We first obtain real-space entanglement spectra for the bosonic integer quantum Hall effect at $v = 2$, which plays the role of a filled Landau level for the bosonic system. We then show that at $v = 4/3$ the system is described by a Jain state which is the particle-hole conjugate of the Halperin (221) state at $v = 2/3$. We will show a similar relationship between non-singlet states at $v = 1/2$ and $v = 3/2$. We will also study the case of $v = 1$, providing unambiguous evidence that the ground state is a composite Fermi liquid. Taken together our results demonstrate that there is indeed an emergent particle-hole symmetry in bosonic quantum Hall systems.

Dynamical signatures of quantum spin liquids

Frank Pollmann

Technical University of Munich
E-mail: frank.pollmann@tum.de

Condensed matter is found in a variety of phases, the vast majority of which are characterized in terms of symmetry breaking. However, the last few decades have yielded a plethora of theoretically proposed quantum phases of matter which fall outside this paradigm. Recent focus lies on the search for concrete realizations of quantum spin liquids. These are notoriously difficult to identify experimentally because of the lack of local order parameters. In my talk, I will discuss universal properties found in dynamical response functions that are useful to characterize these exotic states of matter.

First, we will show that the anyonic statistics of fractionalized excitations display characteristic signatures in threshold spectroscopic measurements. The low energy onset of associated correlation functions near the threshold shows universal behaviour depending on the statistics of the anyons. This explains some recent theoretical results in spin systems and also provides a route towards detecting statistics in experiments such as neutron scattering and tunneling spectroscopy.

Second, we introduce a matrix-product state based method to efficiently obtain dynamical response functions for two-dimensional microscopic Hamiltonians, which we apply to different phases of the Kitaev-Heisenberg model. We find significant broad high energy features beyond spin-wave theory even in the ordered phases proximate to spin liquids. This includes the phase with zig-zag order of the type observed in α-RuCl$_3$, where we find high energy features like those seen in inelastic neutron scattering experiments.

Getting the jump in the Kosterlitz–Thouless transition

Christopher Lobb

University of Maryland
E-mail: lobb@physics.umd.edu

Kosterlitz and Thouless (KT) predicted that their eponymous phase transition should occur in a variety of two-dimensional systems. They correctly excluded two-dimensional superconductors from their list because the vortex-vortex interaction does not have the necessary logarithmic form [1]. Fortunately for experimenters, later theory [2–5] showed that in the right samples vortex-vortex interactions are close enough to logarithmic to cause the KT transition to occur. Experimental results from two different systems, one with [6] and one without [7] the transition, illustrate when the transition occurs, and when it doesn't.

References

1. J. M. Kosterlitz and D. J. Thouless, *J. Phys. C: Solid State Phys.* **6** (1973).
2. M. R. Beasley, J. E. Mooij, and T. P. Orlando, *Phys. Rev. Lett.* **42**, 1165 (1979).
3. S. Doniach and B. A. Huberman, *Phys. Rev. Lett.* **42**, 1169 (1979).
4. B. I. Halperin and D. R. Nelson, *J. Low Temp. Phys.* **36**, 1165 (1979).
5. C. J. Lobb *et al.*, *Phys. Rev. B* **27**, 150 (1983).
6. D. W. Abraham *et al.*, *Phys. Rev. B* **26**, 5268(R) (1982).
7. J. M. Repaci *et al.*, *Phys. Rev. B* **54**, R9674 (1996).

Phase transitions: From Josephson junction arrays to flowing granular matter

Stephen Teitel

University of Rochester
E-mail: stte@pas.rochester.edu

I will review the connection between Josephson junction arrays in a uniform applied magnetic field and the two-dimensional uniformly frustrated XY model. I will then focus on the particular case of the fully frustrated XY model which possesses both a continuous $U(1)$ and a discrete $Z(2)$ symmetry. The interaction between excitations of these two different symmetries leads to the sequence of phase transitions in which first the $U(1)$ order is lost and then the $Z(2)$ order is lost. Topological excitations will be shown to play a key role in understanding the sequence of the transitions. I will then turn to more recent work looking at transitions in the rheological behaviour of sheared granular fluids.

Topological phase transitions in photonic lattices

Yidong Chong

Nanyang Technological University
E-mail: yidong@ntu.edu.sg

Topologically non-trivial bandstructures are not limited to condensed-matter systems, but they can also be realized in a variety of specially-designed photonic systems, such as photonic crystals and arrays of coupled optical waveguides. In some of these photonic systems, it is exceptionally easy to tune system parameters to drive the bandstructure through a topological phase transition, which can produce various interesting effects. In this talk, I will discuss an optical waveguide array that exhibits a particularly accessible transition between a 2D conventional insulator and a 2D topological insulator phase. The transition point corresponds in the full 3D bandstructure to a Type-II Weyl point, the first to be found experimentally in photonics. Moreover, in the regime of nonlinear optics, I will show that the topological phase transition gives rise to a novel family of "self-induced topological solitons", which inherit the properties of gap solitons as well as topological edge states.

Instabilities and solitary waves of light and atoms in photonic crystal fibres*

Mike Gunn

University of Birmingham
E-mail: J.M.F.Gunn@bham.ac.uk

The unexpected experimental discovery of the topologically-ordered Fractional Quantum Hall (FQH) states showed that the powerful diagrammatic perturbation theoretic methods of the time were only useful for a subclass of problems adiabatically related to free-particle problems, and instead, Laughlin's discovery of a model state that describes "flux attachment" to form composite particles has been the source of most subsequent understanding of the effect. In recent years, it has become apparent that "flux attachment" has important sort-distance geometrical properties as well as long-distance topological entanglement properties. I will describe geometric analogies between the unit cell of a solid and the "composite boson" which is the elementary unit of incompressible FQH liquids, and the place for "composite fermions" in their description.

---

*Work performed in collaboration with Jack Gartlan and Nicola Wilkin.

Spin topology architectures in low dimensional magnets

Christos Panagopoulos

Nanyang Technological University
E-mail: christos@ntu.edu.sg

Using particle-like spin structures as a paradigm, I will demonstrate that the states induced by spin orbit coupling and inversion symmetry breaking in magnetic multilayers open a broad perspective with significant impact in the practical technology of spin topology. In particular, I will discuss our effort to modulate interfacial properties for functional skyrmions at room temperature. First, I will introduce a materials recipe to tune the skyrmion's size and density, thermodynamic stability parameter, as well as the crossover between isolated and disordered-lattice configurations. Second, I will present results on the electrical signature of skyrmions using transport and imaging experiments, and their spin dynamics and collective spin excitation modes from high frequency measurements. Finally, I will address their nucleation, stability, current-induced formation and dynamics in design-nanostructures.

Realizing and manipulating topological metals and their exotic properties

Zidan Wang

The University of Hong Kong
E-mail: zwang@hku.hk

Symmetry and topology, as the two fundamentally important concepts in physics and mathematics, have not only manifested themselves in science, but also provided us with profound understanding of various natural phenomena. Recently, topological gapless systems, including Z_2 topological metals/semimetals [1–3], have attracted significant research interests both theoretically and experimentally. In this talk, we will report experimental realization and manipulation of several topological semimetal bands in superconducting quantum circuits as well as the detection of exotic topological characters in these systems [4] and [5].

References

1. Y. X. Zhao and Z. D. Wang, *Phys. Rev. Lett.* **110**, 240404 (2013).
2. Y. X. Zhao and Z. D. Wang, *Phys. Rev. Lett.* **116**, 016401 (2016).
3. Y. X. Zhao, A. P. Schnyder, and Z. D. Wang, *Phys. Rev. Lett.* **116**, 156402 (2016).
4. X. Tan *et al.*, submitted (2017).
5. X. Tan *et al.*, submitted (2017).

Tuning magnetism and topology in topological insulators with broken time reversal symmetry

Yayu Wang

Tsinghua University
E-mail: yayuwang@mail.tsinghua.edu.cn

The interplay between nontrivial topology and broken time reversal symmetry in Topological Insulator (TI) can lead to exotic quantum phenomenon such as the quantum anomalous Hall effect. However, there are still many open questions regarding the mechanism of magnetic order and magneto transport in TI. In this talk, we will present transport studies on magnetically doped TI thin films grown by molecular beam epitaxy. In Cr doped BiSeTe, we observe a magnetic quantum phase transition accompanied by the sign reversal of the anomalous Hall effect induced by Se substitution of Te. ARPES band mapping reveals that the ferromagnetic order is favoured by the nontrivial bulk band topology, revealing a close correlation between the magnetism and topology. More recently, we found a gate-tuned ferromagnetic to paramagnetic phase transition near the topological quantum critical point. We propose that the most likely mechanism is the Stark effect induced electronic energy level shift, which causes a topological quantum phase transition followed by magnetic phase transition. In Mn doped Bi_2Te_3, we observed that the topological Hall effect characteristic of magnetic skyrmions, indicating the coexistence of both real space and momentum space topological structure in magnetic topological insulators.

Anomalous collective modes in topological matter

Justin Song

Nanyang Technological University
E-mail: justinsong@ntu.edu.sg

In the presence of electron interactions, a rich array of topological phases and behaviours are expected to manifest. I will describe in detail how the combined action of Berry curvature and electron interactions dramatically alters the collective behaviour of interacting electron liquids, yielding a new class of collective excitations – Berry plasmons. Berry plasmons manifest as chiral propagating plasmonic modes, which are confined to system boundaries, and appear even in the absence of a magnetic field. They exhibit a rich phenomenology including split energy dispersions for oppositely directed plasmon modes, with splitting that depends directly on interaction strength. Berry plasmons arise generically in anomalous Hall metals, and provide a window into the role of interactions in topological matter.

A second example are collective modes of Fermi-arc carriers in time reversal broken Weyl semimetals. These chiral fermi arc plasmons possess open dispersions, featuring hyperbolic constant frequency contours and group velocity vectors directed along a few specific collimated directions. As a result, a large range of surface plasmon wave vectors can be supported at a given frequency. Both Berry plasmons and Fermi-arc plasmons can be probed via nanophotonic methods, and are parts of an increasingly rich new tool box to manipulate light in a topological matter.

www.ingramcontent.com/pod-product-compliance
Lightning Source LLC
Chambersburg PA
CBHW081513190326
41458CB00015B/5358